Elementary Mathematical Ecology

Elementary Mathematical Ecology

John Vandermeer

Division of Biological Sciences
University of Michigan
Ann Arbor, Michigan

A Wiley-Interscience Publication

JOHN WILEY AND SONS **New York · Chichester · Brisbane · Toronto**

Library of Congress Cataloging in Publication Data

Vandermeer, John H
Elementary mathematical ecology.

"A Wiley-Interscience publication."
Includes bibliographical references and index.
1. Ecology—Mathematics. I. Title.

QH541.15.M34V36 574.5'0151 80–18664
ISBN 0–471–08131–0

Printed in the United States of America

10 9 8 7 6 5 4 3 2 1

FOR HARRY AND ELSIE

Preface

The inspiration for this volume came from teaching a course in mathematical ecology to a group of undergraduate and graduate biology students. I started putting together the notes from which this book evolved after four years of experience in teaching that course, during which I found that the math anxiety of the students was frequently a real handicap to their desire to understand mathematical ecology.

To sidestep this anxiety I developed a problem-solving approach to the subject, starting with extremely straightforward problems and progressing to the more complex as the students grew in understanding and self-confidence. Judging from both student claims and their actual performance on exams, this approach has been extremely effective. Students routinely report that for the first time in their lives a mathematical subject makes sense to them and that the progressive nature of the problems makes it such that the mathematics never seems to get any more difficult.

My original notes have been class-tested now for seven years and revised each year on the basis of the students' comments and criticism. This book now includes the ideas of more than 150 advanced undergraduate and graduate biology students. Particular care has been taken to ensure that when the text is actually worked through, not just skimmed, the difficulty of the problems progresses in easy stages. Thus, in Chapter 1 students encounter an extremely elementary introduction to the nature of the exponential function but by Chapter 8 are asked to compute eigenvalues for systems of differential equations.

The mathematics required for this text is basic calculus and elementary linear algebra. One need not, however, be brilliant in these two fields. A basic gut-level knowledge of calculus (understanding the concept of a derivative and integration and being able to integrate and differentiate simple functions) and a working knowledge of elementary matrix manipulations (matrix addition, subtraction, multiplication, and an intuitive understanding of the inverse of a matrix) are quite sufficient. Much of the necessary mathematics is explained in the text. After completing this book the student should have a working knowledge of the important mathematical techniques needed to appreciate the contemporary ecological literature. It is not, however, intended as a complete survey of the latest developments in theoretical ecology.

As in any book of this sort, the individuals who contributed in one way or another to its formation are too numerous to mention. Two people must be singled out for special thanks: Robert MacArthur who not only introduced me to most of the topics herein but was also the first to point out to me the problem of math anxiety faced by most biology students, and Steve Hubbell who originally taught the course with me and suggested the problem-solving approach.

Finally, a great deal of thanks must go to approximately 150 undergraduate and graduate students at the University of Michigan whose comments and criticisms forced me to revise the text seven times to reach its current level. The book is as much theirs as it is mine.

JOHN VANDERMEER

Ann Arbor, Michigan
January 1981

Contents

1. The Exponential and Logistic Equations 3

The exponential equation, 3
Density dependence, 6
Time lags, 12
Answers to exercises, 14
References, 26

2. Population Projection Matrices 27

Basic formulation, 27
Stable age distribution, 30
The intrinsic rate of natural increase, 31
The stage projection matrix, 33
Density dependence in projection matrices, 35
Answers to Exercises, 38
References, 45

3. Discrete Models of Population Changes 47

Specific forms and their biological interpretations, 47
Equilibrium points and their dynamics, 50
Complicated behavior from simple forms, 54
Answers to exercises, 57
References, 67

4. Life Tables I 69

The survivorship curve, 69
Life tables, 71
Lotka's equation, 72
The intrinsic rate of natural increase, 74
Answers to exercises, 79
References, 94

5. Life Tables II 95

Stable versus Stationary populations, 95
Birth rates, death rates, and mean generation time, 99
Answers to exercises, 102
References, 117

6. The Analysis of Spatial Pattern 119

The Poisson distribution, 121
The methods of Lloyd and Morisita, 132
Nearest Neighbor techniques, 140
Answers to exercises, 142
References, 156

7. Interspecific Competition 158

The Lotka–Volterra equations, 158
Estimating competition coefficients in the real world, 171
Answers to exercises, 173
References, 196

8. Predator–Prey Theory 198

The analysis of general autonomous systems, 199
Predator–prey theory, 204
The method of Rosenzweig and MacArthur, 206
Limit cycles, 212
Answers to exercises, 213
References, 232

9. Species Diversity 234

The lognormal distribution, 234
The measurement of species diversity, 239
The relationship between species number and area, 242
Answers to exercises, 245
References, 264

10. Dynamics of Multiple Species Assemblages 266

The equilibrium theory of island biogeography, 266
The relationship between species and area again, 271
Introduction to population interactions and community structure, 273
The community matrix, 274
The consumer-resource equations, 277
Answers to exercises, 279
References, 289

Index 293

Introduction

Gaining access to the literature in contemporary ecology has become a different task than it was 15 years ago. Today a background in mathematics is a requirement for the serious ecologist. Without a working knowledge of certain mathematical concepts the current ecological literature is virtually meaningless. Although the training of students in ecology usually includes a brief exposure to the more historically important mathematical results, it is only the exceptional student who will make a special effort to develop an understanding of the mathematical concepts that are at the base of the latest in the literature. Yet the ability to use that literature to the full rests on that understanding. The problem is that the average ecologist has a great deal of difficulty keeping up with modern trends. The material herein represents an attempt to fill the void and to provide students with an approach to many of the mathematical techniques that are needed to appreciate contemporary ecological knowledge. It is not intended as a complete, up-to-date exposition of the "latest" developments in theoretical ecology.

This book was designed as a "programmed" learning text. Consequently there is a specific way in which it should be used for maximum effectiveness. It does not contain the sort of material that one can read casually, regardless of how carefully one reads or the quality of one's background. The explanatory material is not meant to stand alone and therefore if one concentrates on it a confusing picture may emerge. Rather, the exercises must be worked out as they are presented. The following outline should be adhered to closely:

1. Read the explanatory material until you come to a set of exercises.
2. *Before reading any further* do the exercises.
3. Each exercise should be attempted without reference to the answer at the end of the chapter. If the way in which the exercise should be done is not evident after a few minutes of thought, look at the answer to determine general drift but not in great detail. Go back and try it again.
4. After all exercises in a section have been completed (and only then) go on to the next section of explanatory material and repeat the process.

As you complete the exercises be sure to check the answers at the end of the chapter to make sure you have done them as I did. It is not that there is only one way of doing them but that frequently the result of a particular exercise is used in subsequent exercises. If your answers do not correspond to mine, you may become confused in later exercises.

Frequently the motivation for an exercise is discussed after the exercise, perhaps a strange concept at first glance. I have found that it is most effective pedagogically to have had experience with the mechanics of solving a problem before the underlying principles are introduced. For some concepts the mechanics become confused with the principles if they are introduced simultaneously. Thus, when you approach a particular exercise and find that you really do not understand why you have been asked to do it in the first place, please have patience. Its rationale will be made evident either in subsequent exercises or in the explanatory text.

It is worth emphasizing that the two parts of this text (explanatory material and exercises) are intimately related. The text was *not* written to allow you to read the explanatory material and skip the exercises. The exercises are an integral part of the whole—indeed probably more important than the other material. Many concepts are introduced *only* in the exercises and often the explanatory material is intimately dependent on these concepts.

The exercises have been developed on the basis of seven years of classroom testing and both their number and the amount of repetition are carefully geared to giving the student enough experience with each problem to grasp it effectively. Some concepts are mastered only after enough practice has been gained in dealing with the mechanics of the computations involved in their application.

By and large I have tried to keep the chapters to the same "effective" length; that is, each chapter should require about the same amount of time for completion. Unfortunately, as judged by student use, I have not been entirely successful. Some of the chapters are harder than others conceptually; others require more repetitive calculations. On the whole you should find yourself spending three to seven hours on each chapter. If you find you need more time, you may not have the proper background for this textbook.

It may be apparent that the notation is not always consistent from chapter to chapter. This is not an oversight. When you go to the literature to study these techniques in recent work, you will find considerable variation in the notation from paper to paper. Indeed such a state seems to be one source of difficulty for students first trying to read the literature in theoretical ecology. For each type of problem I have tried to use the notation most commonly encountered in that topic.

1. The Exponential and Logistic Equations

THE EXPONENTIAL EQUATION. We begin our study of demography by stripping it of all of its complicated details. To make clear the underlying processes that can be objectively and rigorously quantified, we reduce the individuals in a population to particles that do nothing but replicate themselves. So, if we begin with one individual at the present time, we will have two individuals some time in the near future. For the sake of simplicity let us assume that all individuals in the population replicate (produce a baby) after a particular time unit (one day, 30 years, etc.). Suppose, in particular, that each individual replicates once each day (produces one new individual each day). Then, if at day zero we start with one individual, by day 1 we will have two, by day 2 we will have four, by day 3 we will have eight, and so on. If we call the number of individuals in the population at some particular time $N(t)$, we have, for the above example, $N(0) = 1$, $N(1) = 2$, $N(2) = 4$, $N(3) = 8$, and so on.

□ **EXERCISES**

1 If every individual produces one baby per day, how many individuals will be in the population after four days if $N(0) = 15$? If $N(0) = 35$? (Assume that no deaths occur and that a baby produced today does not reproduce until tomorrow.)

2 If every individual produces four babies per day and $N(0) = 5$, what will be the values of $N(1)$, $N(2)$, $N(3)$, $N(4)$? (Assume here, as in the preceding exercise, that no deaths occur and that a baby produced today does not reproduce until tomorrow).

3 Repeat exercise 2 for $N(0) = 10$.

4 Compute $N(t)$, $t = 5, 6, \ldots, 10$, for the example in exercise 2 and plot $N(t)$ against t. □

Thus, if the series we generate $t = 0, 1, 2, 3, 4, \ldots$, is $1, 2, 4, 8, 16, \ldots$, obviously we can represent the relationship between $N(t)$ and t as

$$N(t) = 2^t \tag{1}$$

The number 2 may be written in numerous ways ($\frac{4}{2}, 2 \times 1, 0.5^{-1}$). In particular, it can be written $2 = a^r$, where there are infinite combinations of a and r which, when plugged into a^r, will yield the number 2; but we are concerned with one particular value of the constant a. The value that we allow a to assume, by convention, is Euler's constant: $e = 2.71828 \cdots$. If you have recently had calculus, you probably know why it is convenient to let $a = e$. If you have forgotten your calculus, be sure that you understand that it is *valid* to represent 2 (or, for that matter, any other number) as e^r and that it would be equally valid to represent it in other ways, but that for reasons that are merely convenient and need not concern you we choose e^r, where e is Euler's constant and r is a constant. Specifically, if $e^r = 2$, $r = 0.693$ (recall that by definition $\ln e^r = r$; therefore, if $e^r = 2$, $\ln 2 = r$). Equation 1 then becomes

$$N(t) = e^{(0.693)t} \tag{2}$$

Note that in this exercise, as well as in the rest of this book and in all of the literature, $N(t)$ is variously written as $N(t)$ or N_t or sometimes $X(t)$ or X_t, depending on the author or context. Frequently the functional dependence on t is tacitly assumed; that is, $N(t)$ may be written as N and $N(0)$ is almost always written as N_0.

□ EXERCISES

5 Present the model populations in exercises 2 and 3 in the general form of equation 2.

6 If an individual produces 0.5 offspring per day (on the average) and we begin with a population size of 2, what is $N(15)$?

Equation 2 was written explicitly with the assumption that we began with a single individual; that is, when $t = 0$, $N(t) = 1$. To be more general we must multiply the right-hand side of equation 2 by $N(0)$, the number of individuals we started with, to obtain

$$N(t) = N(0)e^{rt} \tag{3}$$

as the general equation of population growth. It is called the exponential equation. The parameter r is central to population ecology. Mathematically it is the parameter of the exponential equation (equation 3) and biologically it is called the intrinsic rate of natural increase. This concept is discussed at length in later chapters. For now you should have an intuitive feeling for what it means (i.e., the number to which Euler's constant must be raised to obtain the replication (reproductive) rate, as introduced in this chapter).

We note that we may rewrite equation 3 as

$$\ln N(t) = \ln N(0) + rt$$

and differentiate with respect to t:

$$\frac{d \ln N(t)}{dt} = r$$

Recall from basic calculus that in general

$$\frac{d \ln x}{dt} = \frac{d \ln x}{dx}\frac{dx}{dt} = \frac{1}{x}\frac{dx}{dt}$$

Therefore

$$\frac{d \ln N(t)}{dt} = r$$

$$\frac{1}{N(t)}\frac{d\,N(t)}{dt} = r$$

$$\frac{d\,N(t)}{dt} = r\,N(t) \qquad (4)$$

Equation 4 is a differential equation and equation 3 is its integrated form. Both are termed the exponential equation of population growth or simply the exponential equation.

□ EXERCISES

7 If r is 0.69, what is the replication rate (number of offspring per day $+1$; 1 represents the adult doing the reproducing)? What if $r = 1.098$? What if $r = 0.92$? What if $r = 1.39$? What if $r = 4.5$? What if $r = 6.8$?

8 Let $r = 0.83$ and begin with a population size of 2. Plot $N(t)$ against t for $t = 0, 1, \ldots, 6$.

9 Plot $\ln N(t)$ against t for $t = 0, 1, 2, \ldots, 6$, for the data in exercise 8. If you wanted to compute r from this graph, how would you do it?

10 What is the doubling time (how long will it take for the population to double in size) if $r = 0.993$ and $N(0) = 10$? If $r = 0.993$ and $N(0) = 20$?

11 Derive a general equation for doubling time; for tripling time. □

To this point we have examined in the abstract a self-replicating population of particles. Clearly, even at this oversimplified level we have glossed over some fairly universal and important biological facts. Most organisms don't "replicate" like DNA or carbon paper. Most organisms are born, live, and die and we must be concerned with the rate of birth and death, not with the rate of replication.

The usual procedure is to look at the per capita rate of population increase. The per capita rate of increase must be equal to the birth rate minus the death rate. Thus we may write

$$\frac{d\,N(t)}{N(t)\,dt} = b - d \qquad (5)$$

and we then see from equation 4 that $b - d = r$.

Intuitively, nothing changes. Instead of visualizing a population with individuals replicating themselves at some rate, we conceive of them as producing offspring and being subjected to a certain probability of dying such that as a whole the population can be said to have a death rate.

□ EXERCISES

12 In a population of *Paramecium* it is known that under a well-defined set of circumstances each *Paramecium* will divide twice in one day (i.e., in one day a single *Paramecium* turns into four individuals. In terms of the preceding exercises you might think of each individual as producing three babies.) Death is unknown. What is the "instantaneous" birth rate (b in equation 5)? What is the instrinsic rate of increase?

13 Suppose that in the course of dividing (in the example from exercise 12) 50% of the time the individual that was to divide died instead. What are b, d, and r?

14 Repeat exercise 13 but assume that 25% of the attempted divisions resulted in death.

DENSITY DEPENDENCE. The most trivial observations of the most casual observer will reveal a basic inadequacy in the exponential equation, at least insofar as it might be applied to real populations. The equation, if taken as a model that is supposed in some way to represent a natural population, leads to a blatent prediction. Populations grow without limit. It is not necessary to cite experimental evidence to show that the prediction is not borne out in nature.

Two schools of thought emerged from the realization that populations are somehow limited in their growth (i.e., cannot follow the exponential equation forever). One school, typified by the well-known book by Andrewartha and Birch (1954), claimed that most populations did, in fact, follow the exponential equation but that frequently, at more or less random intervals, the population is decimated by some catastrophic event. Thus natural populations behave in a stop-and-start fashion, growing exponentially until some "disaster" forces the population numbers down. The critical feature of this sort of approach is that the factor that knocks the population down is independent of the number of individuals in the population. The other school emphasized the feedback of population numbers on population growth rate. As the number of individuals in the population became larger there was a decrease in the rate at which new individuals were produced and/or an increase in the likelihood of individuals dying. Thus the basic form of population growth was not really exponential because the rate of growth was a decreasing function of the size of the population. (For a summary of the debate see Erlich and Birch [1967] and Slobodkin, Smith, and Hairston [1967].)

Most biologists now agree that many, if not all, natural populations are, at least potentially, subject to these density-dependent constraints (even the strictest adherents to the density-independent school will admit that *some* populations are controlled by density-dependent factors). Thus it makes sense to modify the basic picture of poulation growth as presented above to account for population limitation of the density-dependent type.

In a general way we may say that the per capita rate of change is a function of population size; that is,

$$\frac{dN}{N\,dt} = f(N)$$

and simply note that $f(N)$ gets smaller as N gets larger (i.e., $\partial f/\partial N < 0$) and that f is largest as N approaches zero (i.e., any function that satisfies these two assumptions will provide us with a "reasonable" model of density-dependent population growth). Making a convenient mathematical assumption, we let f take on a simple linear form; that is, we suppose that when N is very small ($N \to 0$) the population grows like an exponential equation (in a density-independent fashion) and for every individual added to the population the per capita growth rate is decreased by a particular amount. Thus the differential equation of population growth becomes

$$\frac{dN}{N\,dt} = r - aN \tag{6}$$

where a is an arbitrary constant that represents the "particular" amount the growth rate is decreased by the addition of a single individual.

☐ EXERCISES

15 For what values of N will the per capita rate be equal to zero (in equation 6)?

16 Call the largest value of N for which $dN/N\,dt = 0$ the carrying capacity (K) and rewrite equation 6 in terms of K, r, and N only.

17 Suppose that the maximum number of individuals sustainable by the environment is K. Suppose that the per capita rate of increase of the population is directly proportional to the fraction of K not yet attained. What would $f(N)$ be? ☐

The equation derived in exercises 16 and 17 is called the logistic equation and is usually written

$$\frac{dN}{dt} = rN\left(\frac{K - N}{K}\right) \tag{7}$$

Equations 6 and 7 are, of course, identical; the form presented in (7) is more common only because it has such obvious biological interpretations. The logistic equation is but one of an infinite number of equations that describe a process of density-dependent population growth. The logistic has the simple property that it describes a population which initially increases at an increasing rate (like an exponential), but as N gets larger the rate of increase becomes smaller until N reaches a maximum value (K), after which the population size no longer changes.

This derivation of the logistic equation is simple and straightforward. Nevertheless, it does not provide much insight into the equation's dynamical meaning (in terms of biological processes). The following derivation is more complex mathematically but it does provide a better understanding of what the logistic means, biologically.

We can ask how many individuals will be in the population at time $t + 1$, given a certain number at time t, the way in which the exponential was introduced. In a perfect density-independent situation we of course have the exponential equation

$$N(t) = N(0)e^{rt}$$

Let $t = 1$ and we obtain

$$N(1) = N(0)e^{r}$$

Let $t = 2$ and we obtain

$$N(2) = N(0)e^{r2}$$

But we see that $e^r = N(1)/N(0)$, so that

$$N(2) = N(0)e^{r}e^{r} = N(0)e^{r}(N(1)/N(0))$$

$$N(2) = N(1)e^{r},$$

and, in general,

$$N(t + 1) = N(t)e^{r} = N(t)\lambda$$

where we have written λ in place of e^r. In this equation, it will be recalled, the only things we have assumed about the population is that it reproduces itself and that the rates of birth and death are invariant with respect to changing population density.

Such are the assumptions of density independence. To modify the exponential equation in this form we must postulate or speculate about the form that the density dependence will take. Just as it was necessary to postulate a specific form for $f(N)$ in the derivation of the differential logistic equation so must we postulate a way in which density will affect the production of $N(t + 1)$ by $N(t)$.

Suppose we have 10 individuals in the population at $t = 0$ and at $t = 1$ we get 20; that is, the population is doubled $[N(1) = 2N(0)]$. Suppose also that instead of going from $N(1) = 20$ to $N(2) = 40$ (as we would expect from the exponential law) the population goes only from $N(1) = 20$ to $N(2) = 30$; that is, it increases by a third $[N(2) = 1.5N(1)]$. The "apparent" value of λ went from 2 to 1.5 by the addition to 10 individuals to the population. We might postulate that each individual introduced into the population reduces the apparent λ by a factor of $(2 - 1.5)/10 = 0.05$. Accepting the postulate that each individual introduced into the population reduces the apparent λ by some particular constant fraction, we find that the apparent value $[\lambda'(N)]$ must be that constant fraction of the density independent value

$$\lambda'(N) = \frac{1}{C}\lambda$$

and that it must be reduced by some amount by every new individual added to the population. Therefore C must equal 1 when $N = 0$ $[\lambda'(N_0) = \lambda]$ and C must increase by some constant factor as $N(t)$ increases (we assume that each individual decreases λ by a constant factor). Thus we have $c = 1 + a\,N(t)$, which, applied to the above equation, yields

$$\lambda'(N) = \frac{1}{1 + a\,N(t)}\lambda$$

and the old exponential equation becomes $N(t + 1) = \lambda'(N)\,N(t)$ or

$$N(t + 1) = \frac{N(t)}{1 + a\,N(t)}\lambda \qquad (8)$$

When the population reaches its carrying capacity, we have

$$K = \frac{\lambda K}{1 + aK}$$

$$1 + aK = \lambda$$

$$a = \frac{\lambda - 1}{K}$$

Equation 8 then becomes

$$N(t + 1) = \frac{\lambda N(t)}{1 + [(\lambda - 1)/K]N(t)} \qquad (9)$$

In fact, equation 9 is equivalent to equation 7, as shown below. Also of note, however, is that equation 9 is frequently important practically in computing projected population histories. Next consider equation 9 with $t = 0$.

$$N(1) = \frac{\lambda N_0}{1 + [(\lambda - 1)/K]N_0} \qquad [N_0 = N(0)]$$

which can be written

$$N(1) = \frac{\lambda K N_0}{\lambda N_0 - N_0 + K}$$

and, finally,

$$N(1) = \frac{K}{1 + [(K - N_0)/N_0]\lambda^{-1}}$$

Making the same derivation for $t = 1$ [i.e., let $N(t) = N(1)$ and $N(t + 1) = N(2)$], we obtain

$$N(2) = \frac{\lambda N(1)}{1 + [(\lambda - 1)/K]N(1)}$$

Substituting for $N(1)$ and rearranging, we have

$$N(2) = \frac{K}{1 + [(K - N_0)/N_0]\lambda^{-2}}$$

and, in general,

$$N(t) = \frac{K}{1 + [(K - N_0)/N_0]\lambda^{-t}} \tag{10}$$

Equation 10 gives $N(t)$ directly as a function of t, and it is the integrated form of the logistic differential equation which can be seen by differentiating. Rewrite equation 10 as

$$N(t) = \frac{K}{1 + ce^{-rt}} = Ke^{rt}(e^{rt} + c)^{-1}$$

where $C = (K - N_0)/N_0$, $e^r = \lambda$. Differentiate directly

$$\frac{d\,N(t)}{dt} = \left[\frac{d(Ke^{rt})}{dt}(e^{rt} + c)^{-1}\right] + \left\{\frac{d[(e^{rt} + c)^{-1}]}{dt} Ke^{rt}\right\}$$

and we obtain

$$\frac{dN}{dt} = r\left[\frac{Ke^{rt}}{e^{rt} + c} - \frac{K^2e^{2rt}}{K(e^{rt} + c)^2}\right]$$

Recalling equation 10, we have

$$\frac{dN}{dt} = r\left[N(t) - \frac{1}{K}N(t)^2\right]$$

which is identical to the logistic differential equation.

We have now derived the logistic in two apparently different ways: first by a direct modification of the differential exponential equation and second by a modification of the integrated form of the exponential. We see that *the increase in the ratio between growth rates at successive time periods is a linear function of the population density* is mathematically identical to *the per capita growth rate is a linear function of the population density.*

There is potential danger in this analysis. Deriving the equation from two points of view helps to clarify its general meaning and behavior. Concomitently, it might tempt the reader to take the equation too seriously, having seen the same equation fall out of different lines of reasoning. In fact, it is a very nice equation, intuitively reasonable, and analytically tractable, but it has not yet proved to be that useful in nature. Much theoretical work is based on it (Levins [1968]; Vandermeer [1970]; MacArthur [1968]; MacArthur and Levins [1967]; Gause [1934]) and small laboratory organisms seem to adhere to it fairly well (Gause [1934]; Vandermeer [1969]), but most populations are too complex to offer situations in which the model could be tested adequately. Indeed, critics of the equation note that natural populations have age structure, exhibit time lags, and respond to stochastic (random) environmental inputs and other such complexities for which the logistic does not account. Such criticisms, although certainly correct, do not get to the heart of the matter and consequently may be misleading. It should be understood that nature's complexity is only one factor to be weighed in examining a model like the logistic equation. If we should judge the logistic equation to be useless because natural populations have age structure, time lags, and stochastic input (say), does this imply that once we shore up the logistic with age structure, time lags, and stochastic inputs it will be useful? If we simply reject the logistic because nature is "too complex," does this imply that we won't reject it when the "complexities" have been added to the basic logistic form? A more significant question, one that has not really been approached adequately, is whether the basic qualitative form of the logistic is a reasonable approximation of nature, even if all the complexities were disregarded.

Perhaps all this will be made clear by a quick methodological review of what has been done so far in this chapter. We began with the exponential equation. In deriving it, only a few assumptions were needed—many, such as instantaneous response and continuity, went unstated. The fundamental biological assumption was simply that organisms reproduce. No more! The assumption is so close to being a definition of life that we can hardly question its validity. Yet it is just that assumption and only that assumption (ignoring the more esoteric mathematical assumptions) that results in the exponential equation.

Next, we treated the subject of density-dependent growth. First, it was noted that a population increases at differing rates, depending on the population density. Most biologists will agree that this, too, is a common, if not universal, premise, but it does not give rise to the logistic equation. It does give rise to

$$\frac{dN}{dt} = f(N)$$

certainly a general equation but perhaps also a trivial one.

Then we began postulating, hypothesizing, or speculating by asking in what way, exactly, the rate of change is related to density. One way is the logistic equation. It is no more correct than any of the other possible choices that could be made.

The point is that a world of difference, methodologically or philosophically, exists between the derivation of the exponential and logistic equations. The first follows directly from a biological universal: living organisms reproduce. It is absolutely correct but rather uninteresting. The second follows from a speculation—the exact form of $f(N)$—the accuracy of which we cannot even begin to assess. It is quite possibly not correct but certainly more interesting than the exponential.

☐ EXERCISES

18 Integrate equation 7 (separate the variables N and t and use a table of integrals).

19 Show that $\ln[(K - N)/N]$ is a linear function of t in the logistic equation.

20 For two populations of some sort of bacteria, grown independently of each other, we have the following data:

t	0	1	2	3	4	5	6	7	8	9	10
$N_1(t)$	10	11	10	20	25	60	110	140	165	175	185
$N_2(t)$	30	50	75	110	145	170	180	185	180	180	180

Plot $N_1(t)$ and $N_2(t)$ versus time. Which population is most like the logistic equation?

21 Begin with a population of two individuals: $r = 0.83$ and $K = 302$. Use equation 9 to compute $N(t)$ for $t = 0, 1, 2, \ldots, 15$. Plot $N(t)$ against t and compare the graph with the one you made in exercise 8.

22 For the data generated in exercise 21 plot $\ln\{[K - N(t)]/N(t)\}$ against t. Draw a line through the points. What is the slope of the line? ☐

TIME LAGS. Among the many complications that should or could be added to this basic formulation, one that seems to come most quickly to the mind of the biologist is associated with the basic assumption of instantaneous response of the population. Most organisms do not respond immediately to environmental cues. Many populations exhibit density dependence with an extreme time lag; for example, the phenomenon of delayed implantation results in part in females producing young as a function of population densities at some time in the past. Numerous other obvious situations exist.

To analyze time lags completely is beyond the scope of this book, but the following discussion will give the reader an intuitive grasp of the effects of time lags.

Recall the derivation of equation 9. We began with the simple exponential equation

$$N(t + 1) = \lambda N(t)$$

We then added density dependence by supposing λ changed as a function of $N(t)$ and eventually wound up with

$$N(t + 1) = N(t) \frac{\lambda}{1 + [(\lambda - 1)/K]N(t)}$$

We could have achieved the same derivation by assuming that the original λ was a function of $N(t - \Delta)$ rather than $N(t)$; that is, instead of presuming that the finite rate of population increase depends on the current population density $N(t)$, we presume that it depends on the population density as it was Δ time units ago, $N(t - \Delta)$. Going through the same derivation that gave us equation 9 but allowing the density-dependent feedback to be effected by $N(t - \Delta)$, we obtain

$$N(t + 1) = N(t) \left\{ \frac{\lambda}{1 + [(\lambda - 1)/K]N(t - \Delta)} \right\} \tag{11}$$

□ EXERCISES

23 Let $r = 0.83$, $K = 300$, $N(0) = 2$, and $N(1) = 4.6$. Using equation 11, let $\Delta = 1$ and compute $N(t)$ for $t = 3, \ldots, 20$.

24 Let $r = 0.83$, $K = 300$, $N(0) = 2$, $N(1) = 4.6$, and $N(2) = 10.3$. Using equation 11, let $\Delta = 2$ and compute $N(t)$ for $t = 3, \ldots, 15$.

25 Plot the results of exercises 21, 23, and 24 on the same graph of $N(t)$ against t. □

Exercises 23, 24, and 25, although only examples, demonstrate what is apparently a general result. As time lags are added, density-dependent population models tend to exhibit damped oscillations. Furthermore, the greater the lag in general, the larger the oscillations and the more slowly they converge. An excellent summary of the effects of time lags on population models can be found in May et al. [1974].

ANSWERS TO EXERCISES

1 $N(4) = 240$ (if $N(0) = 15$).

$N(4) = 560$ (if $N(0) = 35$).

2 $N(0) = 5$; $N(1) = (5)(5) = 25$. (*Note.* Each day each individual in the population produces four *new* individuals. Thus for each individual in the population today there will be five tomorrow, four new ones plus the old one. $N(2) = 25(5) = 125$; $N(3) = 125(5) = 625$; $N(4) = 625(5) = 3125$.

3 $N(0) = 10$; $N(1) = 10(5) = 50$; $N(2) = 50(5) = 250$; $N(3) = 250(5) = 1250$; $N(4) = 1250(5) = 6250$.

4

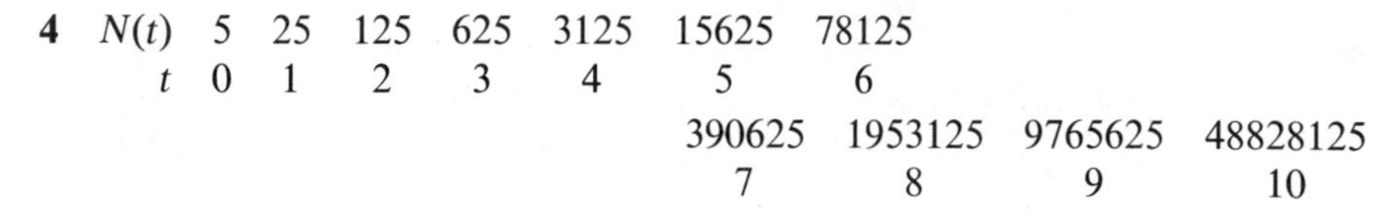

$N(t)$	5	25	125	625	3125	15625	78125
t	0	1	2	3	4	5	6

390625	1953125	9765625	48828125
7	8	9	10

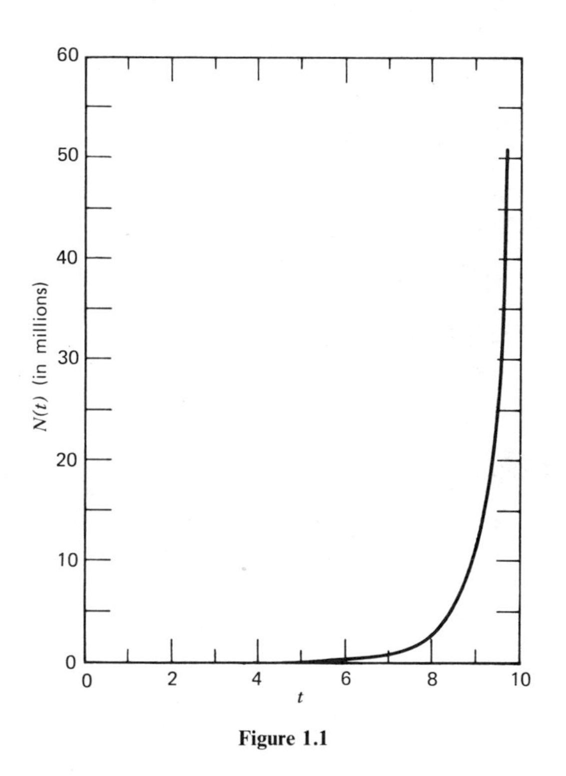

Figure 1.1

5 For exercise 2 each day we obtain five individuals for every individual in the population. Therefore we would have to write $N(t) = 5^t$ if we began with only a single individual at $t = 0$. At $t = 0$, however, we had five individuals and we must write $N(t) = (5)5^t$. Putting the replication rate (reproductive rate) in terms of Euler's constant, we have $N(t) = 5e^{(1.61)t}$. For exercise 3 we obtain $N(t) = 10e^{(1.61)t}$.

6
$$N(15) = (2)(1.5)^t.$$
$$N(15) = 2e^{(0.405)(15)}.$$
$$\begin{aligned}\ln N(15) &= \ln 2 + 0.405(15)\\ &\ 0.69 + 6.075\\ &\ 6.765.\end{aligned}$$
$$N(15) = \text{antilog}\ (6.765) = 867.$$

7

r	0.69	1.10	0.92	1.39	4.5	6.8
Replication rate (e^r)	2	3	2.5	4	90	898

8
$$N(t) = N(0)e^{rt} = 2e^{0.83t}.$$
$$e^{0.83} = 2.3$$

$N(t)$	2	4.6	10.58	24.33	55.97	128.72	296.07
t	0	1	2	3	4	5	6

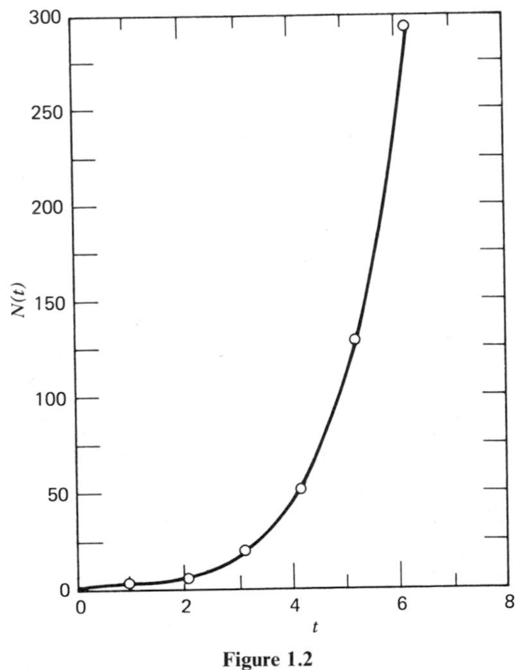

Figure 1.2

9

$N(t)$	2	4.6	10.58	24.33	55.97	128.72	296.07
$\ln N(t)$	0.69	1.53	2.36	3.19	4.02	4.85	5.69

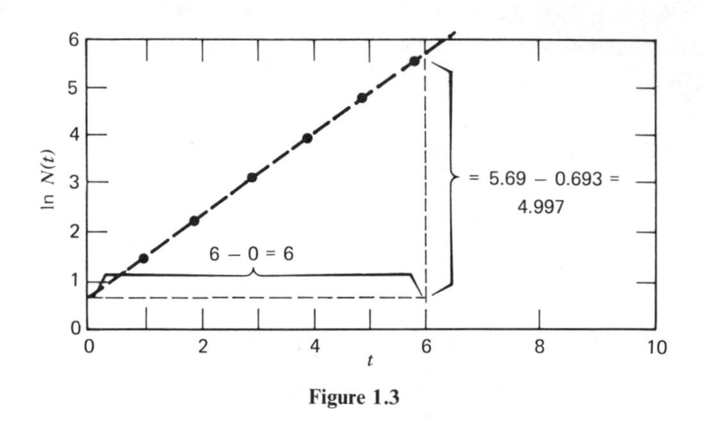

Figure 1.3

Because the equation that describes these data is a straight line [$\ln N(t) = \ln N(0) + rt$], the slope of the line will be equal to r. Thus

$$r = \frac{4.997}{6} = 0.833$$

which agrees with our original r, as given.

10 $N(t) = 10e^{0.993t}$. We want to know how long it will take the population to reach 20; that is, we need to solve for t^* in the equation $20 = 10e^{0.993t^*}$. Thus

$$t^* = \frac{\ln 20 - \ln 10}{0.993} = \frac{2.896 - 2.302}{0.993} = 0.698$$

Therefore the doubling time = 0.698 (days, years, or whatever the unit time is). If $N(0) = 20$,

$$40 = 20e^{0.993t^*}$$

$$t^* = \frac{\ln 40 - \ln 20}{0.993} = \frac{3.689 - 2.996}{0.993} = 0.698$$

Therefore the doubling time is the same as when we started with 10.

11 For doubling time we want to know the value of t that gives $N(t + 1) = 2N(0)$. So

$$2N(0) = N(0)e^{rt^*}$$

$$2 = e^{rt^*}$$

$$\ln 2 = rt^*$$

$$t^* = \frac{\ln 2}{r} = \frac{0.693}{r}$$

For tripling time, similarly,

$$3N(0) = N(0)e^{rt^*} \qquad 3 = e^{rt^*}$$

$$\ln 3 = rt^*$$

$$t^* = \frac{\ln 3}{r} = \frac{1.094}{r}$$

12 A single individual dividing twice a day shows a pattern like Figure 1.4:

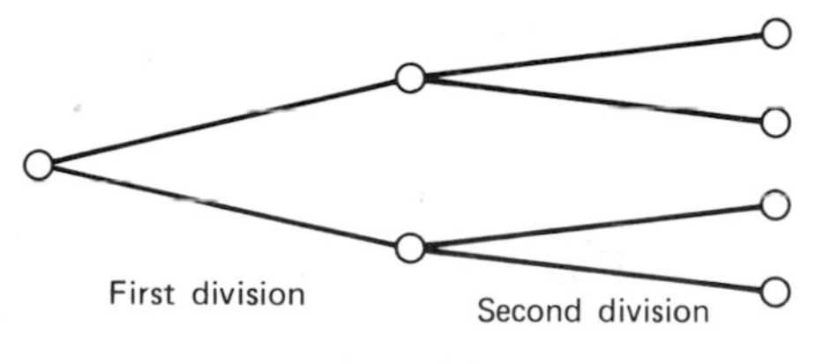

Figure 1.4

In general it will produce the following series of numbers: 1, 4, 16, 64, etc.; that is,

$$N(t) = 4^t = e^{(1.39)t}$$

Thus $r = 1.39$. Because it was stipulated that $d = 0$ and $r = b - d$, it must be that $b = 1.39$. More approximately, but perhaps a bit more intuitively, we note that in one day we went from one to four paramecia. That was a change of three. In going from one individual to four individuals, what was the average population density during that day? Clearly, that average could be approximated (*only* approximated) by $(1 + 4)/2$, which is 2.5. Thus during that day the population was at an average density (more or less) of 2.5 and the number of new individuals produced was three; therefore the per capita rate is $3/2.5 = 1.2$, which is close but of course not equal to the correct figure of 1.39.

13 Instead of the pattern of one to four let us start with two individuals:

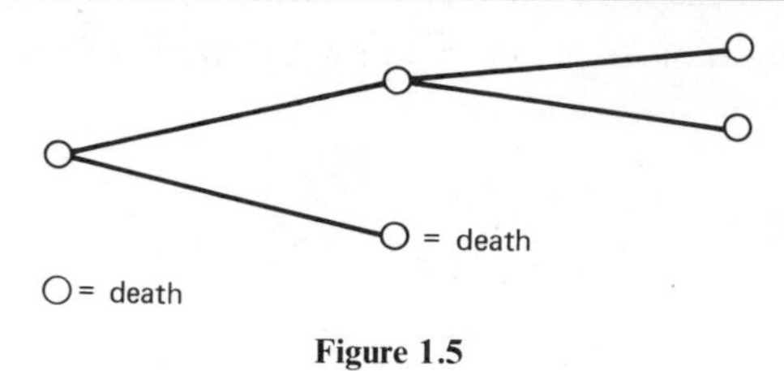

Figure 1.5

In one day two individuals will produce two individuals:

$$N(t) = N(0)e^r$$

$$2 = 2e^r$$

$$1 = e^r$$

$$\ln 1 = r$$

$$r = 0$$

We already know that the birthrate is 1.39; therefore the death rate must also be 1.39 (because $r = b - d$).

14 Begin with eight individuals (you could begin with any number);

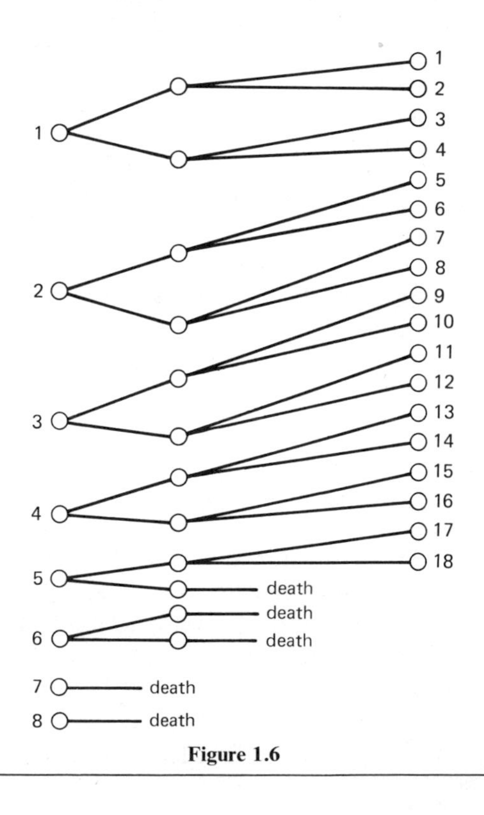

Figure 1.6

Eight will produce 18 in one day and we have

$$18 = 8e^r$$

$$\ln\left(\frac{18}{8}\right) = r$$

$$\ln(2.25) = r$$

$$r = 0.81093$$

We know that $b = 1.39$. Therefore

$$d = b - r = 1.39 - 0.81093 = 0.579$$

As in exercise 12, we can compute the death rate approximately by noting that five individuals died during a time when the population ranged from eight to 18. Therefore $N(t) \approx (8 + 18)/2 = 13$, and $d \approx \frac{5}{13} = 0.385$, again close but not equal to 0.579.

15 $dN/dt = N(r - aN) = Nr - aN^2$; $0 = Nr - aN^2$. Clearly this is a quadratic equation. One root is obviously $N = 0$. The other is $r - aN = 0$; $N = r/a$.

16 $K = r/a$ (from exercise 15).

$$\frac{dN}{dt} = N(r - aN) = Na\left(\frac{r}{a} - N\right) = Na(K - N)$$

but $a = r/K$; therefore

$$\frac{dN}{dt} = N\frac{r}{K}(K - N) = rN\left(\frac{K - N}{K}\right)$$

17 If there are N individuals in the population, the number of individuals not realized is $K - N$. The fraction not realized is then $(K - N)/K$. If that fraction is to be directly proportional to the per capita growth rate (the assumption given in the statement of the problem), we obtain

$$\frac{dN}{N\,dt} = r\left(\frac{K - N}{K}\right)$$

or

$$\frac{dN}{dt} = rN\left(\frac{K - N}{K}\right)$$

Therefore

$$f(N) = r\left(\frac{K - N}{K}\right)$$

18
$$\frac{dN}{dt} = rN\left(\frac{K - N}{K}\right) = r\left(N - \frac{1}{K}N^2\right)$$

$$\int \frac{dN}{N[1 - (1/K)N]} = \int r\,dt$$

From the table of integrals we have

$$\int \frac{dx}{x(a + bx)} = \frac{-1}{a}\ln\frac{a + bx}{x}$$

If $x = N, a = 1, b = -1/K$,

$$\int \frac{dN}{N[1 - (1/K)N]} = -\frac{1}{1}\ln\frac{1(1/K)N}{N} = rt + c$$

$$\frac{1 - (1/K)N}{N} = ce^{-rt}$$

19 Begin with equation 10;

$$N(t) = \frac{K}{1 + [(K - N_0)/N_0]\lambda^{-t}}$$

$$N(t)\left(1 + \frac{K - N_0}{N_0}\lambda^{-t}\right) = K$$

$$K - N(t) = N(t)\frac{K - N_0}{N_0}\lambda^{-t}$$

$$\ln\left[\left(\frac{K - N(t)}{N(t)}\right)\right] = \ln\left[\left(\frac{K - N_0}{N_0}\right)\right] - t\ln\lambda$$

and because $\lambda = e^r$ we have

$$\ln\left[\frac{K - N(t)}{N(t)}\right] = \ln\left[\frac{K - N_0}{N_0}\right] - rt$$

20

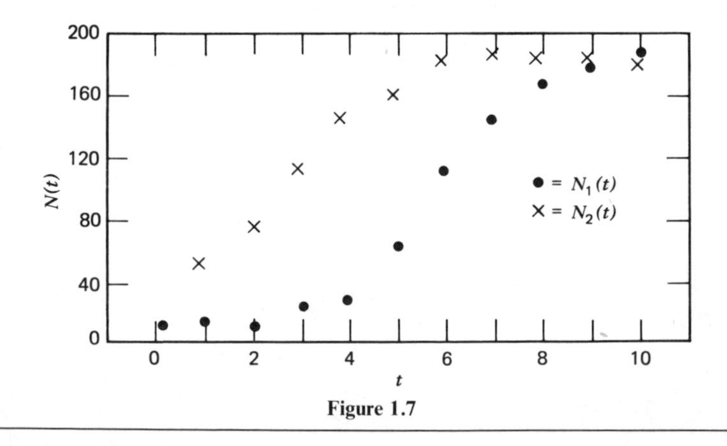

Figure 1.7

It is difficult to determine which population behaves more like the logistic equation from an examination of Figure 1.7. We know from exercise 19 that $\ln[(K - N)/N]$ is linearly related to t. Thus, if we plot $\ln[(K - N)/N]$ against t for successive values of $N(t)$ and t, we should obtain a straight line with slope $-r$. It is most convenient to set up the following table:

t	$N_1(t)$	$(K-N)/N$*	$\ln(K-N)/N$	t	$N_2(t)$	$(K-N)/N$	$\ln(K-N)/N$
0	10	17.0	2.83	0	30	5.0	1.61
1	11	15.4	2.73	1	50	2.6	0.96
2	10	17.0	2.83	2	75	1.4	0.34
3	20	8.0	2.08	3	110	0.64	−0.45
4	25	6.2	1.82	4	145	0.24	−1.43
5	60	2.0	0.69	5	170	0.06	−2.83
6	110	0.64	−0.45	6	180	0	
7	140	0.29	−1.25	7	185	−0.03	
8	165	0.09	−2.40	8	180	0	
9	175	0.03	−3.56	9	180	0	
10	185	0.03		10	180		

* It is necessary to estimate K, which in this case is 180 for both populations, from a simple extrapolation on the graph.

Graphing $\ln[(K - N)/N]$ against t (on the table), we obtain

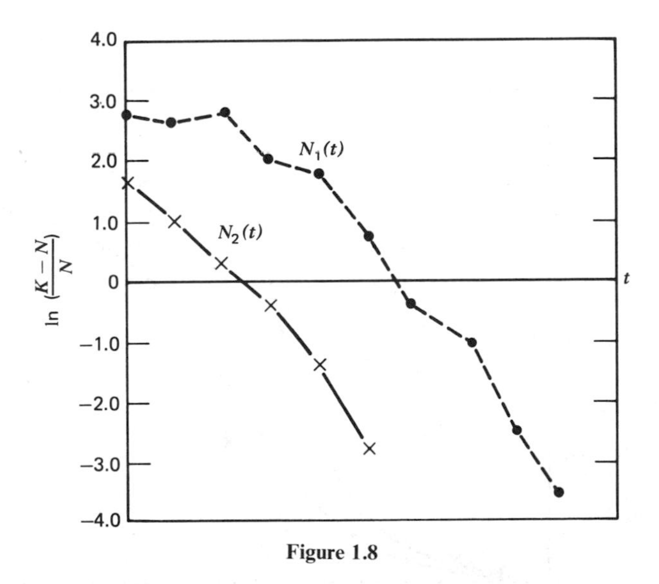

Figure 1.8

Clearly both are approximately linear, but $N_2(t)$ is much closer than $n_1(t)$ to being perfectly linear.

21 $e^{0.83}$ = antilog 0.83 = 2.3 = λ.

$$\frac{\lambda - 1}{K} = \frac{2.3 - 1}{302} = \frac{1.3}{302} = 0.0043$$

Set up the following table:

t	$N(t)$	$\lambda N(t)$	$1 + (0.0043)\,N(t)$	$N(t+1)$
0	2	4.6	1.0086	4.56
1	4.56	10.48	1.0196	10.28
2	10.28	23.64	1.0442	22.64
3	22.64	52.07	1.0973	47.45
4	47.45	109.14	1.2040	90.65
5	90.65	208.50	1.3898	150.02
6	150.02	345.05	1.6451	209.74
7	209.74	482.40	1.9019	253.64
8	253.64	583.37	2.0906	279.04
9	279.04	641.79	2.1999	291.74
10	291.74	671.00	2.2545	297.63
11	297.63	684.55	2.2798	300.27
12	300.27	690.62	2.2912	301.42
13	301.42	693.27	2.2961	301.93
14	301.93	694.44	2.2983	302.15
15	302.15			

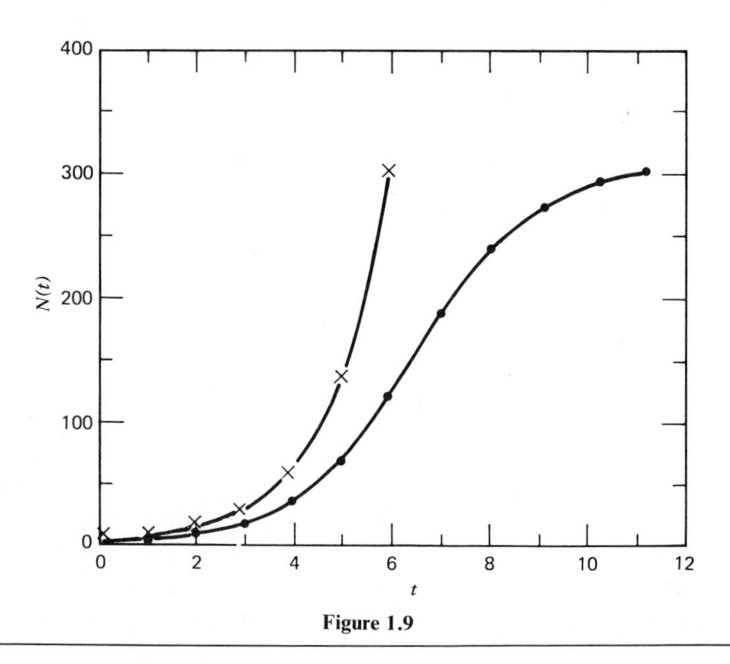

Figure 1.9

22 As in exercise 20, we set up the following table ($K = 300$):

t	$N(t)$	$\frac{K-N}{N}$	$\ln \frac{K-N}{N}$
0	2	149	5.00
1	4.56	64.79	4.17
2	10.28	28.18	3.34
3	22.64	12.25	2.50
4	47.45	5.32	1.67
5	90.65	2.31	0.84
6	150.02	1.00	0.00
7	209.74	0.43	−0.84
8	253.64	0.18	−1.70
9	279.04	0.08	−2.59
10	291.74	0.03	−3.56

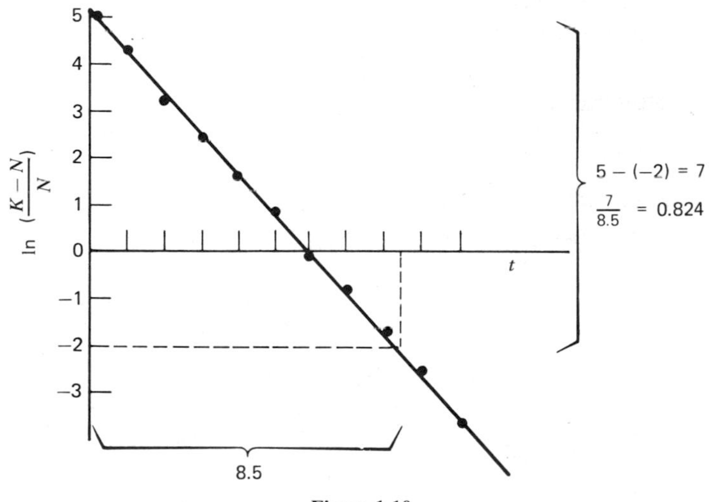

Figure 1.10

23

t	$N(t)$	$\lambda N(t)$	$N(t-1)$	$1 + (0.0043)\,N(t-1)$	$N(t+1)$
0	2				
1	4.6	10.58	2	1.0086	10.49
2	10.49	24.13	4.6	1.0198	23.66
3	23.66	54.42	10.49	1.0451	52.07
4	52.07	119.76	23.66	1.1017	108.71
5	108.71	248.63	52.07	1.2239	203.15
6	203.15	467.24	108.71	1.4675	318.40
7	318.40	732.32	203.15	1.8735	390.88
8	390.88	899.02	318.40	2.3691	379.48
9	379.48	872.80	390.88	2.6808	325.58
10	325.58	748.83	379.48	2.6318	284.53
11	284.53	654.42	325.58	2.4000	272.67
12	272.67	627.14	284.53	2.2235	282.05
13	282.05	648.72	272.67	2.1725	298.60
14	298.60	686.78	282.05	2.2128	310.37
15	310.37	713.85	298.60	2.2840	312.54
16	312.54	718.84	310.37	2.3346	307.91
17	307.91	708.19	312.54	2.3439	302.14
18	302.14	694.92	307.91	2.3240	299.02
19	299.02	687.75	302.14	2.2992	299.12
20	299.12				

24

	$\lambda = 2.3$				
t	$N(t)$	$\lambda N(t)$	$N(t-\Delta)$	$1 + (0.0043)\,N(t-2)$	$N(t+1)$
0	2				
1	4.6				
2	10.3	23.69	2	1.0086	23.48
3	23.48	54.00	4.6	1.0198	52.95
4	52.95	121.78	10.3	1.0443	116.62
5	116.62	268.23	23.48	1.1010	243.62
6	243.62	560.33	52.95	1.2277	456.40
7	456.40	1049.72	116.62	1.5015	699.11
8	699.11	1607.95	243.62	2.0476	785.28
9	785.28	1806.14	456.40	2.9625	609.67
10	609.67	1402.24	699.11	4.0062	350.02
11	350.02	805.05	785.28	4.3767	183.94
12	183.94	423.06	609.67	3.6216	116.82
13	116.82	268.69	350.02	2.5051	107.26
14	107.26	246.70	183.94	1.7909	137.75
15	137.75				

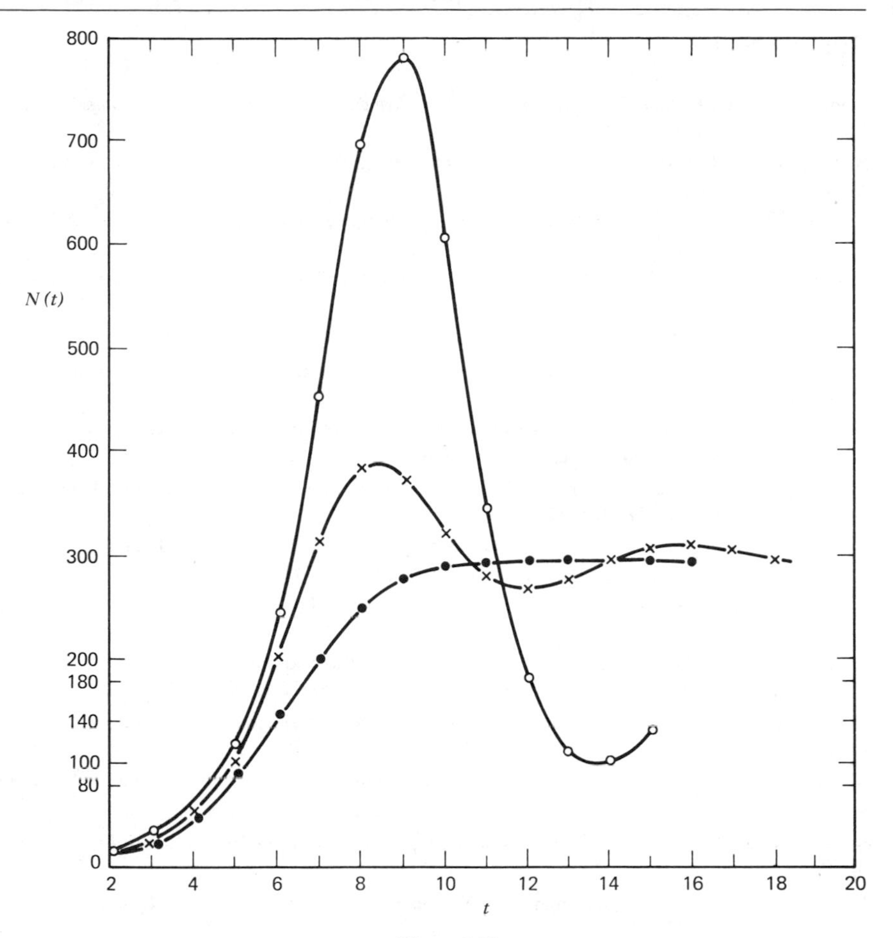
800
700
600
N (t)
500
400
300
200
180
140
100
80
0
2
4
6
8
10
12
14
16
18
20
t

Figure 1.11

REFERENCES

Andrewartha, H. G. and L. C. Birch. 1954. *The distribution and abundance of animals.* 782 pp. Chicago: The University of Chicago Press.

Beddington, J. R. and R. M. May. 1975. Time delays are not necessarily destabilizing. *Math. Biosci.* **27**:109–117.

Cunningham, W. J. 1954. A nonlinear differential-difference equation of growth. *Proc. Nat. Acad. Sci. USA* **40**:708–713.

Ehrlich, P. R. and L. C. Birch. 1967. The "balance of nature" and "population control." *Am. Nat.* **101**:97–107.

Gause, G. F. 1934. *The struggle for existence.* Baltimore: Williams and Wilkins.

Levins, R. 1968. *Evolution in changing environments. Some theoretical explorations.* Princeton, NJ: Princeton University Press.

MacArthur, R. H. 1968. Theory of the niche. R. C. Lewontin, Ed. *Population Biology and Evolution.* Syracuse, NY: Syracuse University Press, pp. 159–176.

MacArthur, R. H. and R. Levins. 1967. The limiting similarity, convergence and divergence of coexisting species. *Am. Nat.* **101**:377–385.

May, R. M., G. R. Conway, M. P. Hassell, and T. R. E. Southwood. 1974. Time delays, density dependence, and single species oscillations. *J. Anim. Ecol.* **43**:747–770.

Nicholson, A. J. and V. A. Bailey. 1935. The balance of animal populations. Part I. *Proc. Zool. Soc. London.* 551–598.

Pearl, R. and L. J. Reed. 1920. On the rate of growth of the population of the United States since 1790 and its mathematical representation. *Proc. Nat. Acad. Sci. USA* **6**:275–288.

Slobodkin, L. B., F. E. Smith, and N. G. Hairston. 1967. Regulation in terrestrial ecosystems and the implied balance of nature. *Am. Nat.* **101**:109–124.

Smith, F. E. 1952. Experimental methods in population dynamics: a critique. *Ecology* **33**:441–450.

Solomon, M. E. 1949. The natural control of animal populations. *J. Anim. Ecol.* **18**:1–35.

Southwood, T. R. E. 1975. The dynamics of insect populations. In D. Pimentel, Ed. *Insects, Science and Society.* New York: Academic, pp. 151–199.

Wangersky, P. J. and W. J. Cunningham. 1957. Time lag in population models, *Cold Spring Harbor Symp. Quant. Biol.* **22**:329–338.

Utida, S. 1967. Damped oscillation of population density at equilibrium. *Res. Popl. Ecol.* **9**:1–9.

Vandermeer, J. H. 1969. The competitive structure of communities: an experimental approach using protozoa. *Ecology* **50**(3):362–371.

Vandermeer, J. H. 1970. The community matrix and the number of species in a community. *Am. Nat.* **104** (935):73–83.

2. Population Projection Matrices

In Chapter 1 we considered the process of population growth from a restricted and biologically oversimplified point of view. In particular, it was assumed that all the individuals in the population were identical. Careful documentation is hardly necessary to verify the restrictiveness of this assumption.

In pursuit of a more reasonable set of assumptions let us ask in what way do individuals in a population differ most typically from one another in terms of mortality and fecundity. If our prejudice is zoological rather than botanical, we recognize age as an important variable. A one-year-old infant has a different likelihood of dying or reproducing—in the next five years (say)—than a 25-year-old. Age is not the only correlate of varying mortalities and fecundities, but in many animal populations it accounts for a great deal of the individual-to-individual variation.

Thus it is natural to expand the approach used in Chapter 1 to include the effects of age on mortality and fecundity. This is not to imply that, in fact, age is the only important determinant of these parameters. Rather, because age is *frequently* the most important variable, it makes sense to consider the basic processes of population growth in which the population is composed of individuals that differ in their mortalities and fecundities and in which such differences are totally manifest in age differences. In other words, every female x years old produces exactly $g(x)$ offspring during an arbitrarily specified time period [e.g., every female between 20 and 25 will have produced (say) 1.5 newborns between 1975 and 1980].

A small amount of matrix algebra, which can be obtained from any book on elementary matrices (e.g., Searle, 1966), is necessary to an understanding of the material in this chapter. I assume in what follows that the reader is familiar with the matrix operations of addition, subtraction, and multiplication and that he or she has an intuitive notion of the meaning of the inverse of a matrix.

The classic papers on the population-projection matrix are one by Lewis (1942) and two by Leslie (1945, 1948). A later paper by Goodman (1968), which this chapter largely follows, presents an elementary view of the projection matrix. Good accounts are also provided by Keyfitz (1968) and Pielou (1969).

BASIC FORMULATION. Consider only the female part of the population. Let $N(x, t)$ be the number of females between ages x and $x + 1$ who are alive at time t (e.g., if the number of females between ages 20 and 21 years at time 1976 is equal to 500, $N(20, 1976) = 500$). Let $a(x)$ be the *proportion* of females between age x and age $x + 1$ that survive through one time interval to enter the class $x + 1$ to $x + 2$; that is,

$$a(x) = \frac{N(x + 1, t + 1)}{N(x, t)}$$

For reasons that will become obvious later $a(x)$ is frequently referred to as a "transition probability." Let $g(x)$ be equal to the number of daughters per live female in the x to $x + 1$ age class, born during the time interval t to $t + 1$, who will be alive in the zeroth age interval ($x = 0$) at time $t + 1$ (e.g., if a female betweeen 20 and 25 at time 1975 produces 1.5 offspring during the years 1975 to 1980, $g(20) = 1.5$: remember that all females of a given age produce the same number of offspring). Assume, to begin with, that $a(x)$ and $g(x)$ are invarying through time and independent of population size.

We may now write the two basic equations from which almost everything in this chapter follows. The number of females aged between 0 and 1 extant in the population at time t must be,

$$N(0, t) = \sum_{x=0}^{x_{max}} g(x)\, N(x, t - 1) \tag{1}$$

that is, the total number of daughters born one time unit ago, in which x_{max} is the oldest an individual can get. The number of females in each age category is given as

$$N(x + 1, t) = a(x)\, N(x, t - 1). \tag{2}$$

Equations 1 and 2 are crucial to the rest of this chapter. Be sure you understand them at a gut level before proceeding.

☐ EXERCISES

1. A certain species of snouter has a maximum life span of four years (snouters are parthenogenetic). Two-year-old snouters (snouters $2 \leq x < 3$ in age) produce two babies per year, and three-year-old snouters produce three babies per year. In 1918 there were 50 baby snouters (babies are between ages 0 and 1), 10 snouters between ages 1 and 2, five between ages 2 and 3, and two between ages 3 and 4. What are the values of $N(x, t)$ for $x = 0$, 1, 2, 3, and $t = 1918$? What are the values of $g(x)$ for $x = 0, 1, 2, 3$?
2. If one were to say that $a(x)$ for $x = 0$ for the example in exercise 1 was equal to 10/50, one would not necessarily be correct. Right? Explain.
3. For the above population of snouters the number of babies in 1919 was 16, the number of one-year-olds was 15, the number of two-year-olds was five, and the number of three-year-olds was two. What are the values of $a(x)$ for $x = 0, 1, 2, 3$?

4 How many snouters in each age category would you expect in 1920?

5 Consider a time interval of two years instead of one year; that is, each increment in t and x is two years (if $x = 0, x + 1 = 2$ years). For the same snouter population what are the values of $N(x, 1918)$ for $x = 0, 1$?

6 For a two-year time interval what would the value of $a(0)$ and $a(1)$ be for the same snouter population?

7 In a human population the number of 20-year-old females at time 0 (arbitrarily call some year zero) is 5000, the number of 21-year-olds is 4900, the number of 22-year-olds is 4800, and the number of 23-year-olds is 4600. At time 2 the number of 22-year-olds is 4700, the number of 23-year-olds is 4500, the number of 24-year-olds is 4100, and the number of 25-year-olds is 4000. At time 4 the number of 24-year-olds is 4000, the number of 25-year-olds is 3800, the number of 26-year-olds is 3700, the number of 27-year-olds is 3500, and the number of 28-year-olds is 3300. What is $a(20)$ for a two-year projection interval? What is $a(20)$ for a four-year projection interval? □

Equations 1 and 2 can be written in matrix form as

$$n_t = Mn_{t-1} \tag{3}$$

where n_t is a column vector, called the "age-distribution vector,"

$$n_t = \begin{vmatrix} N(0, t) \\ N(1, t) \\ \vdots \\ N(m, t) \end{vmatrix}$$

and M is the "population-projection matrix"

$$M = \begin{vmatrix} g(0) & g(1) & \cdots & g(m-1) & g(m) \\ a(0) & 0 & \cdots & 0 & 0 \\ 0 & a(1) & \cdots & 0 & 0 \\ 0 & \cdot & & \cdot & \cdot \\ \cdot & \cdot & & \cdot & \cdot \\ 0 & 0 & \cdots & a(m-1) & 0 \end{vmatrix}$$

Equation 3 is nothing more than another way of writing equations 1 and 2. M is called the population projection matrix because it projects the population from one point in time to the next point. Note that we project the population into the future, or forward. It is also possible to derive equations to ask what the population looked like in the past or to "project" the population backward (see Goodman, 1968).

□ EXERCISES

8 Consider a population with three age categories. Show that $n_t = Mn_{t-1}$ is the same as equations 1 and 2.

9 Show the projection matrix for the snouter example in the preceding set of exercises.

10 For the snouters in the examples given what are n_{1918}, n_{1919}, and n_{1920}?

11 Use the projection matrix to obtain n_{1921} for the snouters. □

We have already seen that to project a population one time interval into the future we must perform the operation $n_{t+1} = Mn_t$. Specifically, if we begin at $t = 0$, $n_1 = Mn_0$. If we begin at $t = 1$, $n_2 = Mn_1$. If we begin at $t = 2$, $n_3 = Mn_2$. If the n vector represents the same population, we can begin substituting. So, recalling that $n_2 = Mn_1$, we may write $n_3 = M(Mn_1)$. Further, recalling that $n_1 = Mn_0$, we may write $n_3 = M(M(Mn_0))$, which may be shortened to read $n_3 = M^3 n_0$. In general, we have $n_t = M^t n_0$.

STABLE AGE DISTRIBUTION. We now wish to examine the behavior of a model population as it is projected into the future. In particular, we wish to determine how the proportional representation in each age class changes, the percentage of the total population that exists in a given class, and how that percentage changes over time. If we let $\hat{N}(t)$ symbolize the total number of individuals in the entire population $[\hat{N}(t) = \sum_{x=0}^{m} N(x, t)]$ at time t, we can examine $B_x(t)$, where $B_x(t) = [N(x, t)/\hat{N}(t)]$. Exactly how will $B_x(t)$ behave as we apply the projection matrix repeatedly to the age-distribution vector?

□ EXERCISES

12 Consider a model population in which $g(0) = 0$, $g(1) = 2$, $g(2) = 3$, $a(0) = 0.5$, and $a(1) = 0.2$. Begin with the following age-distribution vector

$$\begin{vmatrix} 5 \\ 10 \\ 20 \end{vmatrix}$$

and project the population 10 times. Compute the values of $B_x(t)$ for values of $x = 0$, 1, 2 and values of $t = 0 - 10$. Plot $B_0(t)$ against time, $B_1(t)$ against time, and $B_2(t)$ against time. Plot the total number of individuals $\hat{N}(t)$ against time. □

As long as the elements of $M(g(x), a(x)$ for all x) remain constant, the proportional representation of any age class will tend to be constant as the projection matrix is repeatedly applied. (This is not *strictly* true in a mathematical sense, but for most biological systems it is true; that is, $B_x(t) \approx B_x(t+1)$ if t is very large in the equation $n_t = M^t n_0$.) The remarkable fact is that it doesn't matter at all what the initial age-distribution vector is. Any initial vector will generate exactly the same $B_x(t)$ if the projection matrix is applied enough times. When $B_x(t)$ is equal to $B_x(t+1)$ (and, of course, $B_x(t) = B_x(t+\varepsilon)$, where ε is any positive integer), we give it a special name and symbol; that is $B_x = c_x$ and c_x is called the *stable age distribution.* This is an extremely important concept in demography. I have tried to show it intuitively (in the last exercise), but it is a provable theorem [see Pielou (1969) or Leslie (1945)]. Rather than labor on an exact proof, I ask the reader to accept as fact that, given a constant projection matrix, any initial-age distribution vector (n_t) will tend toward stable age distribution (the stable vector). It must be emphasized that it is the proportional representation of each age category that remains constant. The total numbers may be increasing, remaining constant, or decreasing. The stable age distribution refers only to the constancy of the proportion of the total population to be found in a given age category.

THE INTRINSIC RATE OF NATURAL INCREASE. We know that after the stable age distribution has been attained the proportional representation of each age category remains constant through time; that is, if the population has attained a stable age distribution,

$$B_x(t) = B_x(t+1)$$

which can be rewritten

$$\frac{N(x, t)}{\hat{N}(t)} = \frac{N(x, t+1)}{\hat{N}(t+1)}$$

By algebraic rearrangement we obtain,

$$\frac{\hat{N}(t+1)}{\hat{N}(t)} = \frac{N(x, t+1)}{N(x, t)}$$

where every age category is increasing (or decreasing) in numbers at a rate that is constant and equal to the rate at which the entire population is increasing (or decreasing). Let that constant be symbolized by λ; that is,

$$\frac{\hat{N}(t+1)}{\hat{N}(t)} = \frac{N(x, t+1)}{N(x, t)} = \lambda$$

Thus we may write

$$N(x, t+1) = \lambda N(x, t)$$

or in matrix notation

$$n_{t+1} = Mn_t = \lambda n_t \tag{4}$$

(remember λ is a scalar). In a sense, after the stable age distribution has been reached, the constant λ takes the place of the projection matrix.

□ EXERCISES

13 Repeat exercise 12, beginning with the age-distribution vector

$$\begin{vmatrix} 23 \\ 10 \\ 2 \end{vmatrix}.$$

14 Is the age-distribution vector in exercise 13 the stable age distribution?

15 Estimate the value of λ, using only that which has been developed in the text so far. □

It is important to realize that because the stable age distribution will be reached regardless of the initial age-distribution vector λ is determined *only* by the projection matrix. The constant λ is called the dominant eigenvalue of the matrix M (or the dominant latent root of the matrix M).

Going back to equation 4, we may write the following series of equations (assuming that we begin at a stable age distribution):

$$n_1 = \lambda n_0$$

$$n_2 = \lambda n_1$$

$$n_3 = \lambda n_2$$

which may be combined algebraically to give

$$n_3 = \lambda^3 n_0$$

Generalizing, we have

$$n_t = \lambda^t n_0 \tag{5}$$

again, assuming that we begin with a stable age distribution. If equation 5 is valid, we can write,

$$\hat{N}(t) = \lambda^t \hat{N}(0) \tag{6}$$

Equation 6 is identical to an equation in Chapter 1. Because λ is a constant, we can express it however we wish as virtually any combination of constants. In particular, let us express λ as Euler's constant raised to a power (here, as in

Chapter 1, Euler's constant is chosen for convenience—any other constant could have been chosen as well); that is, let $\lambda = e^r$. Then (6) becomes

$$\hat{N}(t) = e^{rt}\hat{N}(0)$$

which will be recognized as the integrated form of the simple differential equation

$$\frac{d\hat{N}}{dt} = r\hat{N}$$

We discussed this equation in detail in Chapter 1 as the exponential equation.

Thus we obtain a fundamental result. Any population that has achieved a stable age distribution—which must happen to any population that has a projection matrix with constant elements—will behave according to the exponential equation. The parameter of that equation (r) is called (as before) the intrinsic rate of natural increase.

□ EXERCISES

16 For the values of $\hat{N}(t)$ from exercise 12 plot $\ln\hat{N}(t)$ against t.

17 For the values of $\hat{N}(t)$ from exercise 13 plot $\ln\hat{N}(t)$ against t.

18 Write the linearized form of equation 6 which describes the data as plotted in exercise 17. Estimate the value of λ for the data in exercise 17. □

THE STAGE PROJECTION MATRIX. The development just discussed hinged on the supposition that all variability in fecundity and survival is attributable to interindividual age differences. For long-lived organisms this approach has frequently been impossible; for example, an oak tree, which as an adult may be 300 years old, achieves most of its significant morality between the ages of zero and one year (the seed). Consequently, to capture the better part of the organism's biology in a population model one must choose age categories of one year length if the standard demographic approach is used. Thus a model of the oak population would be a 300 × 300 population projection matrix. Such a matrix is unwieldy.

When the individuals in the population cannot be aged in the first place, the standard approach is similarly useless. Largely in response to this problem of not being able to age individuals in most populations Lefkovitch (1965a) proposed an alternate model, analytically similar to the Lewis–Leslie model but applicable to a broader spectrum of biological situations. Rather than categorizing individuals by age, this approach arranges them according to some arbitrary morphological or physiological feature; for example, instead of defining stage 1 as composed of all individuals between the ages of 0 and 5 (say), we define stage 1 as all first instar larvae or all seedlings between 0 and 20 cm in height. The Lefkovich approach, hereafter referred to as the stage projection matrix, is particularly useful for populations of long-lived individuals such as most perennial plant species.

Therefore choose an arbitrary categorization variable (e.g., height of plant, instar, or weight) and divide the individuals in the population into categories based on that variable. Like the age distributed population, we assume that all variability in survivorship and fecundity in the "stage" distributed population is due solely to the categorization variable.

Let the number of individuals in the xth stage category at time t be $N(x, t)$. We are concerned with the likelihood that an individual found in the xth stage now will be found in some other stage at the beginning of the next time interval. Let the probability that an individual will change from stage y to stage x during a time interval be P_{xy}. Let the number of offspring produced by an individual in stage x be $g(x)$. Assuming that all newborns are born into the 0th stage category, we have the following equations which parallel equations 1 and 2 for the age-distributed population:

$$N(0, t + 1) = \sum_{x=0}^{m} g(x)\, N(x, t) \qquad (7)$$

and

$$N(x, t + 1) = \sum_{i=0}^{m} P_{xi} N(i, t) \qquad (8)$$

☐ EXERCISES

19 Bristleberry trees occur in four distinguishable forms: seed, seedling, small tree, large tree. In 1974 there were 5000 viable seeds, 500 seedlings, 100 small trees, and 10 large trees. In 1975 there were 6000 seeds, 500 seedlings, 100 small trees, and 11 large trees. Let $x = 0$ represent seeds, $x = 1$, seedlings, $x = 2$, small trees, and $x = 3$, large trees. What are the values for $N(x, 1975)$, $x = 0, 1, 2, 3$? What are the values for $N(x, 1974)$, $x = 0, 1, 2, 3$.

20 Can you estimate any of the P_{xy} values from the data given in exercise 19? If so, which? What are their values? If not, why not?

21 In 1974 in a random sample of 2000 seeds from this bristleberry population each seed was marked. By repeatedly sampling the seeds we know that 20 of them germinated by 1975; the rest rotted. Also 100 randomly selected seedlings were marked. By 1975 20 had died, five had become small trees, and the rest were still seedlings. In addition 50 small trees were marked and it was found that two had become large trees, five had died, and the rest were still small. All 11 large trees were marked; one died and the rest were still alive. Estimate P_{11}, P_{00}, P_{10}, P_{20}, P_{21}, P_{22}, P_{31}, P_{32}, and P_{33}. ☐

Equations 7 and 8, like equations 1 and 2, may be put in matrix form. Thus we have a stage-distribution vector

$$n'_t = \begin{vmatrix} N(0, t) \\ N(1, t) \\ \vdots \\ N(m, t) \end{vmatrix}$$

where the value of x in $N(x, t)$ refers to a stage category rather than an age. Further, we have a stage-projection matrix:

$$M' = \begin{vmatrix} g(0) & g(1) & \cdots & g(m-1) & g(m) \\ P_{21} & P_{22} & & P_{2,m-1} & P_{2m} \\ P_{31} & P_{32} & \cdots & P_{3,m-1} & P_{3m} \\ P_{m} & P_{m2} & & P_{m,m-1} & P_{mm} \end{vmatrix}$$

As in the age-projection matrix, we have

$$n'_t = M'n'_{t-1}$$

and on repeated application we obtain

$$n'_t = M'n'_{t-1} = \lambda n'_{t-1}$$

where n'_t is a stable *stage* distribution.

DENSITY DEPENDENCE IN PROJECTION MATRICES. To this point the analysis has followed closely that of Chapter 1, albeit with some seemingly simple modifications. We have allowed variability in survivorship and fecundity; that is, not all individuals in the population necessarily reproduce and die at the same rate. We first supposed that age was the variable that dictated the variability. We then generalized and allowed an arbitrary categorization variable to dictate the variability, but in both cases we merely added some complications to the general analysis in Chapter 1. In the end we still wind up with an exponentially growing population. A certain realism has been added in that we no longer assume that all individuals in a population are identical and we have introduced a new and perhaps unexpected concept, the stable-age distribution. The fundamental picture of an exponential pattern of growth remains the single important qualitative result of a density-*independent* model.

Thus it makes certain sense to follow the reasoning that introduced this chapter and assign density dependence to the age-distributed or stage-distributed model.

Qualitatively, at least three considerations must be taken into account when adding density dependence to an age- or stage-distributed population. First, does the density dependence act only on reproduction or survivorship or on some combination of the two (see Leslie, 1948)? Second, is density dependence mediated simply by total population density $N(t)$ or is the effect distributed differentially over the various age (stage) categories? Third, are density-dependent effects felt instantaneously or is there a time lag of some sort? We have already considered time lags, but the first two considerations were not even relevant in a population of constant individuals (Chapter 1).

In a general way we can write,

$$g(x) = f_x(n_t, n_{t-1}, \ldots, n_{t-z})$$

$$P_{ij} = h_{ij}(n_t, n_{t-1}, \ldots, n_{t-z})$$

which says that survivorships, growths, and fecundities are functions of everything between now and z time units in the past. Such a generalized formulation might be useful in a complete statement of the problems of density dependence or in a simulation of a real population (f and h are totally unspecified).

The two functions f and h are similar to the function $f(N)$ in the equation $dN/N\,dt = f(N)$ (Chapter 1). Just as $f(N)$ is a statement of the way in which population density feeds back on per capita birthrate, so f and h are statements of population density feedback on survivorship and reproduction. Recall that a particular form of $f(N)$ (Chapter 1) gave the logistic equation. Other forms were possible. Similarly, many forms are possible for f and h in this chapter.

☐ EXERCISES

22 It has been found that three-year-old snouters (see exercises 1–6) exert a competitive effect on all reproductive snouters. It turns out that each new three-year-old snouter depresses by a factor of $\frac{1}{2}$ the reproduction of both 2- and 3-year-olds. Express this biological condition as a mathematical model; that is, what is the explicit form of f_2 and f_3? (Use the figures in exercise 1.)

23 Using the survivorships in exercise 3, begin with the following age distribution vector

$$\begin{vmatrix} 10 \\ 5 \\ 1 \\ 0 \end{vmatrix}$$

and project the population 10 time intervals, using $g(2) = f_2$ and $g(3) = f_3$ as computed in exercise 22. At each projection round off the elements of the age-distribution vector to the nearest integer.

24 In the bristleberry population of exercises 19–21 it turns out that the large trees effectively shaded the small trees and seedlings. At a population level it has been determined that the probability of growing from a seedling to a small tree is reduced to one-third its previous value for each adult tree added to the population. Also, the probability of growing from a small tree to a large tree is reduced to one-half its previous value for each adult tree added to the population. We know from independent data that the probability of growing from seedling to small tree in the absence of any adult trees at all is 0.8 and the probability of growing from small tree to adult tree under those conditions is 0.5. Express this in a mathematical model.

25 Express f_2 and f_3 of exercise 22 and h_{21} and h_{32} of exercise 24, using Euler's constant.

26 Suppose that survivorship becomes reduced as a consequence of rising population density in an *age*-distributed population. $P_{ij} = a(j)$ is the survival probability at a low population density (theoretically $\hat{N} = 0$) and $P'_{ij}(t)$ is the new probability, adjusted for $\hat{N}(t)$. Express the difference between P and P' as a fraction of P'. Suppose that fraction is directly proportional to the total population density. Construct a mathematical model (i.e., express P'_{ij} as a function of P_{ij} and $\hat{N}$). □

Most highly specific and realistic formulations will be impossible to deal with analytically (i.e., the generalized formulations of f_x and h_{ij}). Usually one must go to computer simulation to study such models, but for heuristic purposes it is useful here to derive a somewhat less realistic, yet tractable model of density dependence in an age-distributed population.

Let us suppose that each element in the projection matrix changes as a function of the current size of the population and of the population as it existed at the beginning of the time interval in which individuals currently of age x were born; for example, the survivorship factor for animals in the third age class (between 3 and 4 years old) at time 5 (say) should be some function of the total population density at time 5 and the total population density at time 1 (the beginning of the second time interval). Making the simplest mathematical assumption, we suppose that the xth survivorship and fecundity factor $a(x)$ and $g(x)$ are multiplied by $1/q(x, t)$ where $q(x, t) = 1 + bN(t - x - 1) + aN(t)$, where a and b are constants independent of age. This formulation is simply an extension of that derived in exercise 26. Thus the projection matrix becomes a function of time, $M_t = MQ^{-1}$, where

$$Q_t = \begin{vmatrix} q(0, t) & 0 & \cdots & 0 \\ 0 & q(1, t) & & \\ \vdots & & & \\ 0 & & & q(m, t) \end{vmatrix}$$

[Note that the inverse of Q_t is a diagonal matrix with elements $1/q(x, t)$.] The basic matrix equation becomes $MQ_t^{-1}n_t = n_{t+1}$. It can be shown that repeated application of MQ_t^{-1} to the vector n will eventually reach a point at which the relative proportions of the elements in the n vector remain constant. At this point once again we say that the age distribution is stable. However, the addition of the multiplier Q_t^{-1} also means that eventually the population will reach a constant density from which it will no longer deviate; that is, eventually the total number of individuals in the population will reach a level K as in the logistic equation. At this point—constant population size and stable-age distribution—it is said that the population is at a "stationary age distribution."

Recalling the preceding analysis, we find that the eigenvalue of the matrix M is equal to the antilog of the intrinsic rate of natural increase ($\lambda = e^r$). From the density-dependent analysis, when $MO_t^{-1}n_t = n_t$, λ is equal to 1 because the population is at a stationary state.

Pielou (1969) presents an excellent analysis of this particular formulation [devised by Leslie (1959)]. In particular, after a stable age distribution is reached the population as a whole will grow according to the logistic equation [Leslie (1959)]. We can see how this model works by recalling that in general $Mn_t = \lambda n_t$ when we have a stable age distribution. We might also write $\lambda^{-1}Mn_t = n_t$. In a comparison of this equation with $MQ_t^{-1}n_t = n_t$ it is easy to see that, if the elements of Q are all λ, $MQ_t^{-1} = \lambda^{-1}M$ (remember Q is a diagonal matrix). (The λ we are referring to is the dominant eigenvalue of the matrix M, not of the matrix MQ_t^{-1}). Thus the population will be at a stable age distribution when all the elements of Q are equal to the dominant eigenvalue of the matrix M. Furthermore, the total size of the population remains constant through time; that is, if $MQ_t^{-1}n_t = n_t - \lambda^{-1}Mn_t$, $\hat{N}(t) = \hat{N}(t+1) = K$, where K is the carrying capacity, as defined in Chapter 2. Recall that $q_x = 1 + a\,N(t) + b\,N(t - x - 1)$. When $MQ_t^{-1} = \lambda^{-1}M, q_x = \lambda$. Therefore $\lambda = 1 + aK + bK$. On rearrangement we obtain

$$K = \frac{\lambda - 1}{a + b}$$

□ EXERCISES

27 If the survivorship and fecundity of individuals in age class x in an age-distributed population decreases in proportion to $\hat{N}$, that is, for each increment in $\hat{N}$ the survivorship and fecundity values are reduced by the factor α, what will be the value of K? (Use reasoning similar to that given in the paragraph immediately preceding this exercise.) □

ANSWERS TO EXERCISES

1 $N(0, 1918) = 50$, $N(1, 1918) = 10$, $N(2, 1918) = 5$, $N(3, 1918) = 2$, $g(0) = 0$, $g(1) = 0$, $g(2) = 2$, $g(3) = 3$.

2 Yes. The 10 snouters between ages 1 and 2 did not necessarily derive from 50 newborns.

3 $N(0, 1919) = 16, N(1, 1919) = 15, N(2, 1919) = 5, N(3, 1919) = 2.$

$$a(0) = \frac{N(1, 1919)}{N(0, 1918)} = \frac{15}{50} = 0.3$$

$$a(1) = \frac{N(2, 1919)}{N(1, 1918)} = \frac{5}{10} = 0.5$$

$$a(2) = \frac{N(3, 1919)}{N(2, 1918)} = \frac{2}{5} = 0.4$$

$$a(3) = \frac{N(4, 1919)}{N(3, 1918)} = \frac{0}{2} = 0$$ (because we have already stipulated that snouters never live beyond 4 years).

4 Equation 1 says $N(0, t) = \sum_{x=0} g(x)\, N(x, t-1)$. Therefore $N(0, 1920) = g(0)\, N(0, 1919) + g(1)\, N(1, 1919) + g(2)\, N(2, 1919) + g(3)\, N(3, 1919) = (0)(16) + (0)(15) + (2)(5) + (3)(2) = 16.$
Equation 2 says,

$$N(x + 1, t) = a(x)\, N(x, t - 1)$$

and for $x = 0, 1, 2$

$$N(1, 1920) = a(0)\, N(0, 1919) = (0.3)(16) = 4.8$$

$$N(2, 1920) = a(1)\, N(1, 1919) = (0.5)(15) = 7.5$$

$$N(3, 1920) = a(2)\, N(2, 1919) = (0.4)(5) = 2.0$$

5 $N(0, 1918) = 50 + 10 = 60, N(1, 1918) = 5 + 2 = 7.$

6 $a(0) = N(1, 1920) N(0, 1918) = (7.5 + 2.0) 60 = 0.158.$ $a(1) = N(2, 1920)$ $N(1, 1918) = 0.$

7 For a two-year projection $a(20) = (4700 + 4500)(5000 + 4900) = 0.929$. For a four-year projection $a(20) = (4000 + 3800 + 3700 + 3500)(5000 + 4900 + 4800 + 4600) = 0.777$.

8

$$n_t = \begin{vmatrix} N(0, t) \\ N(1, t) \\ N(2, t) \end{vmatrix} \qquad M = \begin{vmatrix} g(0) & g(1) & g(2) \\ a(0) & 0 & 0 \\ 0 & a(1) & 0 \end{vmatrix}$$

and

$$\begin{vmatrix} N(0, t) \\ N(1, t) \\ N(2, t) \end{vmatrix} = \begin{vmatrix} g(0) & g(1) & g(2) \\ a(0) & 0 & 0 \\ 0 & a(1) & 0 \end{vmatrix} \begin{vmatrix} N(0, t-1) \\ N(1, t-1) \\ N(2, t-1) \end{vmatrix}$$

Carrying out the multiplication on the right-hand side of the equation, we have for the first category

$$N(0, t) = g(0)\, N(0, t-1) + g(1)\, N(1, t-1) + g(2)\, N(2, t-1)$$
$$= \sum_{x=0} g(x)\, N(x, t-1)$$

which is equation 1. Similarly for the second and third categories

$$N(1, t) = a(0)\, N(0, t-1)$$
$$N(2, t) = a(1)\, N(1, t-1)$$

which is, in general,

$$N(x+1, t) = a(x)\, N(x, t-1) \text{ for } x = 0, 1, \ldots$$

which is equation 2.

9

$$\begin{vmatrix} 0 & 0 & 2 & 3 \\ 0.3 & 0 & 0 & 0 \\ 0 & 0.5 & 0 & 0 \\ 0 & 0 & 0.4 & 0 \end{vmatrix} = M$$

10

$$n_{1918} = \begin{vmatrix} 50 \\ 10 \\ 5 \\ 2 \end{vmatrix} \qquad n_{1919} = \begin{vmatrix} 16 \\ 15 \\ 5 \\ 2 \end{vmatrix} \qquad n_{1920} = \begin{vmatrix} 16.0 \\ 4.8 \\ 7.5 \\ 2.0 \end{vmatrix}$$

11

$$\begin{vmatrix} 0 & 0 & 2 & 3 \\ 0.3 & 0 & 0 & 0 \\ 0 & 0.5 & 0 & 0 \\ 0 & 0 & 0.4 & 0 \end{vmatrix} \begin{vmatrix} 16.0 \\ 4.8 \\ 7.5 \\ 2.0 \end{vmatrix} = \begin{vmatrix} (2)(7.5) + (3)(2.0) \\ (16)(0.3) \\ (4.8)(0.5) \\ (7.5)(0.4) \end{vmatrix} = \begin{vmatrix} 21.0 \\ 4.8 \\ 2.4 \\ 3.0 \end{vmatrix}$$

12

$$n_t = \underset{t=0}{\begin{vmatrix} 5 \\ 10 \\ 20 \end{vmatrix}} \underset{1}{\begin{vmatrix} 80 \\ 2.5 \\ 2 \end{vmatrix}} \underset{2}{\begin{vmatrix} 11 \\ 40 \\ 0.5 \end{vmatrix}} \underset{3}{\begin{vmatrix} 81.5 \\ 5.5 \\ 8 \end{vmatrix}} \underset{4}{\begin{vmatrix} 35 \\ 40.5 \\ 1.1 \end{vmatrix}} \underset{5}{\begin{vmatrix} 84.8 \\ 17.5 \\ 8.2 \end{vmatrix}} \underset{6}{\begin{vmatrix} 59.4 \\ 42.4 \\ 3.5 \end{vmatrix}} \underset{7}{\begin{vmatrix} 95.3 \\ 29.7 \\ 8.5 \end{vmatrix}} \underset{8}{\begin{vmatrix} 84.95 \\ 47.65 \\ 5.9 \end{vmatrix}}$$

$$\underset{t=9}{\begin{vmatrix} 113.0 \\ 42.5 \\ 9.5 \end{vmatrix}} \underset{t=10}{\begin{vmatrix} 113.5 \\ 56.5 \\ 8.5 \end{vmatrix}}$$

$B_0 =$	0.143	0.95	0.21	0.86	0.46	0.77	0.59	0.71	0.61	0.68	
$B_1 =$	0.29	0.03	0.78	0.06	0.53	0.16	0.42	0.22	0.34	0.26	
$B_2 =$	0.57	0.02	0.01	0.08	0.01	0.07	0.03	0.06	0.04	0.06	
$t =$	0	1	2	3	4	5	6	7	8	9	10
$\hat{N} =$	35	84	52	95	77	110	105	133	138	165	178

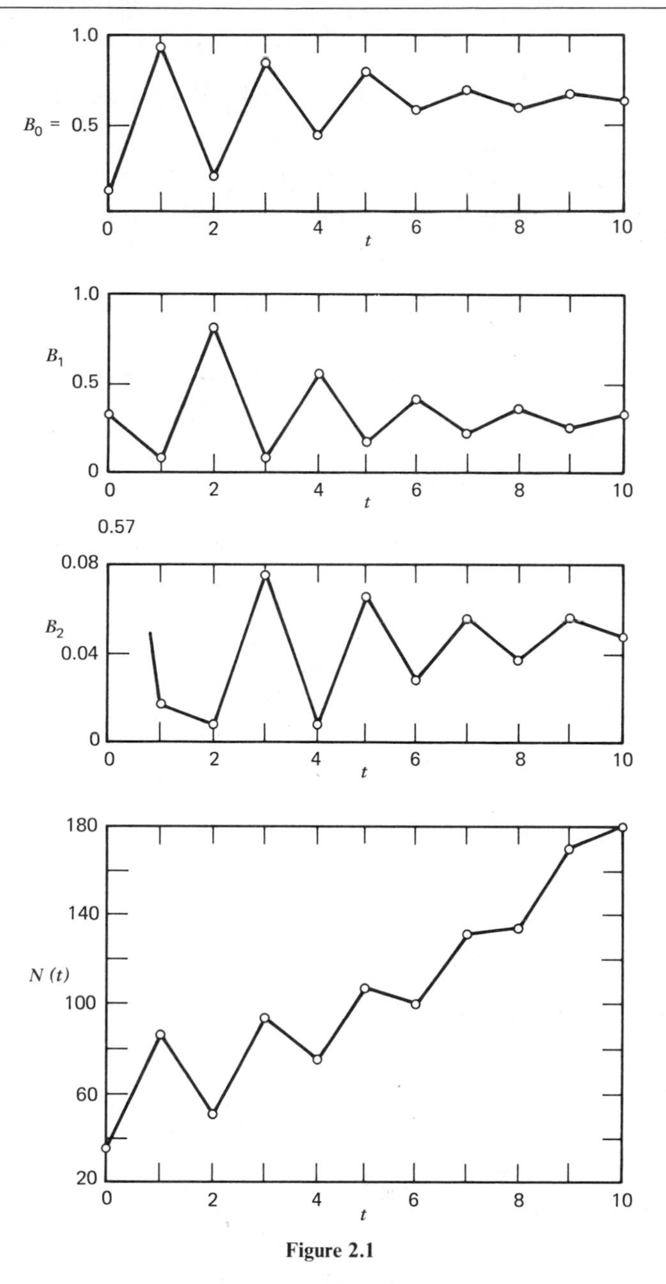

Figure 2.1

13

$$n_t = \begin{vmatrix} 23 \\ 10 \\ 2 \end{vmatrix} \begin{vmatrix} 26 \\ 12 \\ 2 \end{vmatrix} \begin{vmatrix} 30 \\ 13 \\ 2 \end{vmatrix} \begin{vmatrix} 32 \\ 15 \\ 3 \end{vmatrix} \begin{vmatrix} 39 \\ 16 \\ 3 \end{vmatrix} \begin{vmatrix} 41 \\ 20 \\ 3 \end{vmatrix} \begin{vmatrix} 49 \\ 20 \\ 4 \end{vmatrix} \begin{vmatrix} 52 \\ 24 \\ 4 \end{vmatrix} \begin{vmatrix} 60 \\ 26 \\ 5 \end{vmatrix} \begin{vmatrix} 67 \\ 30 \\ 5 \end{vmatrix} \begin{vmatrix} 75 \\ 34 \\ 6 \end{vmatrix}$$

$t =$	0	1	2	3	4	5	6	7	8	9	10
$\hat{N} =$	35	40	45	50	58	64	73	80	91	103	115

t	0	1	2	3	4	5	6	7	8	9	10
P_0	0.66	0.65	0.67	0.64	0.67	0.64	0.67	0.65	0.66	0.65	0.65
P_1	0.29	0.30	0.29	0.30	0.28	0.31	0.27	0.30	0.29	0.29	0.30
P_2	0.06	0.05	0.04	0.06	0.05	0.05	0.06	0.05	0.06	0.05	0.05

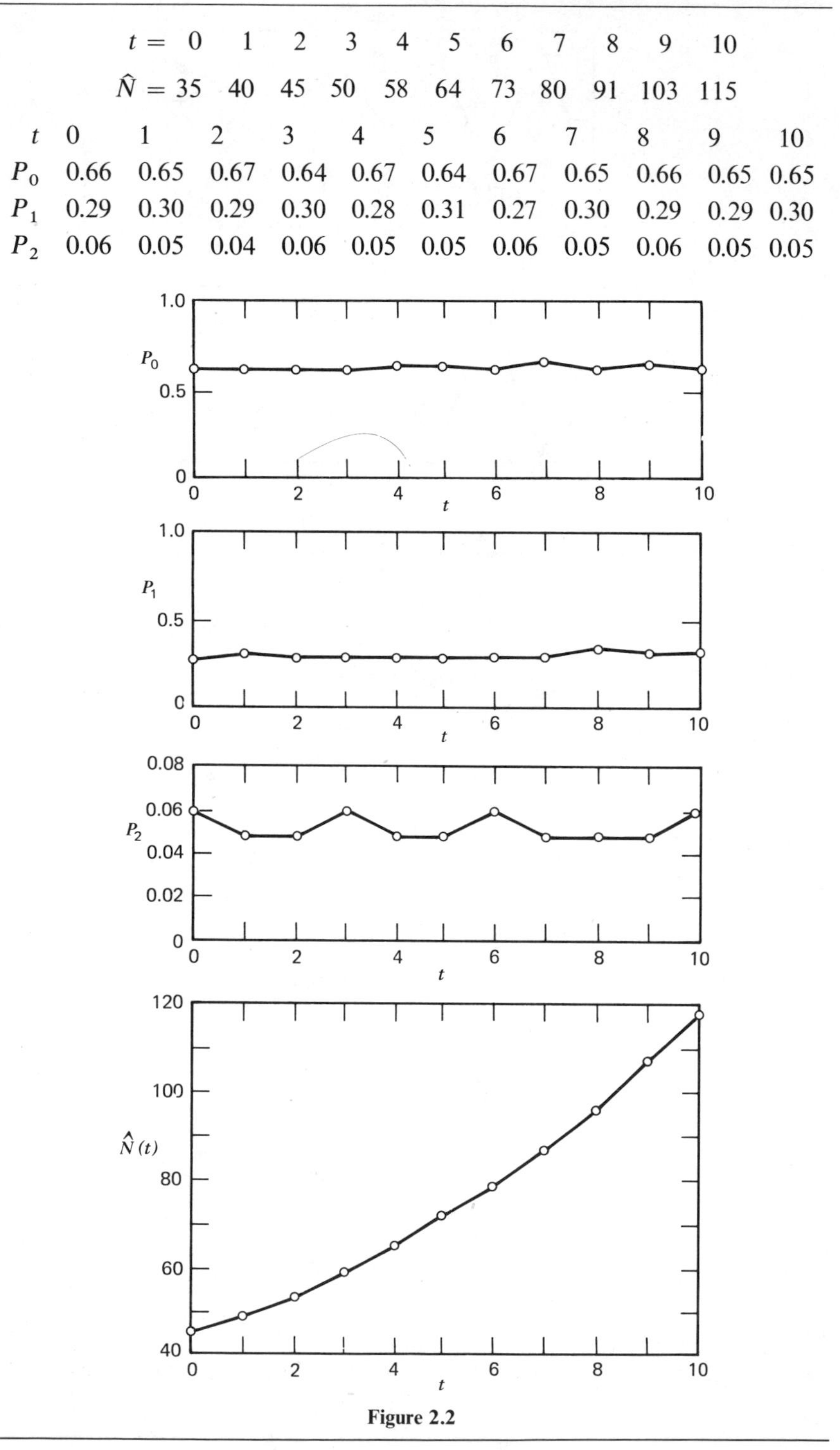

Figure 2.2

14 Yes. The variation in P_x is most likely due to round off error.

15 Because at the stable age distribution $Mn_t = \lambda n_t$ or $n_{t+1} = \lambda n_t$, let us take n_t for $t = 0, 1$ from the answer to exercise 13 and substitute into $n_1 = \lambda n_0$.

$$\begin{vmatrix} 26 \\ 12 \\ 2 \end{vmatrix} = \lambda \begin{vmatrix} 23 \\ 10 \\ 2 \end{vmatrix} \qquad \lambda = \frac{26}{23} = 1.13$$

$$\lambda = \frac{12}{10} = 1.2$$

$$\lambda = \frac{2}{2} = 1.0$$

Because of the round-off, this method of estimating λ is subject to error.

16

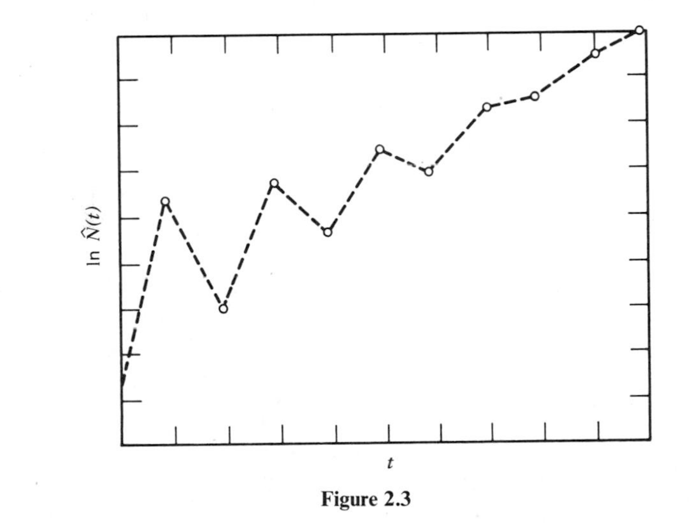

Figure 2.3

17 and **18**

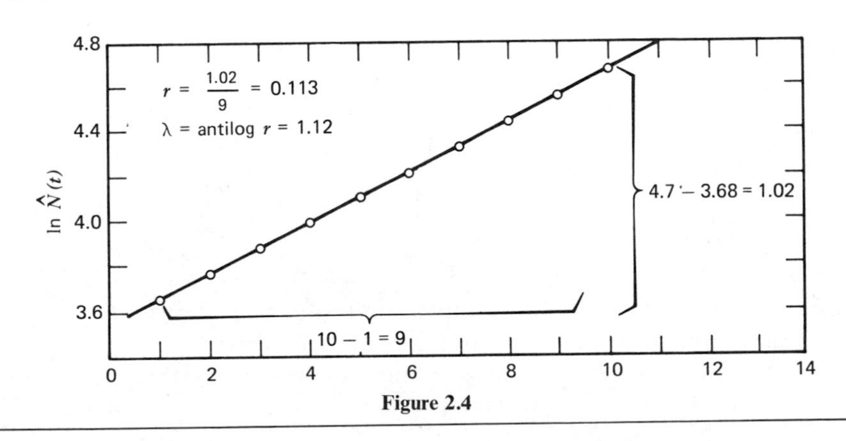

Figure 2.4

19 $N(0, 1975) = 6000$; $N(1, 1975) = 500$; $N(2, 1975) = 100$; $N(3, 1975) = 11$.

$N(0, 1974) = 5000$; $N(1, 1974) = 500$; $N(2, 1974) = 100$; $N(3, 1974) = 10$.

20 No. Impossible to estimate any of the P_{xy} values from these data unless you assume a stationary age distribution.

21 $P_{11} = \frac{75}{100} = 0.75$; $P_{21} = \frac{5}{100} = 0.05$; $P_{22} = \frac{43}{50} = 0.86$; $P_{32} = \frac{2}{50} = 0.04$; $P_{33} = \frac{10}{11} = 0.91$; $P_{10} = \frac{20}{2000} = 0.01$; $P_{31} = P_{20} = P_{00} = 0$.

22 The value of f_2 for $N(3, 1918) = 2$ is 2 [i.e., $g(2) = 2$ from exercise 1]. The value of f_3 for $N(3, 1918) = 2$ is 3 [i.e., $g(3) = 3$ from exercise 1]. Each new individual of age 3 multiplies both f_2 and f_3 by 0.5. Therefore, if we know that for $N(3, t) = 2$, $f_2 = 2$, it must be that for $N(3, t) = 1$, $f_2 = 4$ and for $N(3, t) = 0$, $f_2 = 8$. Similarly, for $N(3, t) = 1$, $f_3 = 6$, and for $N(3, t) = 0$, $f_3 = 12$. Clearly, in a population with zero individuals of age 3 $f_2 = 8$ and $f_3 = 12$. When $N(3, t) = 1$, $f_2 = 8(0.5)$ and $f_3 = 12(0.5)$. When $N(3, t) = 2$, $f_2 = 8(0.5)^2$ and $f_3 = 12(0.5)^2$, etc. In general, we have $f_2 = 8(0.5)^{N(3,t)}$ and $f_3 = 12(0.5)^{N(3,t)}$.

23 Because $f_2 = 8(0.5)^{N(3,t)}$ and $f_3 = 12(0.5)^{N(3,t)}$, we begin with $f_2 = 8$ and $f_3 = 12$ because $N(3, t) = 0$.

$$\begin{vmatrix} 0 & 0 & 8 & 12 \\ 0.3 & 0 & 0 & 0 \\ 0 & 0.5 & 0 & 0 \\ 0 & 0 & 0.4 & 0 \end{vmatrix} \begin{vmatrix} 10 \\ 5 \\ 1 \\ 0 \end{vmatrix}_0 = \begin{vmatrix} 8 \\ 3 \\ 2 \\ 0 \end{vmatrix}_1 = \begin{vmatrix} 16 \\ 2 \\ 2 \\ 1 \end{vmatrix}_2$$

At $t = 2$, $N(3, t) = 1$, so $f_2 = 4$ and $f_3 = 6$.

$$\begin{vmatrix} 0 & 0 & 4 & 6 \\ 0.3 & 0 & 0 & 0 \\ 0 & 0.5 & 0 & 0 \\ 0 & 0 & 0.4 & 0 \end{vmatrix} \begin{vmatrix} 16 \\ 2 \\ 2 \\ 1 \end{vmatrix}_2 = \begin{vmatrix} 14 \\ 5 \\ 1 \\ 1 \end{vmatrix}_3 = \begin{vmatrix} 10 \\ 4 \\ 2 \\ 0 \end{vmatrix}_4$$

$$\begin{vmatrix} 0 & 0 & 8 & 12 \\ 0.3 & 0 & 0 & 0 \\ 0 & 0.5 & 0 & 0 \\ 0 & 0 & 0.4 & 0 \end{vmatrix} \begin{vmatrix} 10 \\ 4 \\ 2 \\ 0 \end{vmatrix}_4 = \begin{vmatrix} 16 \\ 3 \\ 2 \\ 1 \end{vmatrix}_5$$

$$\begin{vmatrix} 0 & 0 & 4 & 6 \\ 0.3 & 0 & 0 & 0 \\ 0 & 0.5 & 0 & 0 \\ 0 & 0 & 0.4 & 0 \end{vmatrix} \begin{vmatrix} 16 \\ 3 \\ 2 \\ 1 \end{vmatrix}_5 = \begin{vmatrix} 14 \\ 5 \\ 2 \\ 1 \end{vmatrix}_6 = \begin{vmatrix} 14 \\ 4 \\ 2 \\ 1 \end{vmatrix}_7 = \begin{vmatrix} 14 \\ 4 \\ 2 \\ 1 \end{vmatrix}_8 = \begin{vmatrix} 14 \\ 4 \\ 2 \\ 1 \end{vmatrix}_9 = \begin{vmatrix} 14 \\ 4 \\ 2 \\ 1 \end{vmatrix}_{10}$$

24 The probability of growing from a seedling to a small tree is $P_{21} = h_{21}$ and the probability of growing from a small tree to an adult tree is $P_{32} = h_{32}$. From the independent data we know that $h_{21}(0) = 0.8$ and $h_{32}(0) = 0.5$ when $N(3, t) = 0$. If we add one more adult tree, $h_{21}(1) = 0.8(0.3)$ and

$h_{32}(1) = 0.5(0.5)$. In general, as in exercise 22, $h_{21}(N(3, t)) = 0.8(0.3)^{N(3,t)}$ and $h_{32}(N(3, t)) = 0.5(0.5)^{N(3,t)}$.

25 $f_2 = 8(0.5)^{N(3,t)} = 8e^{-\alpha N(3,t)}$, where $e^{-\alpha} = 0.5$ or $\ln(0.5) = -\alpha, -0.693 = -\alpha \therefore f_2 = 8e^{-0.693[N(3,t)]}$.

26 The fraction of the "adjusted" probability that changes is $(P - P')/P'$, where P is the probability under no density-dependent effects and P' is the adjusted probability; therefore $(P - P')/P' = b\hat{N}$, where b is an arbitrary constant and $P' + a\,\hat{N}P' = P$, or

$$P' = \frac{P}{1 + b\hat{N}} = \frac{a_j}{1 + b\hat{N}}$$

27 We have $a(x) = a_0(x)e^{-\beta\hat{N}(t)}$, where $e^{-\beta} = \alpha$ and a_0 refers to the transition probability we expect as $\hat{N}(t)$ approaches zero. To make the basic process density dependent we write,

$$MQ_t n_t = n_{t+1}$$

where Q is a diagonal matrix with elements $e^{-\beta\hat{N}(t)}$.

Following the analysis in the text the population will be stable and stationary if $MQ_t^{-1} = \lambda^{-1}M$, which will be true when $e^{\beta\hat{N}(t)} = \lambda$. At that time $\hat{N}(t) = K$. Therefore we have

$$e^{\beta K} = \lambda$$

$$\ln \lambda = \beta K$$

$$K = \frac{r}{\beta}$$

REFERENCES

Baltensweiler, W. 1971. The relevance of changes in the composition of larch bud moth populations for the dynamics of its numbers. In P. J. den Boer and G. R. Gradwell, Eds., Dynamics of populations. Wageningen: Center for Agricultural Publishing, pp. 208–219.

Barclay, G. W. 1958. *Techniques of population analysis.* New York: Wiley.

Bosch, C. A. 1971. Redwoods: A population model. *Science* **172**:345–349.

Dorn, H. F. 1950. Pitfalls in population forecasts and projections. *J. Am. Stat. Assoc.* **45**:311–334.

Goodman, L. A. 1968. An elementary approach to the population projection matrix and to the mathematical theory of population growth. *Demography.* **V**.

Harper. J. L. and J. White. 1972. The demography of plants. *Annu. Rev. Ecol. Syst.* **5**:419–463.

Keyfitz, N. 1968. *Introduction to the mathematics of Population.* 450 pp. Addison-Wesley.

Kitagawa, E. M. 1964. Standardization comparisons in population research. *Demography.* **I**:296–315.

Lefkovitch, L. P. 1965a. An extension of the use of matrices in population methematics. *Biometrics* **22**:1–18.

Lefkovitch, L. P. 1965b. The effects of adult emigration on populations of *Lasioderma serricorne*. Oikos 15:200–210.

Leslie, P. H. 1945. On the use of matrices in certain population mathematics. *Biometrika* **33**: 183–212.

Leslie, P. H. 1948. Some further notes on the use of matrices in population mathematics. *Biometrika* **35**: 213–245.

Leslie, P. H. 1959. The properties of a certain lag type of population growth and the influence of an external random factor on a number of such populations. *Physiol. Zool.* **32**: 151–159.

Lewis, E. G. 1942. On the generation and growth of a population. *Sankhya* **6**: 93–96.

Park, T., P. H. Leslie, and D. B. Mertz. 1964. Genetic strains and competition in populations of *Tribolium*. *Physiol. Zool.* **37**: 97–162.

Pielou, E. C. 1969. *An Introduction to mathematical ecology*. New York: Wiley-Interscience, 286 pp.

Rogers, A. 1968. *Matrix analysis of interregional population growth and distribution*. Berkeley: University of California Press.

Searl, S. R. 1966. *Matrix algebra for the biological sciences, including applications in statistics*. 296 pp. NY: Wiley.

United Nations. 1955. Age and sex patterns of morality: Model life tables for underdeveloped countries. New York.

Vandermeer, J. 1975. On the construction of the population projection matrix for a population grouped in unequal stages. *Biometrics* **31**: 239–242.

Vandermeer, J. H. 1978. Choosing category size in a stage projection matrix. *Oecologia, Berlin.* **32**: 79–84.

3. Discrete Models of Population Changes

SPECIFIC FORMS AND THEIR BIOLOGICAL INTERPRETATIONS. In Chapter 2 the projection-matrix technique was applied to population description. This technique contains the basic formulation

$$N(t + 1) = f(N(t)) \tag{1}$$

where $N(t)$ is the population density at time t and f is an arbitrary function. Equation 1 is a difference equation. The matrix equation which describes population growth (Chapter 2) is a special application of equation 1. In this chapter we shall study the more general form. Equation 1 can exhibit some rather remarkable behavior, a fact of considerable importance because it has seen, and will continue to see, a great deal of practical service.

□ EXERCISES

1 Recall from Chapter 1 that exponential growth is described by

$$N(t) = N(0)e^{rt}$$

(the integrated form of $dN/dt = rN$). Express $N(t + 1)$ as a function of $N(t)$ for this equation.

2 Recall from Chapter 1 that logistic growth is described by the equation

$$N(t) = \frac{K}{1 + \left[\dfrac{K - N(0)}{N(0)}\right]e^{-rt}}$$

(the integrated form of $dN/dt = [rN(K - N)/K]$. Express $N(1)$ as a function of $N(0)$ for this equation. Express $N(2)$ as a function of $N(1)$. Express $N(t + 1)$ as a function of $N(t)$. □

The equations derived in exercises 1 and 2 are, of course, equivalent to the exponential and logistic equations, except that they refer only to discrete points of time. Therefore we have

$$N(t + 1) = N(t)e^{r} \tag{2}$$

and

$$N(t+1) = \frac{e^r K N(t)}{N(t)(e^r - 1) + K} \tag{3}$$

as difference equations that correspond to the exponential and logistic equations, respectively. In many ways the discrete form of the model is easier to work with than the continuous form.

□ EXERCISES

3 Let $r = 0.69$ and $K = 100$. Plot equations 2 and 3 in $[N(t+1), N(t)]$ space.

4 Using the graph for equation 2 plotted in exercise 3, trace graphically the population's history, beginning with $N(0) = 1$. Project $N(0) \rightarrow N(1)$, $N(1) \rightarrow 45°$ line $\rightarrow$ absissa, $N(1) \rightarrow N(2)$ etc. Plot the projection in $[N(t), t]$ space.

5 Using the graph for equation 3 plotted in exercise 3, trace graphically the population's history, beginning with $N = 1$. Plot the projection in $[N(t), t]$ space. □

Besides being easier to work with (at least sometimes), discrete formulations are occasionally more easily interpretable biologically; for example, from equation 2 we can see that *the population at any time is a constant multiple of the former population* is a description of the process of exponential growth. On the other hand, equation 3 is not obviously interpretable in its discrete form. Yet despite its apparent complexity, the discrete form, when graphed, lends itself to an obvious biological interpretation. Whereas exponential growth appears to be linear in $[N(t+1), N(t)]$ space, simple density dependence (such as would be represented by the logistic equation) appears as a "bending" of the exponential extreme, as shown in Figure 3.1:

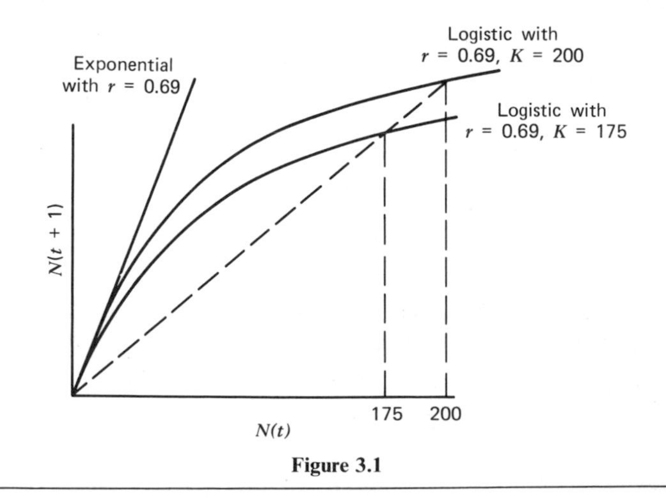

Figure 3.1

Thus frequently the discrete form of the equation has an obvious biological interpretation and, more importantly, the graph in $[N(t+1), N(t)]$ space is almost always suggestive of biological interpretations.

□ EXERCISES

6 Suppose a population behaves according to the equation $N(t+1) = f(N(t))$. Suppose the qualitative description of the population is *the population from now on will* decline *at an exponential rate—will exhibit exponential decay*. Graph f in $[N(t+1), N(t)]$ space qualitatively and, beginning with a relatively large population, project into the future.

7 Suppose the basic form of population dynamics is $N(t+1) = N(t) + g(N(t))$, where we might call the function g the "increment." Suppose the increment is proportional to the fraction of the environment still available, $[(K - N(t))/K]$. Write the complete discrete equation and plot in $N(t+1)$, $N(t)$ space.

8 Suppose the logarithm of the population at $t + 1$ is the result of adding a constant fraction of the space available, $[(K - N(t))/K]$, to the logarithm of the population at t. Write the difference equation (let the fraction $= r$). Plot the equation for $r = 1.0$, $K = 100$, and $r = 3.0$, $K = 100$.

9 Suppose a population increases exponentially when its density is below some value (say $N = \varepsilon$) and decreases exponentially when its density is above that density. Write an equation for this situation and construct its graph.

10 Suppose a population increases exponentially when its density is below some value (say $N = \varepsilon$). Suppose that above that value the individuals are not capable of reproducing at all (but retain a *perfect* survivorship). Construct a graph for this situation.

11 Suppose a particular environment contains exactly 100 "hidey holes" in which the individuals of a particular population live. (Assume that each hidey hole can contain only one individual and that the individuals must obtain hidey holes or they will die). Every individual in the population produces one other individual in each time period, but the new individual will not survive if a hidey hole is not available. Write an equation and construct a graph in $[N(t+1), N(t)]$ space for this situation.

12 Assume the rules of population dynamics outlined in exercise 11. When the number of individuals in the population is greater than 100, however, the individuals will attempt to enter hidey holes already occupied by others. When this takes place, both individuals will die. Construct a graph for this situation. □

These exercises make it clear, I hope, how the underlying biological dynamics of a particular equation are easily understood when the graph in $[N(t+1), N(t)]$

space is examined. It is also true that the general qualitative behavior of the equation is made evident by such a graph, as described in exercises 4 and 5.

□ EXERCISES

13 Using the method in exercises 4 and 5, project the population for $N(0) = 1$ according to the graph you constructed in exercise 12.

14 Perhaps the simplest equation to capture the qualitative dynamics that we expect from biological populations is

$$N(t+1) = a\,N(t)[1 - N(t)]$$

Plot this equation in $[N(t+1), N(t)]$ space for $a = 1.5$ and $a = 2.5$. Using these two graphs, project the populations from $N(0)$ for 10 time periods. (Begin with $N < 1$.) □

EQUILIBRIUM POINTS AND THEIR DYNAMICS. It should be clear from exercises 13 and 14 that the discrete models we are discussing are similar to the continuous models presented in Chapter 1 in that they usually exhibit the basic processes of population growth, followed by some sort of regulation or limitation, and eventually stabilize at some equilibrium point, the carrying capacity. In addition, these discrete forms may exhibit oscillations similar to those that occurred when time lags were added to the continuous forms. In a sense, merely by writing the equation in a discrete form, one has added an implicit time lag; therefore the existence of oscillations is not really surprising. As we shall see in a later section, however, these oscillations may take on an extremely complicated form.

Before going on to an analysis of the more complicated dynamics we must first consider the determination of equilibrium points and their dynamics.

□ EXERCISES

15 For the equation $N(t+1) = a\,N(t)(1 - N(t))$ determine the equilibrium point (the point for which $N(t+1) = N(t)$ or the point at which $f(N)$ crosses the 45° line) for $a = 1.5$, $a = 2.5$, $a = 3.5$.

16 For the equation

$$N(t+1) = \frac{e^r N(t) K}{N(t)(e^r - 1) + K}$$

compute the equilibrium point for $e^r = 2$, $K = 100$ and $e^r = 3$, $K = 100$.

17 For the equation

$$N(t+1) = N(t)e^{r\{[K-N(t)]/K\}}$$

compute the equilibrium point for $r = 1.5$, $K = 100$.

18 Consider the general form,

$$N(t+1) = F(N(t))$$

Compute $\partial F/\partial N(t)$ for the three equations in exercises 15, 16, and 17. □

Equilibrium points exist wherever $F(N(t))$ crosses the 45° line in $[N(t+1), N(t)]$ space. Among the examples considered so far only one nontrivial (not equal to zero) equilibrium point exists. Of course, nothing stands in the way of complicated functions having numerous equilibrium points; for example, we could imagine a population that exhibits an Allee effect below some critical density (say $< N(t) = A$). (An Allee effect exists when the population cannot grow if it is below some critical population density.) Above that critical density the population grows exponentially and is eventually subjected to density dependence and stabilization at some equilibrium density (K). The graph of this population would look like Figure 3.2:

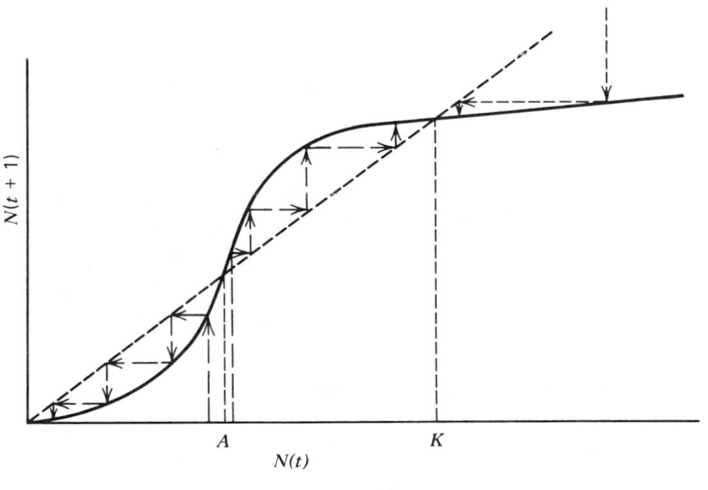

Figure 3.2

Two equilibrium points exist—one at $N(t) = A$, the other at $N(t) = K$. It is obvious that with a little bit of imagination one could construct hypothetical populations in which numerous equilibrium points would exist.

Not all points of equilibrium are the same, however; for example, the equilibrium labled A in the graph is quite different qualitatively than the equilibrium labeled K. As shown in the illustrated trajectories, if the population is located above or below point A, it necessarily diverges away from that point. Indeed, if the population were located *exactly* at point A, it would remain exactly at that point. But if it were just slightly smaller or larger than A, it would diverge toward zero or toward K, the other equilibrium. The equilibrium at K is qualitatively different in that trajectories that begin away from K tend to converge toward K. The equilibrium at point A is called an "asymptotically unstable equilibrium" and at point K is called an "asymptotically stable equilibrium."

The following two graphs illustrate the two other possible types of equilibrium:

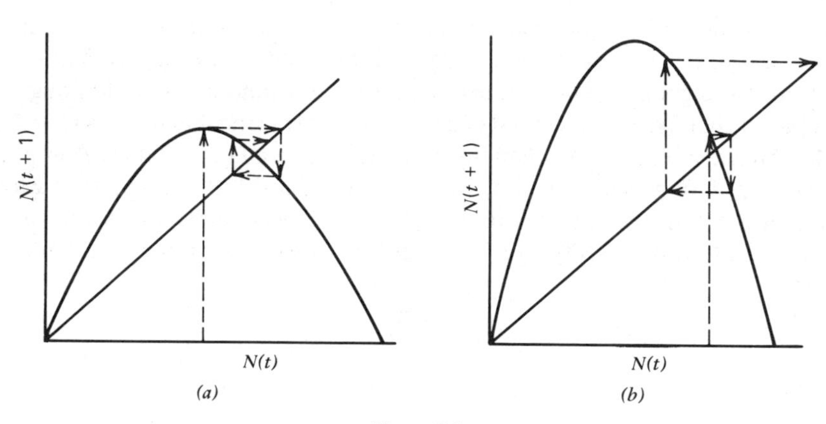

Figure 3.3

In Figure 3.3*a* the population oscillates but with decreasing amplitude until it is eventually located at the equilibrium point (as in exercises 13 and 14). This type of equilibrium point is called a "stable oscillatory equilibrium" or simply "attracting." The equilibrium in Figure 3.3*b* is qualitatively distinct in that any point only slightly removed from the equilibrium point will lead to oscillations that diverge farther and farther from the equilibrium. This equilibrium is referred to as an "unstable oscillatory equilibrium" or simply "repelling."

☐ EXERCISES

19 For each of the following graphs project the population, beginning with the population density indicated by the arrow. Plot the projected population (qualitatively) on a graph of $N(t)$ versus t and indicate the kind of equilibrium that exists for each nontrivial equilibrium.

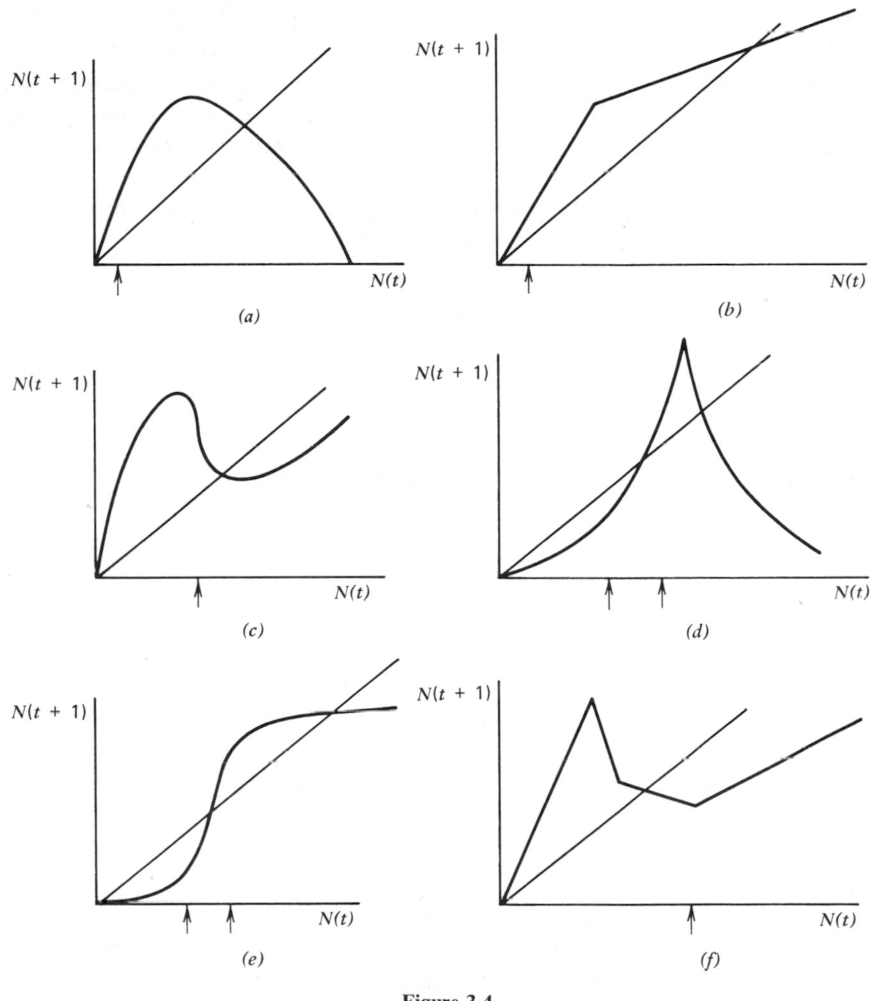

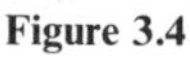

Figure 3.4

20 For each of the equilibrium points in the graphs of exercise 19 draw a tangent to F at the point of intersection with the 45° line. Call the slope of the line λ and determine into which of the following sets the λ for each equilibrium point falls:

(1) $\lambda < -1$
(2) $-1 < \lambda < 0$
(3) $0 < \lambda < 1$
(4) $1 < \lambda$

Describe qualitatively the behavior of each equilibrium point next to the set determination. □

From these exercises we see that the slope of the function F at the point at which it crosses the 45° line determines what the dynamical behavior of the equation will be *near that equilibrium point*. The slope of the function is frequently called the eigenvalue in this context. Obviously, if the eigenvalue (slope of the tangent to F at the point of equilibrium) is less than -1, the system is unstable oscillatory. If the eigenvalue is greater than -1 but less than zero, the system is stable oscillatory. If the eigenvalue is between zero and $+1$, the system is stable asymptotic. If the eigenvalue is greater than $+1$, the system is unstable asymptotic. Thus the dynamics of an equilibrium can be analytically determined by finding the eigenvalue (slope of F at the equilibrium).

□ EXERCISES

21 In exercise 15 you determined the value of the equilibrium of the equation

$$N(t + 1) = a\,N(t)(1 - N(t)).$$

In exercise 18 you computed the derivative of the function F (the right-hand side of the equation). What is the value of the eigenvalue at the point of equilibrium?

22 For what values of a will the equation

$$N(t + 1) = a\,N(t)(1 - N(t))$$

yield a stable asymptote, stable oscillations, unstable oscillations?

23 Show that the discrete form of the logistic equation

$$N(t + 1) = \frac{e^r K\,N(t)}{N(t)(e^r - 1) + K}$$

can never exhibit oscillations, presuming a population with a potential for growth. (Consider the necessary conditions on the derivative).

24 For the equation

$$N(t + 1) = N(t)e^{r\{[K - N(t)]/K\}}$$

find the values of r for which the equation will exhibit oscillations. □

COMPLICATED BEHAVIOR FROM SIMPLE FORMS. The preceding discussion treated particular equilibrium points and their dynamics. Although the treatment of such fixed points is the backbone of dynamic analysis, frequently the most interesting aspects of particular equations are never revealed if the analysis is restricted to specific equilibrium points. In this section we shall examine some rather remarkable behavior, remarkable both in its complexity and in the fact that it derives from simple equations.

☐ EXERCISES

25 Given the equation $N(t+1) = a\,N(t)(1 - N(t))$, let $a = 3.1$, begin with $N(t) = 0.64$, and project the population 10 times into the future (using the equation, *not* the graph of the equation). How would you describe this population behavior (round to nearest hundredth).

26 Using the same equation, let $a = 3.52$ and begin with $N(t) = 0.88$ (always round to two significant figures). Project the population 10 times. How would you describe this population behavior?

These exercises illustrate not only that permanent oscillations can be produced in the model $N(t+1) = f(N(t))$ but also that the number of points in the cycle can vary (e.g., a two-point cycle in exercise 25 and a four-point cycle in exercise 26). Thus in characterizing the dynamics of discrete population models we are faced not only with characterizing equilibrium points as described in the preceding sections but also with characterizing permanent oscillations. The existence of permanent oscillations (or cycles) cannot be inferred directly from an analysis of particular equilibrium points. Indeed, the direct detection of permanent cycles requires somewhat more sophisticated techniques than are offered here. Therefore the reader is referred to Clark (1976). In this presentation we are more concerned with the general qualitative behavior of these models.

Analogous to the simple oscillatory equilibria, permanent cycles can be stable or unstable. A stable permanent cycle exists when particular trajectories converge on the cycle, no matter where they start. An unstable permanent cycle exists when any slight perturbation away from the permanent cycle produces a trajectory that never returns to that permanent cycle. The two basic forms of dynamical behavior are illustrated in $[N(t+1), N(t)]$ space in Figure 3.5 and in $[N(t), t]$ space in Figure 3.6.

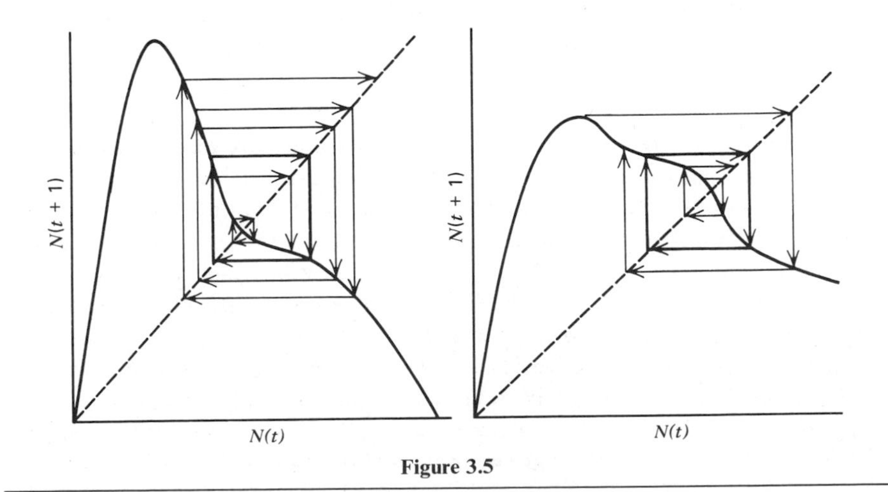

Figure 3.5

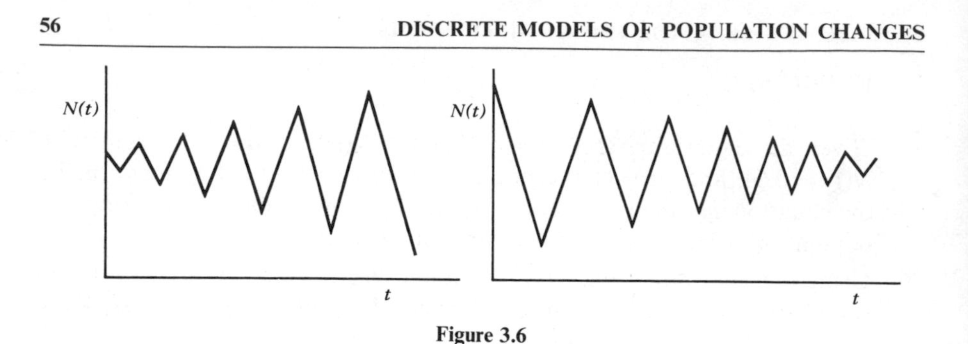

Figure 3.6

For most purposes the only permanent cycles of much interest to an ecologist are the stable varieties. Furthermore, for most situations the existence of a stable permanent cycle has associated with it as a necessary (but obviously not sufficient) condition an *unstable* oscillatory equilibrium point. If it can be shown that a particular model does *not* have an unstable oscillatory equilibrium point it is a reasonably safe bet that it will exhibit no permanent cycles in the stable form.

□ EXERCISES

27 Repeat exercise 26 for $a = 4.2$, beginning with $N(t) = 0.77$ (for convention, allow any population less than or equal to zero to be exactly zero). Plot against time (round to hundredths).

28 (a) Using the equation $N(t + 1) = N(t)e^{r\{[K - N(t)]K\}}$, let $K = 1$ and $r = 3.0$. Begin with a population equal to 1.01 (always round to the second decimal place) and project the population 12 times (using the equation, *not* the graph). Explain the pattern. (b) Using the same equation with the same parameters, begin with a population of 1.86 individuals and project the population 10 times (always round to the second decimal place). □

At this point it should be obvious that the models we are dealing with in this chapter are capable of generating extremely complicated behavior. It should also be recalled that the biological assumptions that go into making these models are simple. It is remarkable that such complicated behavior can be generated with such simple models. One can only wonder what complications will develop if the models themselves are made to be more complex (which naturally must be done if the models are to correspond closely to nature).

To understand the qualitative dynamics of these models we must analyze them at two levels. First, in a local analysis (or neighborhood analysis) we ask how many fixed equilibria are there, where they are located, and what the dynamic characteristics of the equation are near each of the equilibria. As exercises 25 to

28 show, much more can go on dynamically than will be understood from a mere analysis of fixed equilibrium points; that is, the second phase of the analysis must be a study of the permanent cycles that the model is capable of generating. From exercises 25 and 26 we find that permanent cycles can have different numbers of points. From exercise 28 we find that more than one cycle can exist within the same equation with the same parameters (exercise 28 also shows that the number of points in a cycle can be large). Furthermore, a permanent cycle can be stable or unstable, as described above. Thus in the study of permanent cycles for a given model we must ask (1) how many permanent cycles exist? (2) How many points are in each cycle? (3) Is each cycle stable or unstable?

A general qualitative result that has recently come to the attention of ecologists is the existence of population behavior *so* complicated that it defies description by any normal statistical means. This sort of behavior has been termed *chaos*. It exists when the number of different permanent oscillations is infinite and there are cycles with all numbers of points (i.e., a 2-point cycle, a 3-point cycle, a 4-point cycle, a 100-point cycle). Furthermore, the number of aperiodic trajectories is infinite (i.e., a large number of $N(0)$'s for which, no matter how long the population is projected according to $N(t+1) - f(N(t))$, $N(0)$ is never reached again nor is any number *ever* repeated). Such behavior is hopelessly complicated. The truly remarkable thing is that such complications result from extremely simple assumptions; for example, the equation $N(t+1) = N(t)e^{r\{[K-N(t)]/K\}}$ exhibits chaotic behavior whenever $r > 2.69$. An excellent summary of this new and difficult topic is found in Guckenheimer *et al.* (1976).

ANSWERS TO EXERCISES

1

$$\begin{aligned} N(t+1) &= N(0)e^{r(t+1)} \\ &= [N(0)e^{rt}]e^{r} \end{aligned}$$

and because the term in brackets is equal to $N(t)$ [recall $N(t) = N(0)e^{rt}$] we have $N(t+1) = N(t)e^{r}$.

2

$$\begin{aligned} N(t+1) &= \frac{K}{1 + \dfrac{K - N_0}{N_0} e^{-r(t+1)}} \\ &= \frac{K}{e^{-r}\left[e^{r} + \dfrac{K - N_0}{N_0} e^{-rt}\right]} \\ &= \frac{K}{e^{-r}\left[e^{r} - 1 + 1 + \dfrac{K - N_0}{N_0} e^{-rt}\right]} \end{aligned}$$

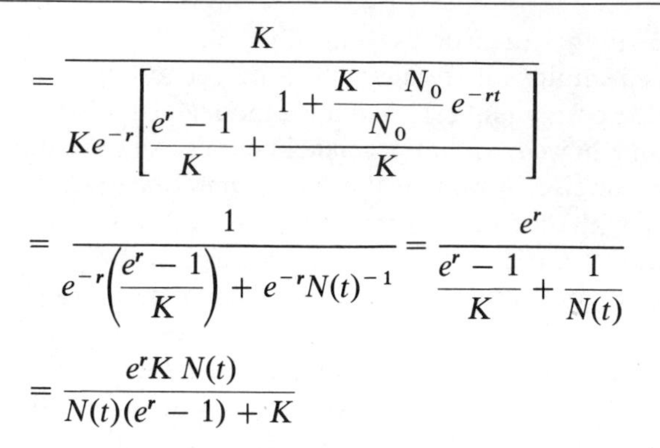

$$= \frac{K}{Ke^{-r}\left[\dfrac{e^r - 1}{K} + \dfrac{1 + \dfrac{K - N_0}{N_0} e^{-rt}}{K}\right]}$$

$$= \frac{1}{e^{-r}\left(\dfrac{e^r - 1}{K}\right) + e^{-r}N(t)^{-1}} = \frac{e^r}{\dfrac{e^r - 1}{K} + \dfrac{1}{N(t)}}$$

$$= \frac{e^r K\, N(t)}{N(t)(e^r - 1) + K}$$

3 For equation 2 we note that it is linear in $N(t + 1)$, $N(t)$ space, with intercept 0 and a slope equal to e^r. Thus the graph is simply

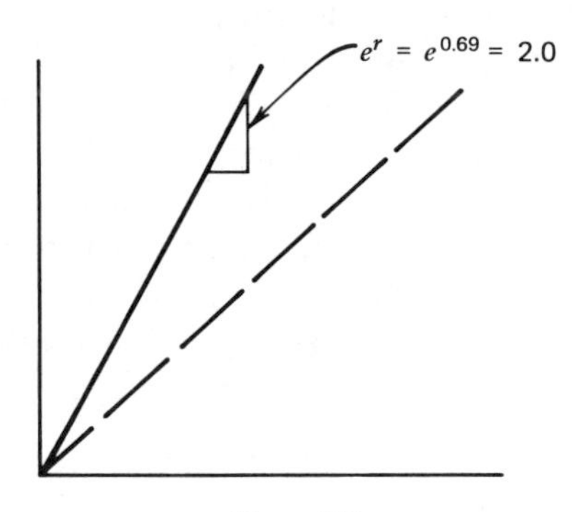

Figure 3.7

For equation 3 we have,

$$N(t + 1) = \frac{200N(t)}{N(t) + 100}$$

Note that $[N(t + 1)]/N(t) = 1$ when $N(t) = 100$ and $[N(t + 1)]/N(t) = 2$ when $N(t) = 0$. Therefore

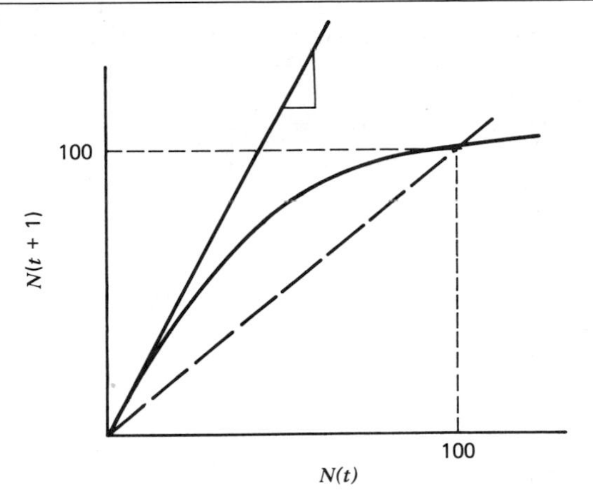

Figure 3.8

4

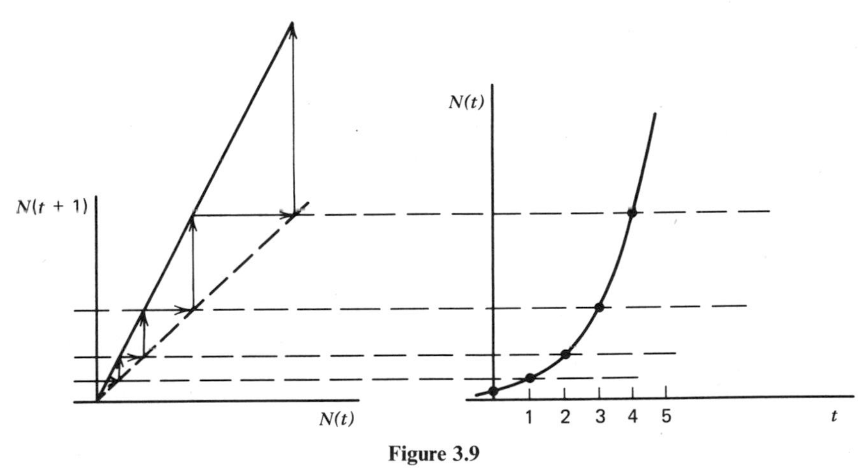

Figure 3.9

5

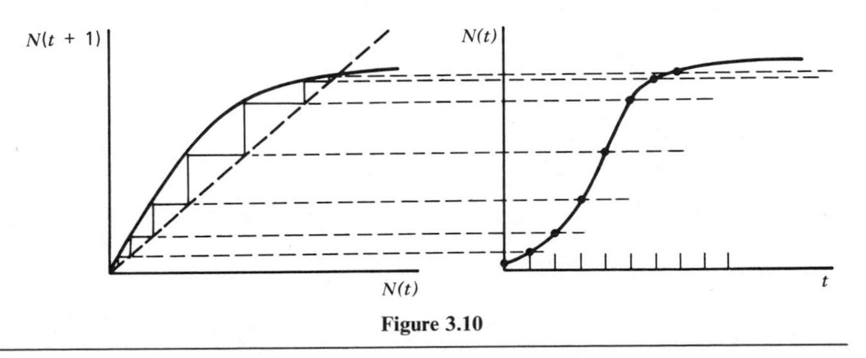

Figure 3.10

6 Exponential decay means that $r < 0$, which means that $e^r < 1$. Thus

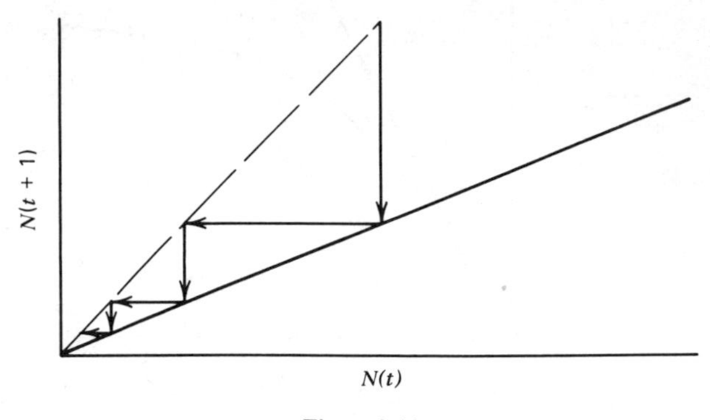

Figure 3.11

7 $g(N(t)) = a\{[K - N(t)]/K\}$, where a is a constant, Thus

$$N(t+1) = N(t)\left\{1 + b\left[\frac{K - N(t)}{N(t)}\right]\right\}$$

where $b = a/K$.

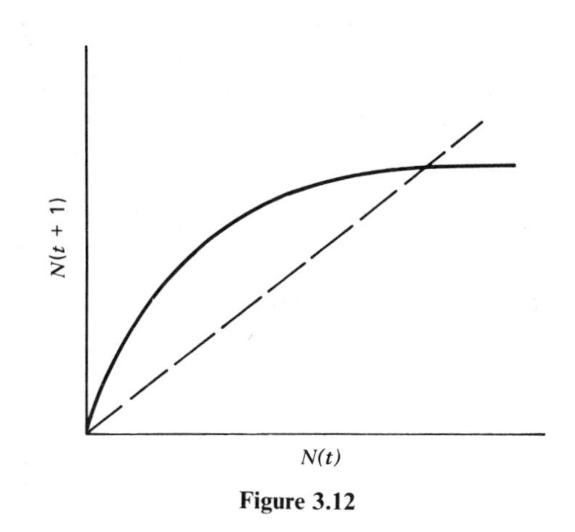

Figure 3.12

8

$$\ln N(t+1) = \ln N(t) + r\left[\frac{K - N(t)}{K}\right]$$

$$N(t+1) = N(t)e^{\{r[K - N(t)]/K\}}$$

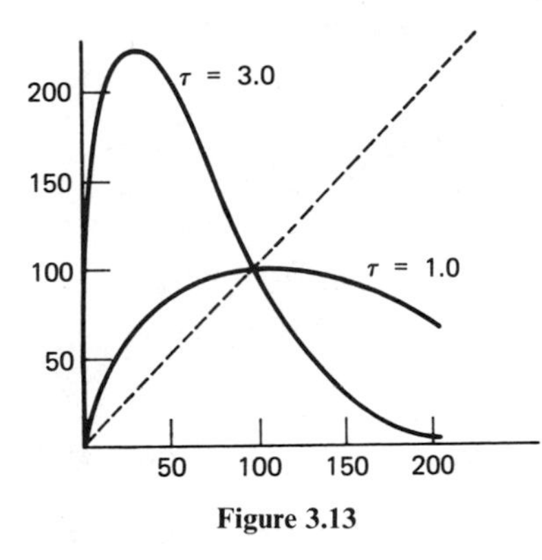

Figure 3.13

9

$$N(t+1) = a_1\,N(t) \quad \text{for} \quad N(t) \leq \varepsilon \text{ (where } a_1 > 1)$$

$$N(t+1) = a_2\,N(t) \quad \text{for} \quad N(t) > \varepsilon \text{ (where } a_2 < 1)$$

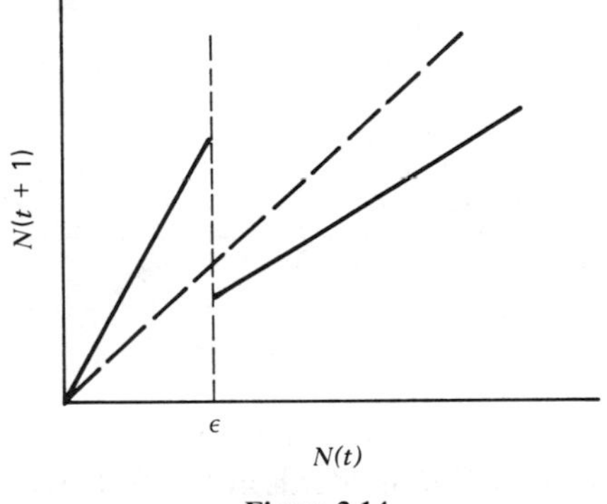

Figure 3.14

10

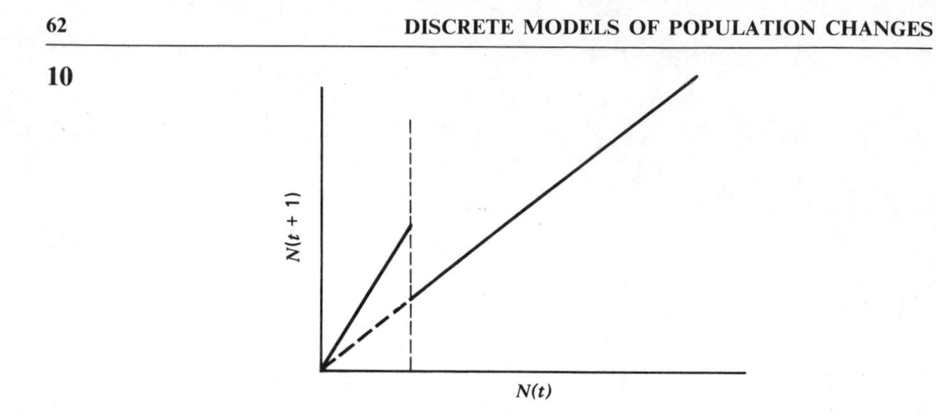

Figure 3.15

11 The equation must be

$$N(t + 1) = 2N(t) \quad \text{for} \quad N < 50$$

$$N(t + 1) = 100 \quad \text{for} \quad N \geq 50$$

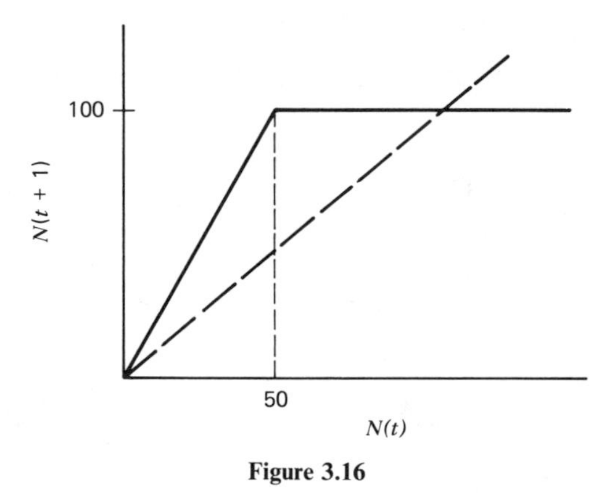

Figure 3.16

12 Suppose we begin with 51 individuals at $t = 0$. These 51 produce 102, which means that two of the hiding places will contain two individuals each. Thus a total of four individuals will die, leaving 98 for the $t + 1$ generation. If by the same reasoning we begin with 52, 96 will be in the $t + 1$ generation, 53 will yield 94, and ultimately 100 will yield 0. Thus the graph is

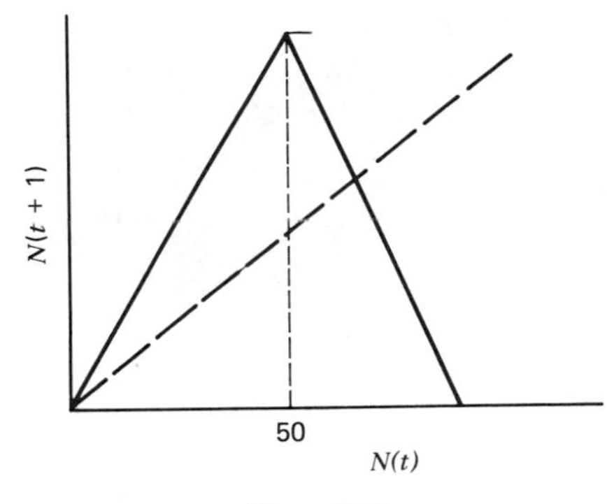

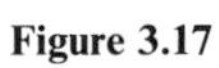

Figure 3.17

13

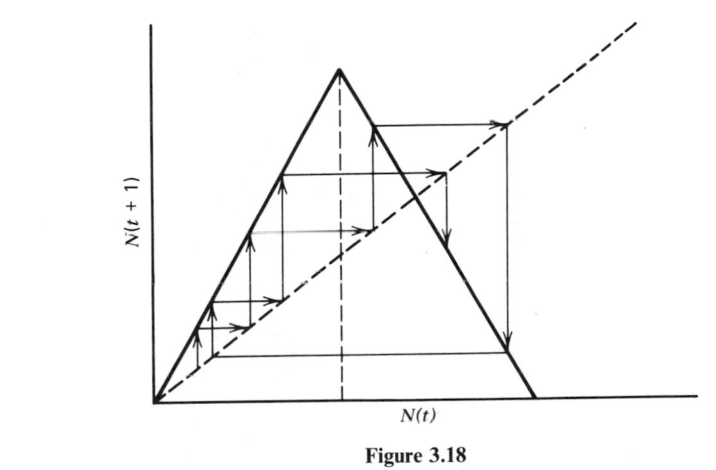

Figure 3.18

14

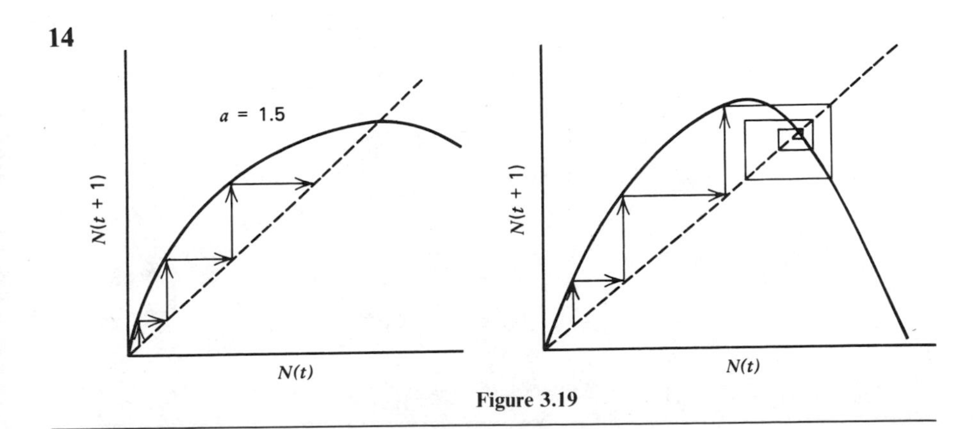

Figure 3.19

15 $N(t+1) = N(t)$ implies $N(t) = aN(t)(1 - N(t))$. Thus $1 = a(1 - N(t)) = a - aN(t)$ and $N(t) = (a-1)/a$. Therefore for $a = 1.5$, $N^* = (a-1)/a = 0.5/1.5 = 0.33$; for $a = 2.5$, $N^* = 1.5/2.5 = 0.60$; for $a = 3.5$, $N^* = 2.5/3.5 = 0.71$.

16

$$N(t) = \frac{\lambda N(t) K}{N(t)(\lambda - 1) + K}$$

$$N^*(t)(\lambda - 1) + K = \lambda K$$

$$N^*(t) = \frac{\lambda K - K}{\lambda - 1} = \frac{(\lambda - 1)K}{\lambda - 1} = K$$

Thus the equilibrium point is always equal to K.

17

$$N(t) = N(t) e^{r\{[K - N(t)]/K\}}$$

$$1 = e^{r\{[K - N(t)]/K\}}$$

$$\ln 1 = \ln[e^{r\{[K - N(t)]/K\}}]$$

$$0 = r\left(\frac{K - N^*(t)}{K}\right) = rK - rN^*(t)$$

$$K = N^*(t)$$

As in exercise 16, the equilibrium point will be K, regardless of the value of r.

18 (a) $N(t+1) = aN(t)(1 - N(t))$

$$F(N(t)) = aN(t)(1 - N(t)) = aN(t) - aN(t)^2$$

$$\frac{\partial F}{\partial N} = a - 2aN(t)$$

(b) $N(t+1) = \dfrac{\lambda K N(t)}{N(t)(\lambda - 1) + K}$ (where $\lambda = e^r$)

$$\frac{\partial F}{\partial N} = \frac{\lambda K}{N(t)(\lambda - 1) + K} - \frac{(\lambda - 1)\lambda K\, N(t)}{(N(t)(\lambda - 1) + K)^2}$$

$$= \frac{e^r K N(e^r - 1) + K^2 e^r - e^{2r} K N + e^r K N}{[N(e^r - 1) + K]^2}$$

$$= \frac{e^r K[N(e^r - 1) + K - e^r N + N]}{[N(e^r - 1) + K]^2} = \frac{e^r K^2}{[N(e^r - 1) + K]^2}$$

(c) $F = N(t) e^{r\{[K - N(t)]/K\}}$

$$\frac{\partial F}{\partial N} = e^{r[(K-N)]/K} + N(t)\,\frac{\partial\{e^{r[(K-N)]/K}\}}{\partial N}$$

let $u = r[(K - N)]/K$, $g = e^{r[(K-N)]/K} = e^u$, Recall that

$$\frac{\partial g}{\partial N} = \frac{\partial g}{\partial u}\frac{\partial u}{\partial N} \quad \text{and} \quad \frac{\partial g}{\partial u} = e^{u}, \frac{\partial u}{\partial N} = -\frac{r}{K}$$

$$\frac{\partial g}{\partial N} = e^{r[(K-N)]/K}\left(-\frac{r}{K}\right)$$

Therefore

$$\frac{\partial F}{\partial N} = e^{r[(K-N)]/K} - \frac{rN}{K}e^{r[(K-N)]/K}$$

$$= e^{r[(K-N)]/K}\left(1 - \frac{rN}{K}\right)$$

19

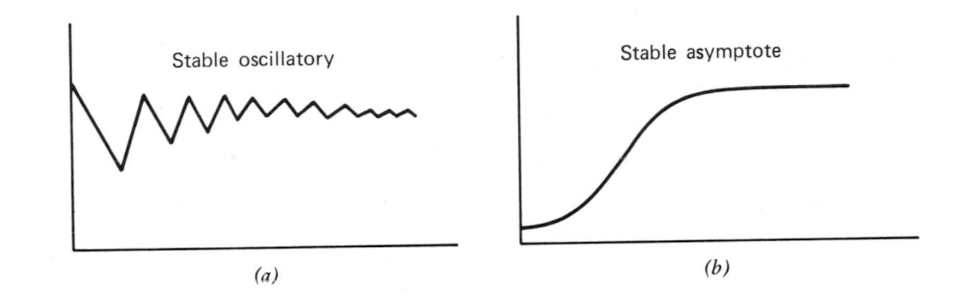

(a) (b)

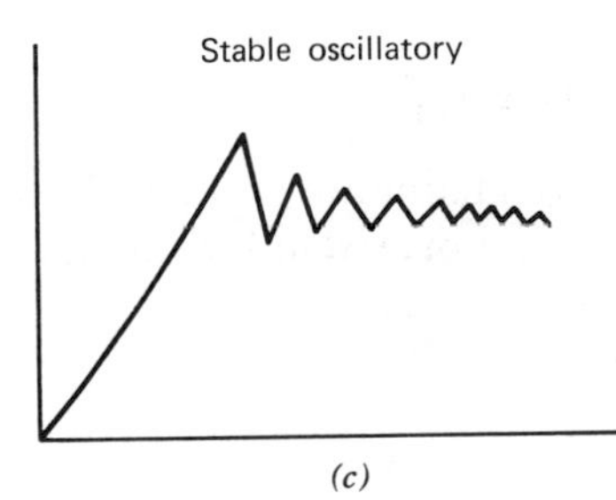

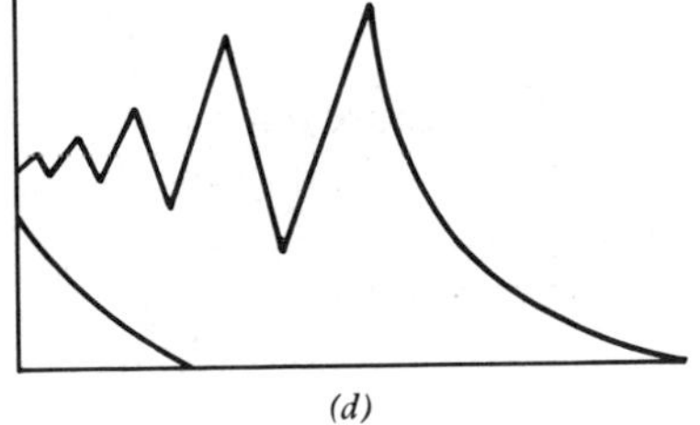

(c) (d)

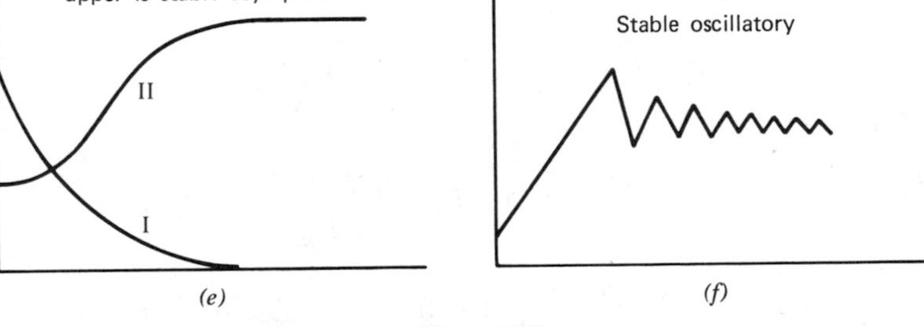

(e) (f)

Figure 3.20

20 (a) $0 > \lambda > -1$ stable oscillatory.
(b) $0 < \lambda < 1$ stable asymptote.
(c) $-1 < \lambda < 0$ stable oscillatory.
(d) $-1 > \lambda$ unstable oscillatory (lower one $= 1 < \lambda$ unstable asymptote.
(e) $\lambda > 1$ unstable oscillatory (upper one $= 0 < \lambda < 1$ stable asymptote).
(f) $-1 < \lambda < 0$ stable oscillatory.

21 Equilibrium value is $(a - 1)/a$
Derivative is $a - 2aN^*$.
Eigenvalue $= \lambda = a - 2a[(a - 1)/a] = a - 2a + 2 = 2 - a.$

22 Stable asymptote if $0 < \lambda < 1$:

$$\left.\begin{aligned} 0 < 2 - a \Rightarrow a < 2 \\ 2 - a < 1 \Rightarrow a > 1 \end{aligned}\right\} \qquad 1 < a < 2$$

Stable oscillatory if $-1 < \lambda < 0$:

$$\left.\begin{aligned} -1 < 2 - a \Rightarrow a < 3 \\ 0 > 2 - a \Rightarrow a > 2 \end{aligned}\right\} \qquad 2 < a < 3$$

Unstable oscillatory if $-1 > \lambda$:

$$-1 > 2 - a \Rightarrow a > 3$$

23 From exercise 18 we know that

$$\frac{dF}{dN} = \frac{e^r K^2}{[N(e^r - 1) + K]^2}$$

Clearly, because both numerator and denominator are positive, the derivative can never be negative, but oscillations cannot exist unless the eigenvalue is negative.

24 From exercise 18

$$\frac{dF}{dN} = e^{r[(K-N)]/K}\left(1 - \frac{rN}{K}\right)$$

For oscillations to exist we require

$$\left.\frac{\partial F}{\partial N}\right|_K < 0$$

Therefore

$$\left. e^{r[(K-N)]/K}\left(1 - \frac{rN}{K}\right)\right|_K < 0$$

$$e^{r[(K-N)]/K}\left(1 - \frac{rK}{K}\right) < 0 \qquad 1 - r < 0 \qquad r > 1$$

25

$N(t+1)$	$N(t+2)$	$N(t+3)$	$N(t+4)$	$N(t+5)$	$N(t+6)$
0.71	0.64	0.71	0.64	0.71	0.64

$N(t+7)$	$N(t+8)$	$N(t+9)$	$N(t+10)$
0.71	0.64	0.71	0.64

26

$N(t+1)$	$N(t+2)$	$N(t+3)$	$N(t+4)$	$N(t+5)$	$N(t+6)$
0.37	0.82	0.52	0.88	0.37	0.82

$N(t+7)$	$N(t+8)$	$N(t+9)$	$N(t+10)$	$N(t+11)$	$N(t+12)$
0.52	0.88	0.37	0.82	0.52	0.88

27

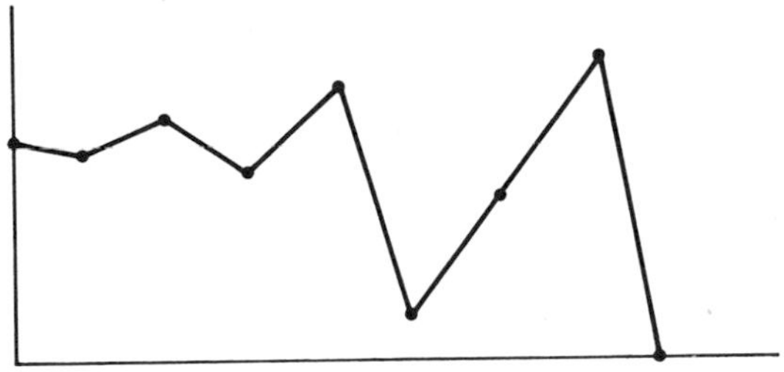

Figure 3.21

By time = 8 the population will become extinct.

28 (a) 1.01, 0.98, 1.04, 0.92, 1.17, 0.70, 1.73, 0.19, 2.20, 0.06, 1.01, 0.98, 1.04
The population exhibits a permanent cycle and the cycle has 10 distinct points.

(b) 1.86, 0.14, 1.86, 0.14, 1.86, 0.14, 1.86, 0.14, 1.86, 0.14
Even though exactly the same equation with exactly the same parameters was used, this time we obtained a cycle with only two points.

REFERENCES

Beddington, J. R., C. A. Free, and J. H. Lawton. 1975. Dynamic complexity in predator-prey models framed in difference equations. *Nature, London.* 255: 58–60.

Beverton, R. J. H. and S. J. Holt. 1957. On the dynamics of exploited fish populations. Fishery investigations, Series II, Vol. XIX, H.M.S.O. London, 1957.

Clark, C. W. 1976. *Mathematical Bioeconomics.* NY: Wiley.

Gordon, H. Scott. 1954. The economic theory of a common-property resource: the fishery. *J. Political Econ.* **62**: 124.

Guckenheimer, J., G. F. Oster, and A. Ipaktchi. 1976. The dynamics of density dependent population models. *Math. Biol.* **4**: 101–149.

Hassell, M. P., J. H. Lawton, and R. M. May. 1976. Patterns of dynamical behaviour in single-species populations. *J. Anim. Ecol.* **45**: 471–486.

Kleoden, P., M. A. B. Deakin, and A. Z. Tirkel. 1976. A precise definition of chaos. *Nature* **264**: 295.

Li, T. Y. and J. A. Yorke. 1975. Period three implies chaos, *Am. Math. Monthly*, **82**: 985–992.

May. R. M. 1974. Biological populations with nonoverlapping generations: stable points, stable cycles, and chaos. *Science* **186**:645–647.

May, R. M. 1975. Biological populations obeying difference equations: stable points stable cycles and chaos. *J. Theor. Biol.* **49**:511–524.

May. R. M. 1976. Simple mathematical models with very complicated dynamics. *Nature* **261**:459–467.

May. R. M. and G. F. Oster. 1976. Bifurcations and dynamic complexity in simple ecological models. *Am. Nat.* **110**:573–599.

Ricker, W. E. 1958. Maximum sustained yields from fluctuating environments and mixed stocks. *J. Fish. Res. Board Can.* **15**:991–1006.

Vandermeer, J. H. 1975. A graphical model of insect seed predation. *Am. Nat.* **109**:147–160.

4. Life Tables I

In the Chapter 3 we developed mathematical models for populations in which individuals were classified into discrete age or stage categories. In this chapter we derive some of the results of the last chapter and some new results by using a slightly different methodology.

THE SURVIVORSHIP CURVE. We begin with a cohort of a particular number of newborn female individuals, say 1000. We then follow them through time, recording the age at death of each. At a given age a certain number of individuals will be alive. Let z_x be the number of individuals still alive at time x. Obviously, all z_x individuals will be age x. If z_0 is the number originally in the cohort, we define the age specific survivorship as

$$l_x = \frac{z_x}{z_0}$$

Clearly $l_0 = 1$. A graph of l_x versus x is called a survivorship curve or survivorship function.

☐ EXERCISES

1 Begin with a population of 50 newborn female hamsters on July 1. By counting the population once a week we find that on July 8 48 hamsters are left; on July 15 47 hamsters are left; on July 22 45 are left, on July 29 44 are left, on August 5 44 are left, on August 12 43 are left, on August 19 40 are left, on August 26, 35 are left, on September 2, 28 are left, on September 9, 19 are left, on September 16, 8 are left, on September 23, 1 is left, and on September 30, all are dead. Graduating x in terms of weeks, draw a survivorship curve (plot l_x against x). Assume no reproduction.

2 At time zero you have 100 female guppies in an aquarium. You count the population once a day and find the following numbers

$t =$	1	2	3	4	5	6	7	8	9	10	11	12	13	14	15	16
$z_x =$	85	72	61	52	44	37	31	26	22	19	16	14	12	10	8	7
$t =$	17	18	19	20	21	22	23	24	25	26	27					
$z_x =$	6	5	4	3	3	2	2	1	1	1	0					

Assuming no reproduction, construct the survivorship curve for this population.

3 In the same aquarium you discover that a common pathogen which operates effectively against undernourished fish was the principal cause of mortality. You repeat the experiment in exercise 2, but this time you provide the fish with abundant food and treat the water with a chemical that retards the growth of the pathogen. Again you start with 100 newborn guppies and count once a day. You obtain the following data:

$t =$	0	1	2	3	4	5	6	7	8	9	10	11	12	13
$z_x =$	100	100	99	99	99	98	98	97	97	97	96	96	96	96

$t =$	14	15	16	17	18	19	20	21	22	23	24	25	26	27	28
$z_x =$	95	95	95	94	94	93	92	91	91	90	89	88	85	70	50

$t =$	29	30
$z_x =$	10	0

Assuming no reproduction, construct a survivorship curve for these numbers.

4 Repeat the above experiment, but this time a new extremely effective pathogen has invaded the aquarium. Your results are as follows.

$t = x =$	0	1	2	3	4	5	6	7	8	9	10	11	12	13
$z_x =$	100	40	10	5	3	2	2	1	1	1	1	1	1	0

Assume that no reproduction has occurred and construct the survivorship curve.

5 Plot the log of the survivorship (ln l_x) against x for the data in exercises 2, 3, and 4, all on the same graph (ignore $l_x = 0$). □

The survivorship curves of exercises 2, 3, and 4 are not typical of what actually occurs in nature, nature being sloppy. Rather they were constructed to illustrate artificially the three types of survivorship curves "classically" noted in the literature. The curve of exercise 3 is usually called a type I curve and is typical of organisms provided with parental care; humans and other primates are, of course the best examples. The curve of exercise 2 is a type II curve which indicates that a constant fraction of the population at any given age will die in each time unit, a condition encountered in many vertebrate populations. One can easily imagine, in theory, how this curve could be generated; for example, if a predator were the principle cause of mortality and did not distinguish among individuals of different ages, a type II curve would result. The curve of exercise 4 is a type III. A host of "lower" organisms and virtually all perennial plants seem to follow it.

It should be noted that a good heuristic notion of the meaning of l_x is *the probability that an individual which was alive at age* 0 *will be alive at age x*. This notation makes obvious the relationship between the approach of this chapter and that of Chapter 3.

The actual calculation of survivorship curves is straightforward if one begins with a cohort of a certain size and follows its members to death. In most situations this is impossible in nature and other techniques, which can easily be misapplied, must be used. This a topic to which we will return.

□ EXERCISES

6 Beginning with a population of 1000 newborn tsetse flies, we count the population at time 1 and find 700 flies, at time 2, 530, at time 4, 300, at time 6, 125, at time 8, 47, at time 9, 5, and at time 10, 0. Graph the survivorship curve and estimate graphically the values of l_3, l_5, l_7, $l_{0.5}$, $l_{5.5}$. □

LIFE TABLES. Data on survivorship are usually summarized in the form of a life table. Frequently demographic parameters of interest to ecologists are computed from the data organized in this fashion, and a familiarity with life-table manipulations is essential to today's ecologist.

Actually, the term life table refers to anything from l_x to virtually all computable demographic aspects of the population. For our purposes we define a life table as a table with seven columns and as many rows as there are age categories. To avoid excessively small numbers and for ease of exposition it is convenient to scale l_x in terms of larger numbers; that is, rather than setting $l_0 = 1.0$, set $l_0 = 1000$ or some other large number. Thus, in essence, l_x becomes equivalent to z_x. The seven columns are defined as follows:

Column 1 Age. Symbolized by x; in a life table x refers to an age interval, specifically the age interval x to $x + 1$.

Column 2 Survivorship, l_x.

Column 3 Age specific death rate, d_x, where $d_x = l_x - l_{x+1}$, or those that die during the time interval.

Column 4 Proportion of those entering the age category that do not survive, q_x, where $q_x = d_x/l_x$.

Column 5 The length of time lived (the number of "organism years") by all the individuals who enter the age category; that is, $L_x = \int_x^{x+1} l_n \, dn$, where n is the variable of integration. Consider how long each individual lives after entering the xth age category and sum the lengths of time. Alternatively, consider how many individuals live to be $x + 0.1$, how many to be $x + 0.2$, etc. Sum these individuals. Because l_x is rarely given in analytical form, we usually have to approximate the integral. A reasonable approximation, if l_x is a type II survivorship curve over the interval $x - (x + 1)$, is $L_x = \frac{1}{2}(l_x + l_{x+1})$.

Column 6 The total lifetime remaining for all individuals attaining the age x, or $T_x = \int_x^\infty L_n\, dn = \int_x^\infty \int_y^{y+1} l_n\, dn\, dy$, where n and y are variables of integration.

Column 7 The life expectancy of an individual aged x, $e_x = T_x/l_x$, that is, how many time units can an individual aged x expect to live?

□ EXERCISES

7 Construct a life table for the data in exercise 2 (first scale the data such that $l_0 = 1000$ and remember that for T_x you sum from the bottom of the column).

8 Let $l_x = 1/(x + 1)^2$. The survivorship values are $l_0 = 1.0$, $l_1 = 0.2500$, $l_2 = 0.1111$, $l_3 = 0.0625$, and $l_4 = 0.0400$.

(a) Compute L_0, using the equation $L_x = \frac{1}{2}(l_x + l_{x+1})$.

(b) Graph l_x against x, draw a smooth curve through the points, and estimate $L_x = \int_x^{x+1} l_n\, dn$ graphically for $x = 0$ and compare it with the value obtained in (a).

9 From the data in exercise 8 compute L_0 directly by integration. Compare this with the values obtained in exercise 8.

10 If l_1 is 500, l_0 is 1000, and l_2 is 250, what is $P_{1,0}$ in the population projection matrix?

11 Derive a general equation for $P_{x+1,x}$ (the projection matrix probability), using the survivorship function.

LOTKA'S EQUATION. Lotka's equation, derived in 1926 is perhaps the most fundamental equation in demography. With the background of Chapters 2 and 3 (so far), we can easily derive this most important equation. First we must define one more parameter, the fecundity function. Note that it is not precisely the same as the fecundity function in the projection matrix. It is usually symbolized as m_x and defined as the mean number of offspring born to individuals aged x in a given unit time period. All the babies born to all the mothers between ages x and $x + dx$ during the time intervals t and $t + dx$, divided by the total number of individuals aged between x and $x + dx$ during that time period, is m_x. Clearly, this is theoretical. To work with fecundity values we are forced to make certain approximations, which are covered later in detail.

Recalling the concept of the stable age distribution from Chapter 3, we let c_x refer to the proportion of the total *stable* population found in a small age category; that is, not in the category $x, x + 1$, but rather in the category $x, x + dx$.

If N_t is the size of the total population at time t, $N_t c_x$ is the number of individuals in the age category $x + dx$ at time t (we assume a stable age distribution—an important fact). Let B_t be the number of individuals born in the entire population over the time interval $t, t + dx$. At time $t - x$ the total number of newborns was B_{t-x}. How many of those newborns do we expect to see at time t? The survivorship function enters here and we can say that the number of individuals in the age category $x + dx$ at time t must be equal to the number born $t - x$ time ago times the probability of surviving over x time units. Thus the number of individuals in the age category $x + dx$ is equal to $B_{t-x} l_x$. Therefore we have

$$N_t C_x = l_x B_{t-x} \tag{1}$$

remembering that the population is at a stable age distribution. Again, recalling Chapter 3, if the population is at a stable age distribution the number of births per capita in the whole population is constant; that is to say B_t/N_t is the same for any value of t. We shall let $B_t/N_t = b$ and call it the instantaneous birth rate.

Recall that by assuming a stable age distribution we can write $N_t = N_0 e^{rt}$ or more generally $N_{t+y} = N_\tau e^{ry}$. Setting $t + y = t$ and $y = x$ we can also write,

$$N_t = N_{t-x} e^{rx} \tag{2}$$

Now we recall that $b = B_t/N_t$ for any value of t. Specifically, we may write

$$B_{t-x} = bN_{t-x}$$

but then we can substitute from equation 2 for N_{t-x} to get

$$B_{t-x} = bN_t e^{-rx} \tag{3}$$

From equation 1 we obtain

$$C_x = \frac{l_x B_{t-x}}{N_t}$$

and, substituting for B_{t-x} from equation 3,

$$C_x = l_x b e^{-rx} \tag{4}$$

Recalling the fecundity function, we calculate the total number of individuals born at time t as

$$B_t = \int_0^\infty N_t c_x m_x \, dx$$

because $N_t c_x$ is the number in the $x, x + dx$ age category and m_x is the number of offspring each of them is expected to produce. Substitute for c_x from equation 4 to obtain

$$B_t = \int_0^\infty N_t(l_x b e^{-rx}) m_x \, dx$$

and rearrange algebraically to obtain,

$$\frac{B_t}{N_t}\frac{1}{b} = \int_0^\infty l_x m_x e^{-rx}\,dx$$

But $b = B_t/N_t$; therefore

$$1 = \int_0^\infty l_x m_x e^{-rx}\,dx \tag{5}$$

Equation 5 is Lotka's equation which forms the basis in practice for estimating r from real data.

□ EXERCISES

12 Derive an equation for b in terms of l_x and r only.

13 Derive an equation for c_x in terms of l_x and r only.

14 Suppose that m_x has the following values (expressed in years): $m_{1.0} = 2.5$, $m_{1.2} = 2.5$, $m_{1.4} = 2.5$, $m_{1.6} = 2.2$, $m_{1.8} = 2.1$, and $m_{2.0} = 2.0$. Suppose l_x has the following values (again in years): $l_{1.0} = 0.9$, $l_{1.2} = 0.80$, $l_{1.4} = 0.75$, $l_{1.6} = 0.72$, $l_{1.8} = 0.71$, and $l_{2.0} = 0.70$. If we begin with 90 individuals exactly one year old at time 0, how many babies will they produce before they reach two years? (Assume that all babies are born at exactly $t = 0$, $t = 0.2$, $t = 0.4$, and so on).

15 In exercise 14 what is the total number of offspring born in the time interval $t = 0$ to $t = 1$, who will be alive at time $t = 1$? The following survivorship values hold; $l_0 = 1.0$, $l_{0.2} = 0.95$, $l_{0.4} = 0.93$, $l_{0.6} = 0.92$, and $l_{0.8} = 0.91$. (Assume that the births occur exactly at $t = 0$, $t = 0.2$, $t = 0.4$, and so on).

16 From the data of exercises 14 and 15 what is the average number of offspring per adult individual born in the time interval $t = 0$ to $t = 1$ who will be alive at time $t = 1$?

17 Derive an exact formula that relates $g(x)$ from Chapter 2 to m_x and l_x of this chapter. (Use the particular result from exercise 16, generalize, and let the time increment go to dx). □

THE INTRINSIC RATE OF NATURAL INCREASE. If the age specific survivorship and fecundity (l_x and m_x) are known for a population, Lotka's equation (equation 5) can be used to compute the intrinsic rate of natural increase. Clearly, the equation cannot be solved directly for r. Nevertheless, it is true that for a particular set of l_x and m_x one and only one specific value of r will satisfy equation 5 and that value is the intrinsic rate of natural increase.

Probably the best way to find the value of r is by trial and error; that is begin with any value of r and compute the integral on the right-hand side of equation 5. If the integral is greater than 1, r is too small; if the integral is less than 1, r is too large. Add or subtract some value and recompute the integral. Repeat the procedure until the integral is sufficiently close to unity. Sophisticated techniques have been devised for estimating r numerically, but they are not dealt with here. High-speed computers have made these techniques anachronistic.

Obviously one cannot truly "compute" the integral without functional forms for l_x and m_x. Usually what is given is a series of values for l_x and m_x. What we must do is approximate the integral with a sum; namely,

$$1 = \sum_{x=0}^{M} l_x m_x e^{-rx}$$

where M is the last age for which $l_x > 0$. We assume that all births occur at the midpoint of the age interval $[(x + x + 1)/2]$; therefore the value of x in the term e^{-rx} is actually $(x + x + 1)/2$ or $x + 0.5$.

☐ EXERCISES

18 Given the following values of l_x and m_x,

x	0	1	2	3	4	5
l_x	1.0	0.98	0.95	0.5	0.3	0.0
m_x	0	0	0	3	4	0

compute the intrinsic rate of natural increase to the second decimal place.

19 From the data in exercise 18 plot $\sum l_x m_x e^{-rx}$ for all the trial values of $r(r^*)$ you used.

20 Compute $\sum l_x m_x e^{-rx}$ for $r = 0.125$ and $r = 0.25$ for the life table in exercise 18. Plot $\sum l_x m_x e^{-rx}$ for the values of r between 0.1 and 0.3 (using the figures computed in this exercise and those computed in exercise 18).

21 Compute $\sum l_x m_x e^{-rx}$ for $r = 0.152$ and $r = 0.154$ and plot $\sum l_x m_x e^{-rx}$ for the values of r between 0.15 and 0.16. Estimate r graphically. ☐

The intrinsic rate of natural increase has appeared in every chapter so far. It also permeates the ecological literature. Unfortunately its use is frequently ambiguous. The concept is used to describe at least three distinct population phenomena and to understand most of the current literature one must understand all three of these uses.

First, the intrinsic rate of natural increase refers to the instantaneous rate of population growth of a population growing according to the exponential law

$$\frac{dN}{dt} = rN \quad \text{or} \quad N(t) = N(0)e^{rt}$$

Any population that exhibits constant survivorship and fecundity will eventually approach a stable age distribution and will then grow according to the exponential law. If we estimate survivorships and fecundities in a natural population and compute r, what we are actually computing is the rate at which the population would grow if it were at the stable age distribution or the rate at which it will grow when it does attain the stable age distribution. This is true, assuming that the survivorships and fecundities remain constant; that is, any population (with few exceptions) that exhibits constant survivorship and fecundity will ultimately reach a stable age distribution, at which point it will grow exponentially at the rate r.

Second, if a population exhibits density-dependent regulation—as many populations undoubtedly do—its growth rate will vary over time. Suppose survivorships and fecundities are measured in a population that, in fact, has undergone density-dependent regulation? If the intrinsic rate of natural increase is calculated, exactly what does it mean? Mathematically it means the rate of growth that the population would exhibit *if* the survivorships and fecundities were to remain constant from that time on. By stipulating that the population exhibits density-dependent feedback we have already stated that survivorships and fecundities do *not* remain constant.

These points can be clarified by reference to Figure 4.1 in which the exponential and logistic equations have been plotted:

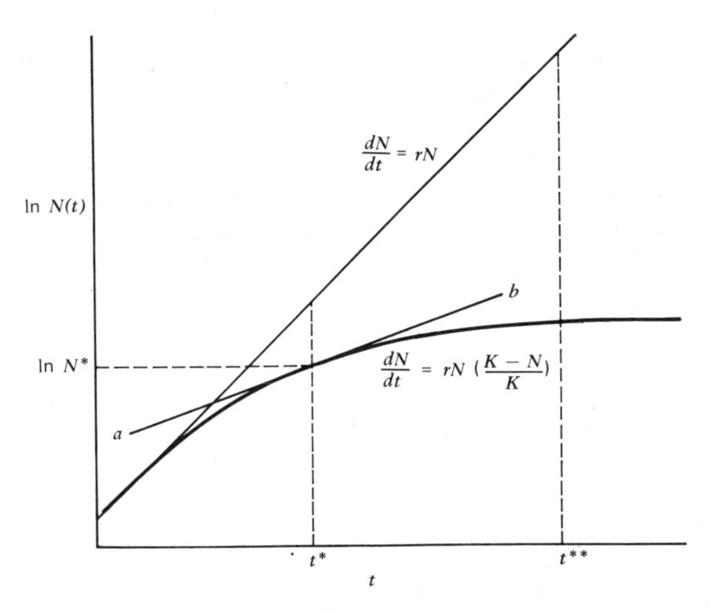

Figure 4.1

These equations have been plotted to equalize the parameter r in both. Suppose that these curves represent two populations and that both populations have attained a stable age distribution. Suppose, further, that we have some method by which we can measure age specific survivorships and fecundities in each of the populations. Consider the survivorship and fecundity data for population I (exponential or density independent) gathered at time t^*. If we plug those data into Lotka's equation and compute the intrinsic rate of natural increase as already described, the value we obtain will be r because its value is defined in the exponential equation. Exactly the same thing would happen if we computed r for population I from survivorship and fecundity data gathered at time t^{**}. Graphically, r is the slope of line I in Figure 4.1.

What if we estimate survivorship and fecundity values at time t^* in population II? If we plug these values into Lotka's equation and compute the value of r, we will not get the r that appears in the logistic or exponential equation in Figure 4.1. In fact, what we will get is the slope of the line a, b in Figure 4.1 or the instantaneous rate of increase of the log of the population at time t^*. Biologically, we compute what the rate of population growth would be if density dependence were to cease suddenly, and the survivorships and fecundities were to stay at exactly the same value they were when the population size was N^* (i.e., its size at t^*). The important point is that the computed value of r would *not* equal the r in either of the equations.

Mathematically what we compute is $[d \ln N(t)]/dt$ which is the same as $dN/N\,dt$. If the population is growing exponentially (density-independent), this quantity is equal to r, but if the population is growing according to the logistic the computed quantity is equal to $r(K - N^*)/K$. What is called the intrinsic rate of natural increase in the literature could be either one. It all depends on the underlying dynamics of the population.

There is a third use of the term intrinsic rate of natural increase which is much less common and usually carries the modifier "maximum" (i.e., maximum intrinsic rate of natural increase or r_{max}). This is a highly theoretical quantity. Its usefulness is questionable. It is defined as the maximum attainable intrinsic rate of natural increase; that is, of all the environmental and biological factors that might impinge on the growth rate of a population, there exists one unique combination of factors that prouces the maximum possible r. That value, not computable in any real sense, is r_{max}.

The literature must be read with caution. Because ecologists are not so busy as physicists, they spend a lot of time calling the same thing different names. At times the r of the logistic equation is referred to as r_{max}, whereas $d \ln N/dt$ is referred to simply as r. Indeed, you can probably find other even more creatively confusing combinations.

☐ EXERCISES

22 Given the following life table

x	l_x	m_x
0	1.0	0
1	0.8	1
2	0.5	2
3	0.0	0

compute r from Lotka's equation.

23 Express the life table as a projection matrix.

24 Begin with an age distribution vector of

$$\begin{pmatrix} 3.86 \\ 2.56 \\ 0.94 \end{pmatrix}$$

Use the projection matrix from exercise 23 and project the population four times. Graph ln $\hat{N}$ ($\hat{N}$ is the total number of individuals in the population) against time and the estimate r. How does this estimate compare with the estimate in exercise 22?

25 Graph the equation $N(t + 1) = \lambda N(t)$ (where $\lambda = e^r$) for the values of r as computed in exercises 22 and 24. (Let $N_0 = 10$.)

ANSWERS TO EXERCISES

1

x	0	1	2	3	4	5	6
z_x	50	48	47	45	44	44	43
l_x	1.0	0.96	0.94	0.90	0.88	0.88	0.86

x	7	8	9	10	11	12	13
z_x	40	35	28	19	8	1	0
l_x	0.80	0.70	0.56	0.38	0.16	0.02	0

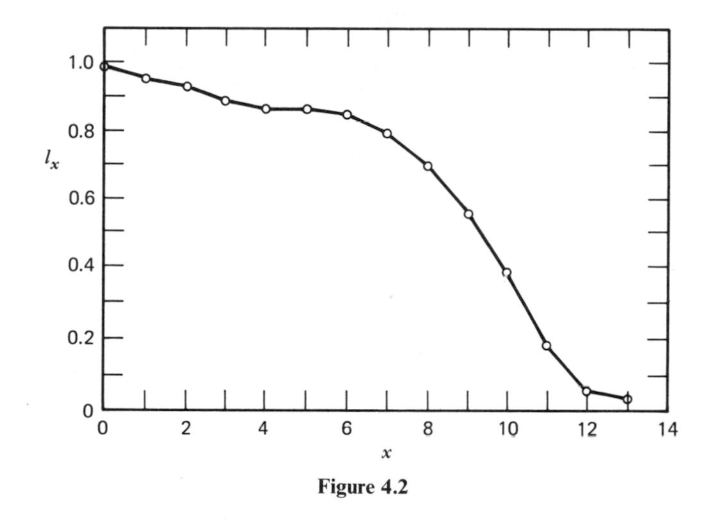

Figure 4.2

2 For our purposes $x = t$ (age = time) because we begin with a population of newborns (age = x = 0) at time zero and age and time are scaled in the same units. We generate the following table:

x	0	1	2	3	4	5	6
z_x	100	85	72	61	52	44	37
l_x	1.0	0.85	0.72	0.61	0.52	0.44	0.37
x	7	8	9	10	11	12	13
z_x	31	26	22	19	16	14	12
l_x	0.31	0.26	0.22	0.19	0.16	0.14	0.12
x	14	15	16	17	18	19	20
z_x	10	8	7	6	5	4	3
l_x	0.10	0.08	0.07	0.06	0.05	0.04	0.03
x	21	22	23	24	25	26	27
z_x	3	2	2	1	1	1	0
l_x	0.03	0.02	0.02	0.01	0.01	0.01	0

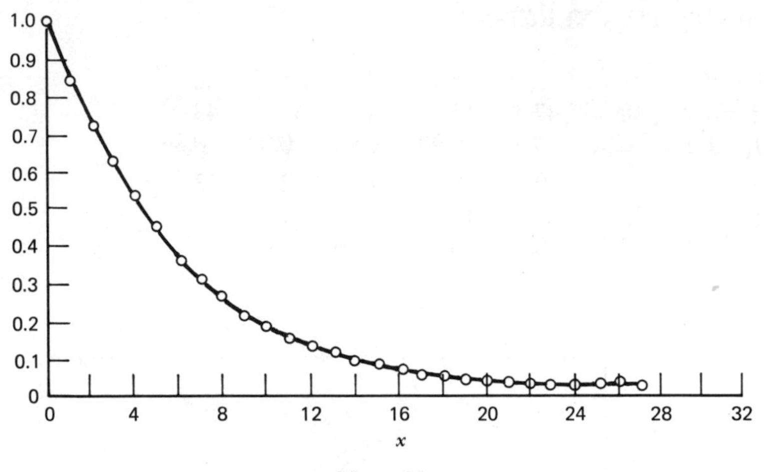

Figure 4.3

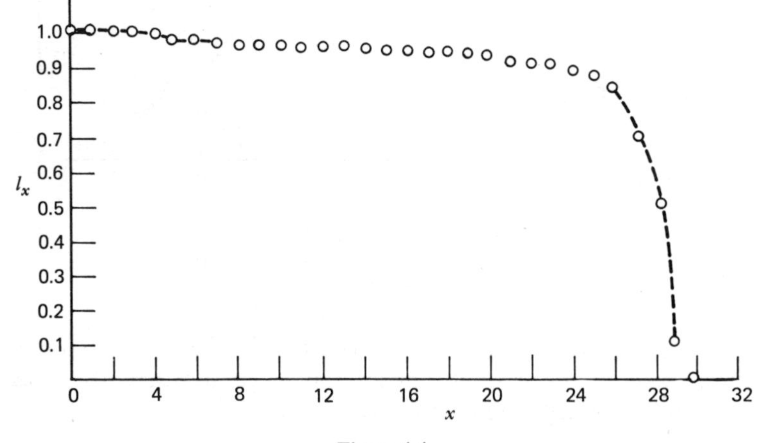

Figure 4.4

4

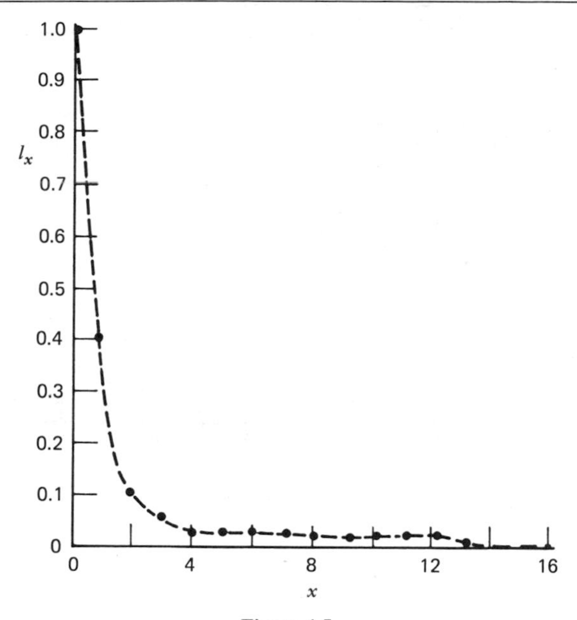

Figure 4.5

5

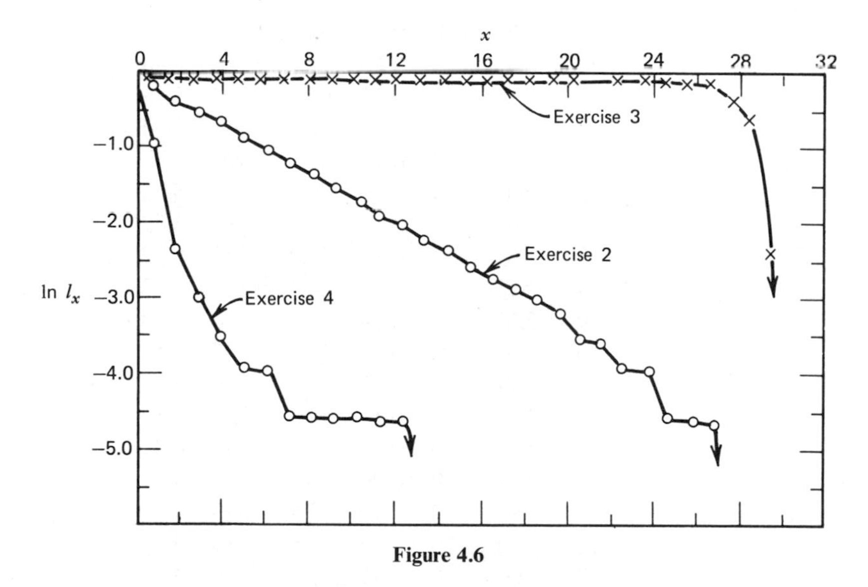

Figure 4.6

6

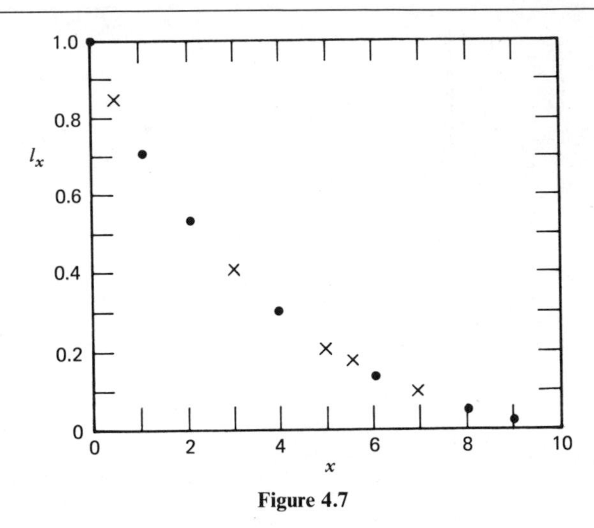

Figure 4.7

The x's are the estimated values; that is, $l_{0\cdot5} = 0.85$, $l_3 = 0.4$, $l_5 = 0.2$, $l_{5\cdot5} = 0.17$, $l_7 = 0.08$

7

x	l_x	d_x	q_x	L_x	T_x	e_x
0	1000	150	0.15	925	5940	5.94
1	850	130	0.15	785	5015	5.90
2	720	110	0.15	665	4230	5.88
3	610	90	0.15	565	3565	5.84
4	520	80	0.15	480	3000	5.77
5	440	70	0.16	405	2520	5.73
6	370	60	0.16	340	2115	5.72
7	310	50	0.16	285	1775	5.72
8	260	40	0.15	240	1490	5.73
9	220	30	0.14	205	1250	5.68
10	190	30	0.16	175	1045	5.5
11	160	20	0.12	150	870	5.4
12	140	20	0.14	130	720	5.1
13	120	20	0.17	110	590	4.9
14	100	20	0.20	90	480	4.8
15	80	10	0.12	75	390	4.9
16	70	10	0.14	65	315	4.5
17	60	10	0.17	55	250	4.2
18	50	10	0.20	45	195	3.9
19	40	10	0.25	35	150	3.8
20	30	0	0	30	115	3.8
21	30	10	0.33	25	85	2.8
22	20	0	0	20	60	3.0

x	l_x	d_x	q_x	L_x	T_x	e_x
23	20	10	0.5	15	40	2.0
24	10	0	0	10	25	2.5
25	10	0	0	10	15	1.5
26	10	10	1.00	5	5	0.5
27	0	0	0	0		

8 (a) $L_0 = \frac{1}{2}(1 + 0.2500) = 0.6250$.

(b) Graph l_x against x, draw a smooth curve, and divide the interval $x = 0$ to $x = 1$ into eight equal parts as in Figure 4.8 (eight was chosen arbitrarily). Measure the height of a rectangle drawn on each of the eight subintervals and sum the products of those heights multiplied by 0.125. Thus $L_0 = (0.90 + 0.75 + 0.62 + 0.53 + 0.44 + 0.37 + 0.32 + 0.27)\ 0.125 = 0.5250$. This value is considerably smaller than 0.6250 from part (a).

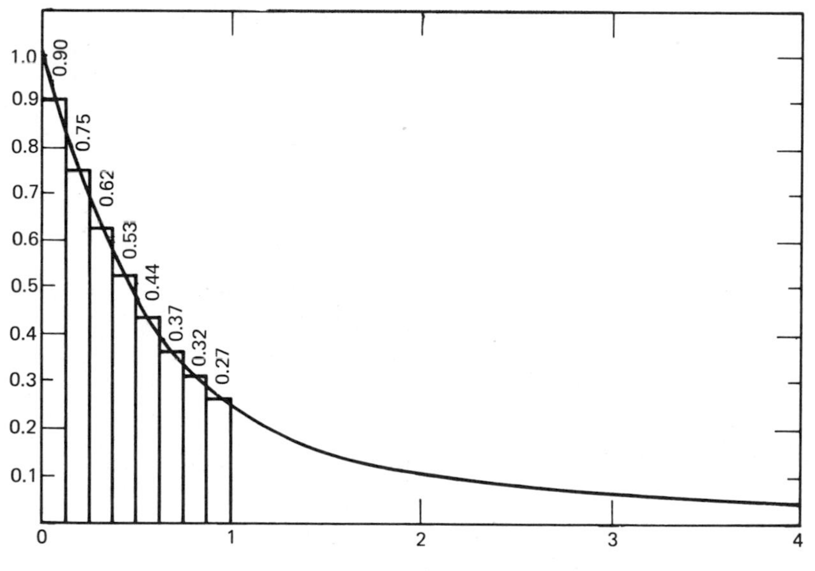

Figure 4.8

9 Because $l_x = 1/(x+1)^2$, we have

$$L_x = \int_x^{x+1} l_n \, dn = \int_x^{x+1} \frac{1}{(n+1)^2} \, dn = -\frac{1}{(1+n)}\Bigg|_x^{x+1}$$
$$= -\frac{1}{1+x+1} + \frac{1}{1+x}$$

For $x = 0$,

$$L_0 = \frac{1}{1+0} - \frac{1}{1+0+1} = 0.5000$$

This value is closer to that obtained graphically than to that obtained by the approximating equation.

10 On the average there were $(1000 + 500)/2 = 750$ individuals in the population during the first time interval ($x = 0 - 1$). On the average there were $(500 + 250)/2 = 375$ individuals in the population during the second time interval ($x = 1 - 2$). Thus $P_{10} = \frac{375}{750} = 0.50$.

11 $P_{x,x+1}$ = the number of individuals between age x and $x+1$ in the population during time t to $t+1$, divided into the number of individuals between age $x+1$ to $x+2$ during time $t+1$ to $t+2$. In general, the number of individuals existing in the population during any particular time unit is

$$L_x = \int_x^{x+1} l_x \, dx$$

Therefore

$$P_{x+1,x} = \frac{\int_{x+1}^{x+2} l_x \, dx}{\int_x^{x+1} l_x \, dx}$$

12 Equation 4 is $C_x = l_x b e^{-rx}$. Integrating, we obtain

$$\int_0^\infty C_x = \int_0^\infty l_x b e^{-rx} \, dx$$

But because C_x is a proportion

$$\int_0^\infty C_x = 1$$

and

$$1 = b \int_0^\infty l_x e^{-rx} \, dx$$

whence

$$b = \frac{1}{\int_0^\infty l_x e^{-rx}\, dx}$$

13 $C_x = l_x b e^{-rx}$ and because

$$b = \frac{1}{\int_0^\infty l_x e^{-rx}\, dx}$$

we have

$$C_x = \frac{l_x e^{-rx}}{\int_0^\infty l_x e^{-rx}\, dx}$$

14 The question is the same as how many babies will be produced by those individuals between 1 and 2 years of age during the time interval $t = 0$ to $t = 1$. During the time interval $t = 0$ to $t = 0.2$ all living individuals produce 2.5 individuals. However, because the 90 individuals alive at $t = 0$ were the result of an original cohort of 100 ($l_1 = 0.9$), we expect that at $t = 0.2$ we shall have only 80 individuals left. If none of the individuals had died during the interval $t = 0$ to $t = 0.2$ (i.e., the 10 individuals died just before $t = 0.2$, perhaps exactly at the point $t = 0.199999\cdots$), there would have been $90(2.5) = 225$ babies. If they had died just after $t = 0$ (say, at $t = 0.000000001$), there would have been $80(2.5) = 200$ babies. Which number should be used to compute the number of babies born between $t = 0$ and $t = 0.2$? Clearly, there is no exact answer unless we know the exact form of l_x, which we do not. We must make some sort of assumption about the way deaths and/or births are occurring during the period $t = 0$ to $t = 0.2$.

In particular, if we assume that all births occur at exactly $t = 0$, $t = 0.2$, $t = 0.4$, $t = 0.6$, $t = 0.8$, and $t = 1.0$, it does not matter how l_x behaves in between these points. We know that at $t = 0$ there are 90 individuals and that each of these individuals produces 2.5 babies. Thus between $t = 0$ and $t = 0.2$, $90(2.5) = 225$ babies are produced. Similarly, for $t = 0.2$ to $t = 0.4$, $(80)(2.5) = 200$. For $t = 0.4$ to $t = 0.6$, $(75)(2.5) = 187.5$. For $t = 0.6$ to $t = 0.8$, $(72)(2.2) = 158.4$. For $t = 0.8$ to $t = 1.0$, $(71)(2.1) = 149.1$. Thus the total number born in the interval $t = 0$ to $t = 1$ (those born at exactly $t = 1$ are considered in the interval $t = 1$ to $t = 2$) is $225 + 200 + 187.5 + 158.4 + 149.1 = 920$.

15 During the period $t = 0$ to $t = 0.2$, 225 babies were born. We assume that the 225 babies are born at exactly $t = 0$. Therefore those babies are subjected to a 0.95 survivorship from the period $t = 0$ to $t = 0.2$. Thus at $t = 0.2$ there will be $(0.95)(225) = 213.75$ babies left. By time $t = 1$ there will be $(0.9)(225) = 202.5$ left. During the period $t = 0.2$ to $t = 0.4$, 200 babies were born. By

$t = 1$ those babies will have gone through 0.8 years of life (we presume that they were born at exactly $t = 0.2$) and there will be $(l_{0.8})(200)$ of them left or $(0.91)(200) = 182.0$. For the 187.5 born during the period $t = 0.4$ to $t = 0.6$, $(187.5)(l_{0.6})$ will have survived or $(187.5)(0.92) = 172.5$. For the period $t = 0.6$ to $t = 0.8$, $(158.5)(0.93) = 147.4$. For the period $t = 0.8$ to $t = 1.0$, $(149.1)(0.95) = 141.6$. Thus the total alive at $t = 1.0$ will be $202.5 + 182.0 + 172.5 + 147.4 + 141.6 = 846.0$.

16 Because m_x refers to the average number of offspring per individual in an interval of length 0.2, we have $(2.5)l_1 + (2.5)l_{0.8} + (2.5)l_{0.6} + (2.2)l_{0.4} + (2.1)l_{0.2} = (2.5)0.9 + (2.5)0.91 + 2.5(0.92) + 2.2(0.93) + 2.1(0.95) = 10.87$. We could also have computed this figure by dividing the total number of babies produced during the period $t = 0$ to $t = 1$ who are alive at $t = 1$ by the average number of adults alive during that time period. The average number of adults, it will be recalled from exercise 14, is one-fifth of $90 + 80 + 75 + 72 + 71 = \frac{388}{5} = 77.6$. Thus from exercise 15 the number of living newborns at $t = 1$ produced during the period $t = 0$ to $t = 1$ is 846. Therefore $846/77.6 = 10.9$.

17 Allowing m_x to refer to the number of births during the period $x + \Delta x$, we see from exercise 16 that if all births occur at the beginning of the time interval each m_x must be multiplied by an appropriate l_x. For exercise 16 we had $m_1 l_1 + m_{1.2} l_{0.8} + m_{1.4} l_{0.6} + m_{1.6} l_{0.4} + m_{1.8} l_{0.2} = g(1)$. If x were equal to (say) 3, $m_3 l_1 + m_{3.2} l_{0.8} + m_{3.4} l_{0.6} + \cdots$. In general, for increments of 0.2 we may write

$$g(x) = \sum_{y=x}^{x+1} m_y l_{x-y+1}$$

where y is incremented by 0.2 in the summation. If we let the increment get very small (approach dx)

$$g(x) = \int_x^{x+1} m_y l_{x-y+1}\, dy$$

18 We presume that all births happen at the midpoint of the time interval. We therefore must use the l_x which is the midpoint of the interval. Begin with $r = 0$. Construct the following table:

x	l_x	m_x	$-rx$	e^{-rx}	$l_x m_x e^{-rx}$
0.5	0.99	0	0	1	0
1.5	0.96	0	0	1	0
2.5	0.72	0	0	1	0
3.5	0.40	3	0	1	1.2
4.5	0.15	4	0	1	0.6

$$\sum l_x m_x e^{-rx} = 1.8$$

The sum is too large. We need a larger value of r. Try $r = 0.5$. (Eliminate the first three age categories from the calculations because $m_x = 0$.)

x	l_x	m_x	$-rx$	e^{-rx}	$l_x m_x e^{-rx}$
3.5	0.4	3	−1.75	0.173	0.2076
4.5	0.15	4	−2.25	0.105	0.0630

$$\sum l_x m_x e^{-rx} = 0.2706$$

That sum is too small. Try $r = 0.1$.

x	l_x	m_x	$-rx$	e^{-rx}	$l_x m_x e^{-rx}$
3.5	0.4	3	−0.35	0.705	0.8460
4.5	0.15	4	−0.45	0.638	0.3828

$$\sum l_x m_x e^{-rx} = 1.2288$$

The sum is too large. We need a larger r. Try $r = 0.3$.

x	l_x	m_x	$-rx$	e^{-rx}	$l_x m_x e^{-rx}$
3.5	0.4	3	−1.05	0.350	0.420
4.5	0.15	4	−1.35	0.259	0.155

$$\sum l_x m_x e^{-rx} = 0.575$$

The sum is too small. We need a smaller r. Try $r = 0.2$.

x	l_x	m_x	$-rx$	e^{-rx}	$l_x m_x e^{-rx}$
3.5	0.4	3	−0.70	0.487	0.5964
4.5	0.15	4	−0.90	0.406	0.2436

$$\sum l_x m_x e^{-rx} = 0.840$$

The sum is too small. Thus we now know that the value of r is between 0.1 and 0.2. Try $r = 0.15$.

x	l_x	m_x	$-rx$	e^{-rx}	$l_x m_x e^{-rx}$
3.5	0.4	3	−0.525	0.591	0.7092
4.5	0.15	4	−0.675	0.509	0.3054

$$\sum l_x m_x e^{-rx} = 1.0146$$

The sum is too large. We need a larger r. Try 0.17.

x	l_x	m_x	$-rx$	e^{-rx}	$l_x m_x e^{-rx}$
3.5	0.4	3	−0.595	0.551	0.6612
4.5	0.15	4	−0.765	0.465	0.2790

$$\sum l_x m_x e^{-rx} = 0.9402$$

We need a smaller r. Try 0.16.

x	l_x	m_x	$-rx$	e^{-rx}	$l_x m_x e^{-rx}$
3.5	0.4	3	−0.560	0.571	0.6852
4.5	0.15	4	−0.720	0.487	0.2922

$$\sum l_x m_x e^{-rx} = 0.9774$$

Clearly, the value of r is between 0.16 and 0.15. Try 0.155.

x	l_x	m_x	$-rx$	e^{-rx}	$l_x m_x e^{-rx}$
3.5	0.4	3	−0.5424	0.582	0.6984
4.5	0.15	4	−0.6975	0.498	0.2988

$$\sum l_x m_x e^{-rx} = 0.99720$$

Thus the value of r is between 0.15 and 0.155, which, after rounding to the second decimal, is 0.15.

19

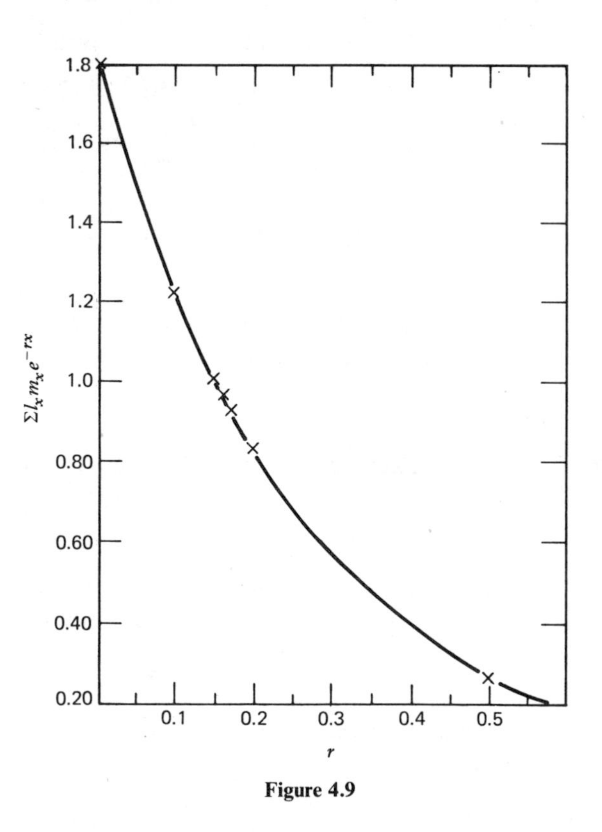

Figure 4.9

20 For $r = 0.125$

x	l_x	m_x	$-rx$	e^{-rx}	$l_x m_x e^{-rx}$
3.5	0.4	3	−0.4375	0.646	0.7752
4.5	0.15	4	−0.562	0.570	0.3420

$$\sum l_x m_x e^{-rx} = 1.1172$$

For $r = 0.25$

x	l_x	m_x	$-rx$	e^{-rx}	$l_x m_x e^{-rx}$
3.5	0.4	3	−0.875	0.417	0.5004
4.5	0.15	4	−1.125	0.325	0.1950

$$\sum l_x m_x e^{-rx} = 0.6954$$

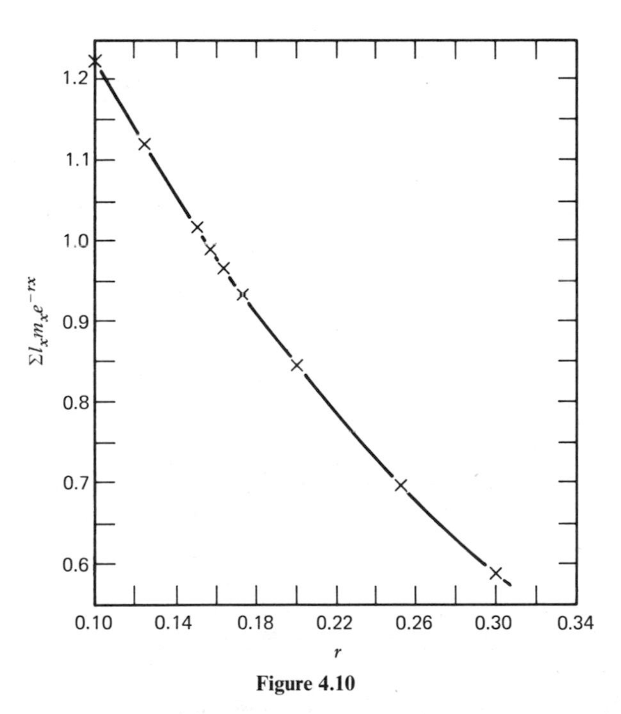

Figure 4.10

21 For $r = 0.152$

x	l_x	m_x	$-rx$	e^{-rx}	$l_x m_x e^{-rx}$
3.5	0.4	3	−0.532	0.587	0.7044
4.5	0.15	4	−0.684	0.504	0.3024

$$\sum l_x m_x e^{-rx} = 1.0068$$

For $r = 0.154$

x	l_x	m_x	$-rx$	e^{-rx}	$l_x m_x e^{-rx}$
3.5	0.4	3	−0.539	0.583	0.6996
4.5	0.15	4	−0.693	0.500	0.3000

$$\sum l_x m_x e^{-rx} = 0.9996$$

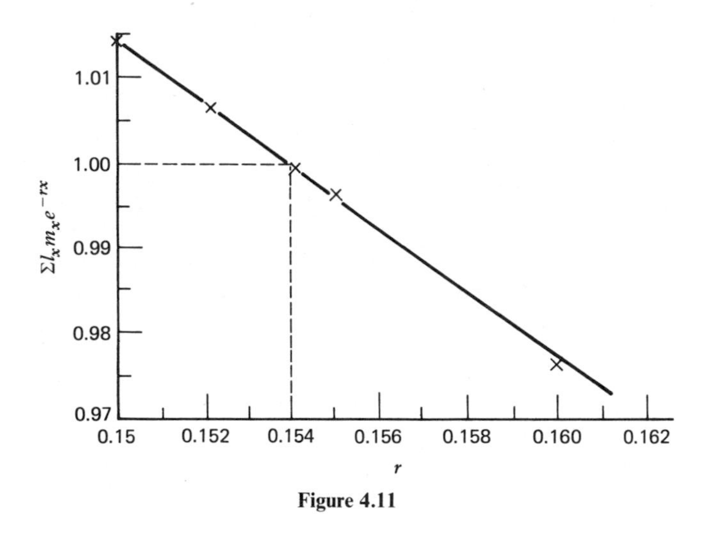

Figure 4.11

By a graphical estimate $r = 0.1539$.
Check

x	l_x	m_x	$-rx$	e^{-rx}	$l_x m_x e^{-rx}$
3.5	0.4	3	−0.5386	0.5835	0.70020
4.5	0.15	4	−0.6925	0.5003	0.30018

$$\sum l_x m_x e^{-rx} = 1.00038$$

Note from the graph that at this level of resolution round-off errors are probably a greater source of error than the basic curvilinear nature of the graph.

22 Clearly, in this case, we must satisfy the equation

$$0.65e^{-r(1.5)} + 0.5e^{-r(2.5)} = 1$$

Try $r = 0.05$.

$$(0.65)(0.928) + (0.5)(0.882) = 1.044$$

Try $r = 0.06$.

$$(0.65)(0.914) + (0.5)(0.861) = 1.025$$

Try $r = 0.07$.

$$(0.65)(0.900) + (0.5)(0.839) = 1.0045$$

Graph

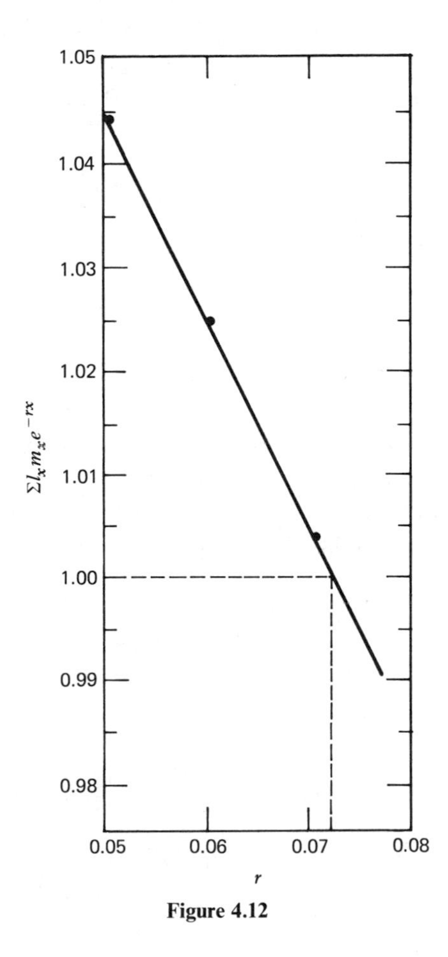

Figure 4.12

By a graphical estimate $r = 0.072$.

23 From exercise 11

$$P_{x+1,x} = \frac{L_{x+1}}{L_x}$$

Approximating L_x as $(l_x + l_{x+1})/2$

$$L_0 = \frac{1.8}{2} = 0.9$$

$$L_1 = \frac{0.8 + 0.5}{2} = 0.65$$

$$L_2 = \frac{0.5}{2} = 0.25$$

whence

$$P_{10} = \frac{0.65}{0.9} = 0.722$$

$$P_{21} = \frac{0.25}{0.65} = 0.385$$

From exercise 17

$$g(x) = \int_x^{x+1} m_y l_{x-y+1}\, dy$$

We approximate the integral by assuming that all births occur equally over the time interval and therefore that approximately one-half of the births occur during the first half of the time interval and one-half during the second half. Thus one-half of the offspring are subjected to a survivorship of l_1 before they enter the next time period; the others, a survivorship of l_0. Thus we can state approximately that

$$g(x) = \frac{l_0 m_x + l_1 m_x}{2}$$

From which we obtain

$$g(0) = 0$$

$$g(1) = \frac{1 + 0.8}{2} = \frac{1.8}{2} = 0.9$$

$$g(2) = \frac{2 + 1.6}{2} = \frac{3.6}{2} = 1.8$$

24

$$\begin{vmatrix} 0 & 0.9 & 1.8 \\ 0.72 & 0 & 0 \\ 0 & 0.38 & 0 \end{vmatrix} \begin{vmatrix} 3.86 \\ 2.56 \\ 0.94 \end{vmatrix} = \begin{vmatrix} 4.00 \\ 2.78 \\ 0.97 \end{vmatrix} = \begin{vmatrix} 4.25 \\ 2.88 \\ 1.06 \end{vmatrix} = \begin{vmatrix} 4.50 \\ 3.06 \\ 1.09 \end{vmatrix} = \begin{vmatrix} 4.72 \\ 3.24 \\ 1.16 \end{vmatrix}$$

t	$\hat{N}$	$\ln \hat{N}$
0	7.36	1.996
1	7.75	2.048
2	8.19	2.102
3	8.65	2.158
4	9.12	2.210

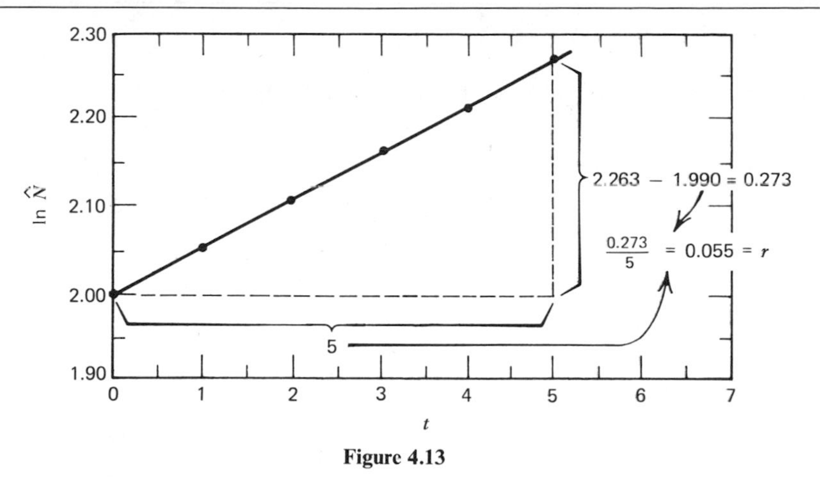

Figure 4.13

Thus $r = 0.055$ from the projection technique and $r = 0.072$ from exercise 22.

25

t	$N(t)_1$	$N(t)_2$
0	10	10
1	10.54	10.78
2	11.11	11.62
3	11.71	12.53
4	12.34	13.50
5	13.01	14.56
6	13.71	15.69
7	14.45	16.92
8	15.23	18.24
9	16.06	19.66
10	16.92	21.19
11	17.84	22.84
12	18.80	24.63
13	19.82	26.55
14	20.88	28.62
15	22.01	30.85

where for population 1 $r = 0.055$ and for population 2 $r = 0.072$. λ for population 1 = 1.056, λ for population 2 = 1.075. Obviously the projections are quite different. In fact, the differences result from slightly different assumptions in computing the intrinsic rate. In both cases we approximate integrals with discrete time intervals. Apparently the approximations are different in the two ways of estimating r. Using the eigenvalue of the projection matrix, we usually obtain a slightly lower value of r than by solving Lotka's equation. The difference is greatly exaggerated when the number of age categories is small.

REFERENCES

References for Chapters 4 and 5 are listed at the end of Chapter 5.

5. Life Tables II

STABLE VERSUS STATIONARY POPULATIONS. In the last two chapters the concept of the stable age distribution was central. In an age- or stage-distributed population once a stable age distribution has been obtained the population as a whole behaves according to the equation

$$\frac{dN}{dt} = rN$$

where r is the intrinsic rate of natural increase. Furthermore, if survivorships and fecundities (and all other transition probabilities in a *stage*-projection matrix) remain constant over time, the population to which those survivorships and fecundities refer will ultimately approach the stable age distribution.

A particularly interesting case is when the intrinsic rate of increase is equal to zero; that is, when the population reaches its stable age distribution it will neither increase nor decrease in numbers from that time on. A population for which $r = 0$ is called a *stationary* population or is said to have a stationary age distribution.

Consider what happens to Lotka's equation in a stationary population. We have

$$1 = \sum_{x=0}^{M} l_x m_x e^{-rx}$$

from Chapter 4. If $r = 0$, $e^{-rx} = 1$, and Lotka's equation becomes

$$1 = \sum_{x=0}^{M} l_x m_x$$

The term $\sum l_x m_x$ is frequently given the symbol R_0.

□ EXERCISES

1 Given the following life tables, determine whether $r = 0$ for each life table (it is not necessary to compute r):

Population 1			Population 2			Population 3		
x	l_x	m_x	x	l_x	m_x	x	l_x	m_x
0.5	0.98	0	0.5	0.98	0	0.5	0.98	0
1.5	0.95	0	1.5	0.90	0	1.5	0.90	0

2.5	0.50	1.40	2.5	0.40	1.40	2.5	0.60	1.40
3.5	0.40	0.75	3.5	0.20	0.75	3.5	0.50	0.75

2 What can you say about R_0 if $r < 0$? If $r > 0$? □

The stable age distribution has a particularly simple form in a stationary population. Recalling the equation derived in exercise 13, Chapter 4, the stable age distribution is

$$C_x = \frac{l_x e^{-rx}}{\int_0^\infty l_x e^{-rx}\, dx}$$

which, if $r = 0$, reduces to

$$C_x = \frac{l_x}{\int_0^\infty l_x\, dx}$$

The integral in the denominator is a constant; therefore

$$C_x \propto l_x$$

Thus in a stationary population the stable age distribution is proportional to the survivorship function. This is an extremely important and useful result. Indeed, if there is reason to assume that a particular population is stationary, its survivorship can be calculated directly from the age distribution.

□ EXERCISES

3 Given the following projection matrix

$$\begin{vmatrix} 0 & 2 & 11 & 16 & 2 \\ 0.2 & 0 & 0 & 0 & 0 \\ 0 & 0.2 & 0 & 0 & 0 \\ 0 & 0 & 0.2 & 0 & 0 \\ 0 & 0 & 0 & 0.2 & 0 \end{vmatrix}$$

compute l_x (using the equation from exercise 11, Chapter 4).

4 Use the first four rows and columns of the matrix in exercise 3 (because $l_4 = 0.002$, we ignore the last age category) and project the following vector until you obtain the stable age distribution (round off to the nearest integer at each projection):

$$\begin{vmatrix} 110 \\ 17 \\ 3 \\ 2 \end{vmatrix}$$

Compare the stable age distribution with the survivorship function in exercise 3.

5 Compute the stable age distribution for all three populations in exercise 1. (Recall exercise 13, Chapter 4). Plot c_x (the stable age distribution) against l_x (scale l_x such that $\sum l_x = 1$, before plotting; this makes c_x and l_x fall on the 45° line in $r = 0$). □

It might frequently be convenient to assume that $r = 0$ and estimate l_x by the actual proportional distribution of individuals into the various age categories. Care must be taken to avoid one particular trap that ecologists occasionally fall into. It is ever so tempting to "assume" a stationary population, estimate l_x by the observed distribution of individuals among ages, and then check on the "assumption" by computing R_0. Remarkably, computing R_0 in this way almost invariably yields a number close to 1!

There are basically two methods of computing life tables from field data. First, the cohort method, or vertical life table, and second the "current" method, or horizontal life table. In a vertical life table a set of newborn individuals, a cohort, is followed through time until all its members have died (e.g., as in exercise 1 of Chapter 4). Those remaining alive x time units after the cohort was first born, divided by the number originally in the cohort, is l_x. The computation is quite direct.

Second, in the current method, or horizontal life table, individuals are censused at one point in time. Typically this method takes two forms. First, we count the extant population, assume it is stationary (at a stable age distribution and $r = 0$), and estimate l_x from the actual distribution of individuals in the age category. Second, we follow individuals of all ages (or stages) through one time unit, observing which individuals die (and change the stage category in a stage-projection matrix). The number that survives during the time interval, divided by the total number in the time interval, is a direct estimate of $P_{x+1,x}$, assuming a stable age distribution. Of course, the transition probabilities can be converted easily to the more traditional survivorship figures. This approach was taken in exercise 21, Chapter 2.

□ EXERCISES

6 Suppose we observe five different cohorts, presumably in identical environments, each beginning at five different times and with different numbers of newborns. The numbers and times are as follows:

t	$N(t)_1$	$N(t)_2$	$N(t)_3$	$N(t)_4$	$N(t)_5$
0	1000				
1	900		600		

2	750	80	540		
3	500	72	450		150
4	100	60	300	200	135
5	0	40	60	180	112
6		8	0	150	75
7		0		100	15
8				20	0
9				0	

Compute l_x for each cohort, using the cohort or vertical life table method.

7 Suppose now that you cannot distinguish among these cohorts, that all five cohorts look like just one large population to you, but that you can tell the exact age of every individual in the population. Thus at time 4 you observe 200 newborns, 135 one-year-olds, 60 two-year-olds, 300 three-year-olds, and 100 four-year-olds. You are tempted to assume the population is at a stationary state. Do so and compute l_x only from the data observed at time 4. Compare with l_x as computed in exercise 6.

8 Suppose that you observe the same population (really the collection of five cohort populations) at $t = 5$, in addition to $t = 4$. Obviously, you observe $N_1 = 180$, $N_2 = 112$, $N_3 = 40$, and $N_4 = 60$. Compute the transition probabilities for the population-projection matrix.

9 Compute l_x from these transition probabilities. Remember, you have complete information; that is, you know the exact ages of all individuals. Furthermore, within one age category all individuals are the same age and were born at the beginning of the time interval. This will be important in the computation of L_x.

10 Suppose we census a population at 10-month intervals. Suppose there are only two age categories and that our census technique can distinguish only 10 month categories; that is, for each individual in the population we can only determine whether it is <10 months old or >10 months old. The survivorships are $l_0 = 1$, $l_{0.2} = 0.9$, $l_{0.4} = 0.8$, $l_{0.6} = 0.7$, $l_{0.8} = 0.6$, $l_{1.0} = 0.5$, $l_{1.2} = 0.4$, $l_{1.4} = 0.3$, $l_{1.6} = 0.2$, $l_{1.8} = 0.1$, $l_{2.0} = 0$. Let us sample the population on July 1, 1976, and May 1, 1977. (a) We don't know it, but the population started on January 1, 1976, with just 100 newborn individuals. How many individuals will be alive when we sample July 1, 1976? What will their ages be? (Remember, we can tell only if an individual is between 0–1 or between 1–2) How many will be alive (and how old) when we sample May 1, 1977? Estimate $P_{1,0}$ for the projection matrix, where the time interval is 10 months. (b) Repeat (a), but suppose that the population started with 100 newborns on May 1, 1976. (c) What can you conclude about how within-category distribution affects the estimate of the elements in the projection matrix? □

BIRTH RATES, DEATH RATES, AND MEAN GENERATION TIME. The number of individuals that die from time t to time $t + 1$, divided by the number of individuals in the population during that same time period, is the "crude" death rate. Similarly, the number of births from time t to time $t + 1$, divided by the number of individuals in the population during that same time period, is the "crude" birth rate. If the population is at the stable age distribution, the crude birth rate is equal to the intrinsic birth rate (frequently called simply the birth rate), and the crude death rate is equal to the intrinsic death rate (frequently called simply the death rate).

☐ EXERCISES

11 Using the projection matrix of exercise 4, project the following population one time unit:

$$\begin{matrix} 110 \\ 17 \\ 3 \\ 2 \end{matrix}$$

What is the crude birth and crude death rate?

12 Using the same matrix, project the following population one time unit:

$$\begin{matrix} 100 \\ 20 \\ 4 \\ 1 \end{matrix}$$

What is the crude birth rate and crude death rate?

13 From the life table computed in exercise 3 use the following equation to compute the intrinsic birth rate and compare it with that in exercise 12.

$$b = \frac{1}{\int_0^\infty l_x e^{-rx}\, dx}$$

14 Compute the intrinsic death rate and compare it with that in exercise 12. ☐

The total number of babies produced by an individual in its entire life time is $\sum_{x=0}^{M} m_x$ if the individual lives to M (remember we are considering only the female sector of the population). If the individual lives to $M - 2$, the total number of babies will be $\sum_{x=0}^{M-2} m_x$. Clearly, on the average, the number of babies produced by a female in the population is $\sum l_x m_x$ because m_x is the number produced at age x and l_x is the probability of reaching age x. Thus for each individual living now we expect $\sum l_x m_x$ individuals to exist in the "next generation." Clearly, if T is the average length of a generation, we can say that if the population size now is $N(t)$ the population size T from now will be $(\sum l_x m_x)\, N(t)$; that is,

$$N(t + T) = R_0\, N(t)$$

We also know that if we have a stable population, in general (i.e., for any value of n),

$$N(t + n) = \lambda^n\, N(t) = e^{rn}\, N(t)$$

and, in particular,

$$N(t + T) = e^{rT}\, N(t)$$

Clearly,

$$R_0 = e^{rT}$$

and the mean generation time is given as

$$T = \frac{\ln R_0}{r}$$

Note that to find the mean generation time we must calculate r. Approximate methods are available for computing T without estimating r, but we shall not go into them.

□ EXERCISES

15 For the following projection matrix

$$\begin{vmatrix} 0 & 2.04 & 12.73 & 20.37 \\ 0.20 & 0 & 0 & 0 \\ 0 & 0.20 & 0 & 0 \\ 0 & 0 & 0.20 & 0 \end{vmatrix}$$

compute R_0, r, and T (mean generation time). Remember that l_x for these transition probabilities has already been given (see exercise 3) as

x	l_x
0	1.0
1	0.20
2	0.04
3	0.008

The computation for m_x is somewhat difficult (see exercise 17, Chapter 4). Do not bother to compute them; they are $m(1) = 4$, $m(2) = 25$, $m(3) = 40$.

16 Compute the stable age distribution from the life table in exercise 15.

17 Consider a population of 10,000 individuals in the stable age distribution computed in exercise 16. Project the population seven times. What is the approximate value of r from a graph of $\ln N(t)$ against t? What is the stable age distribution (approximately) by the projection technique?

18 Using a graph of $\ln N(t)$ against t for the data produced in exercise 17, show graphically the value of T, given the value of R_0 computed in exercise 15.

19 Clearly, such deviant estimates of r (computed in exercise 18) must be accounted for. Remembering that the l_x we used for calculating r was $(l_x + l_{x+1})/2$, let us use a more accurate estimate of the midpoint. The following are the midpoint values for l_x, using

$$L_x = \int_x^{x+1} l_y \, dy$$

for the data in exercise 18:

x	l_x
0.5	0.45
1.5	0.09
2.5	0.018
3.5	0.004

Compute R_0. What can you conclude from exercises 18 and 19?

20 Consider the following life table:

x	l_x	m_x
0	1.0	0
1	0.20	40
2	0.04	250
3	0.008	400

The exact midpoint values are

x	l_x
0.5	0.45
1.5	0.09
2.5	0.018
3.5	0.004

Compute R_0, r, T, and c_x.

21 The projection matrix for the life table in exercise 20 is

$$\begin{vmatrix} 0 & 20.4 & 127.3 & 203.7 \\ 0.2 & 0 & 0 & 0 \\ 0 & 0.2 & 0 & 0 \\ 0 & 0 & 0.2 & 0 \end{vmatrix}$$

Begin with a population of 1000, distributed according to the stable age distribution computed in exercise 20. Project the population four times and plot ln $N(t)$ against t. Estimate r and compare with the estimate in exercise 20. On the same graph plot the line that corresponds to r obtained in exercise 20.

ANSWERS TO EXERCISES

1 If $r = 0$, $R_0 = 1$. Thus we need only to compute $R_0 = \sum l_x m_x$. For population 1 $R_0 = (0.5)(1.40) + (0.40)(0.75) = 1.00$. Therefore $r = 0$ for population 1. For population 2, $R_0 = 0.710$. Therefore $r \neq 0$. For population 3, $R_0 = 1.215$. Therefore $r \neq 0$.

2 If $r < 0$, e^{-rx} will be a number greater than 1 for all values of x. Because each term in $\sum l_x m_x$ is multiplied by some value greater than 1 for Lotka's equation to be true, $\sum l_x m_x$ must be less than 1 for $r < 0$. Similar reasoning shows that $R_0 > 1$ if $r > 0$.

3 From exercise 11 of chapter 4

$$P_{x+1,x} = \frac{\int_{x+1}^{x+2} l_x\,dx}{\int_{x}^{x+1} l_x\,dx} = \frac{\frac{1}{2}(l_{x+1} + l_{x+2})}{\frac{1}{2}(l_x + l_{x+1})}$$

In particular, for this 5 × 5 case

$$P_{10} = \frac{l_1 + l_2}{l_0 + l_1} = \frac{l_1 + l_2}{1 + l_1} \qquad (1)$$

$$P_{21} = \frac{l_2 + l_3}{l_1 + l_2} \qquad (2)$$

$$P_{32} = \frac{l_3 + l_4}{l_2 + l_3} \qquad (3)$$

$$P_{43} = \frac{l_4 + l_5}{l_3 + l_4} = \frac{l_4}{l_3 + l_4} \qquad (4)$$

From equation 4

$$P_{43} = 0.2 = \frac{l_4}{l_3 + l_4}$$

$$0.2l_3 + 0.2l_4 = l_4$$

$$l_4(1 - 0.2) = 0.2l_3$$

$$l_4 = \frac{0.2}{0.8} l_3 = 0.25l_3 \qquad (5)$$

Substituting into equation 3 we have

$$0.2 = \frac{l_3 + 0.25l_3}{l_2 + l_3}$$

$$0.2l_2 + 0.2l_3 = l_3 + 0.25l_3$$

$$0.2l_2 = l_3(1 + 0.25 - 0.2) = 1.05l_3$$

$$l_3 = \frac{0.2}{1.05} l_2 = 0.19l_2 \quad (6)$$

Substituting into equation 2 yields

$$0.2 = \frac{l_2 + 0.19l_2}{l_1 + l_2}$$

$$0.2l_1 + 0.2l_2 = l_2 + 0.19l_2$$

$$0.2l_1 = l_2(1 + 0.19 - 0.2) = 0.99l_2$$

$$l_2 = 0.20l_1$$

Substituting into equation 1 gives us

$$0.2 = \frac{l_1 + 0.20l_1}{1 + l_1}$$

$$0.2 + 0.2l_1 = l_1 + 0.2l_1$$

$$l_1 = 0.2 \quad (7)$$

Substitute back into equation 7:

$$l_2 = (0.2)(0.2) = 0.04$$

Substitute back into equation 6:

$$l_3 = (0.19)(0.04) = 0.0076$$

Substitute back into equation 5:

$$l_4 = (0.25)(0.0076) = 0.0019$$

Thus the survivorship table is

x	l_x
0	1.0
1	0.20
2	0.04
3	0.008
4	0.002

4

$$\begin{vmatrix} 0 & 2 & 11 & 16 \\ 0.2 & 0 & 0 & 0 \\ 0 & 0.2 & 0 & 0 \\ 0 & 0 & 0.2 & 0 \end{vmatrix} \begin{vmatrix} 110 \\ 17 \\ 3 \\ 2 \end{vmatrix} = \begin{vmatrix} 99 \\ 22 \\ 3 \\ 1 \end{vmatrix} = \begin{vmatrix} 93 \\ 20 \\ 4 \\ 1 \end{vmatrix} = \begin{vmatrix} 100 \\ 19 \\ 4 \\ 1 \end{vmatrix}$$

$$= \begin{vmatrix} 98 \\ 20 \\ 4 \\ 1 \end{vmatrix} = \begin{vmatrix} 100 \\ 20 \\ 4 \\ 1 \end{vmatrix} = \begin{vmatrix} 100 \\ 20 \\ 4 \\ 1 \end{vmatrix}$$

Clearly, the stable age distribution is identical to the survivorship function.

5 For population 1 we have already determined that, because $R_0 = 1$, $r = 0$, Thus $c_x = l_x$ and the stable vector is proportional to

$$\begin{vmatrix} 98 \\ 95 \\ 50 \\ 40 \end{vmatrix}$$

In pure proportions it is

$$\begin{vmatrix} 0.35 \\ 0.34 \\ 0.18 \\ 0.14 \end{vmatrix}$$

For population 2 we know that $R_0 < 1$. Therefore we expect $r < 0$. Try $r = -0.5$.

x	l_x	m_x	$-rx$	e^{-rx}	$l_x m_x e^{-rx}$
0.5	0.98	0	0.25	1.28	0
1.5	0.90	0	0.75	2.11	0
2.5	0.40	1.4	1.25	3.49	1.954
3.5	0.20	0.75	1.75	5.75	0.862

$$\sum l_x m_x e^{-rx} = 2.816$$

Because the sum is too large, we need a smaller r (absolute value). Try $r = -0.1$

x	l_x	m_x	$-rx$	e^{-rx}	$l_x m_x e^{-rx}$
2.5	0.40	1.4	0.25	1.28	0.7168
3.5	0.20	0.75	0.35	1.42	0.2130

$$\sum l_x m_x e^{-rx} = 0.9298$$

The sum is too small, try $r = -0.15$:

x	l_x	m_x	$-rx$	e^{-rx}	$l_x m_x e^{-rx}$
2.5	0.40	1.4	0.375	1.45	0.8120
3.5	0.20	0.75	0.525	1.69	0.2535

$$\sum l_x m_x e^{-rx} = 1.0655$$

Clearly r is between -0.1 and -0.15. By a graphical estimate $r = -0.127$. Check

x	l_x	m_x	$-rx$	e^{-rx}	$l_x m_x e^{-rx}$
2.5	0.40	1.4	0.3175	1.375	0.7700
3.5	0.20	0.75	0.4445	1.560	0.2340

$$\sum l_x m_x e^{-rx} = 1.0040$$

which is close enough. The stable age distribution is given as

$$c_x = \frac{l_x e^{-rx}}{\int_0^\infty l_x e^{-rx}\, dx}$$

x	l_x	$-rx$	e^{-rx}	$l_x e^{-rx}$	c_x
0.5	0.98	0.0635	1.065	1.044	0.35
1.5	0.90	0.1905	1.210	1.089	0.36
2.5	0.40	0.3175	1.375	0.550	0.18
3.5	0.20	0.4445	1.560	0.312	0.10

$$\sum l_x = 2.48 \qquad \sum l_x e^{-rx} = 2.995$$

To scale such that $l_x = 1.0$ divide all l_x by 2.48 to obtain

$$\begin{vmatrix} 0.39 \\ 0.36 \\ 0.16 \\ 0.08 \end{vmatrix}$$

For population 3 clearly the value of r is greater than zero. Try $r = 0.1$.

x	l_x	m_x	$-rx$	e^{-rx}	$l_x m_x e^{-rx}$
2.5	0.60	1.40	-0.25	0.778	0.6535
3.5	0.50	0.75	-0.35	0.705	0.2644

$$\sum l_x m_x e^{-rx} = 0.9179$$

We need a smaller value, try $r = 0.05$.

x	l_x	m_x	$-rx$	e^{-rx}	$l_x m_x e^{-rx}$
2.5	0.60	1.40	0.125	0.882	0.7409
3.5	0.50	0.75	0.175	0.839	0.3146

$$\sum l_x m_x e^{-rx} = 1.055$$

The value of r is between 0.05 and 0.1. By a graphical estimate $r = 0.070$. Check

x	l_x	m_x	$-rx$	e^{-rx}	$l_x m_x e^{-rx}$
2.5	0.60	1.40	−0.175	0.839	0.7048
3.5	0.50	0.75	−0.245	0.782	0.2932

$$\sum l_x m_x e^{-rx} = 0.9980$$

This is too small, try 0.069:

x	l_x	m_x	$-rx$	e^{-rx}	$l_x m_x e^{-rx}$
2.5	0.60	1.40	−0.1725	0.842	0.7073
3.5	0.50	0.75	−0.2415	0.786	0.2948

$$\sum l_x m_x e^{-rx} = 1.002$$

which is close enough to 1.000. The stable age distribution is given as follows:

x	l_x	$-rx$	e^{-rx}	$l_x e^{-rx}$	c_x
0.5	0.98	−0.0345	0.970	0.9506	0.36
1.5	0.90	−0.1035	0.902	0.8118	0.30
2.5	0.60	−0.1725	0.842	0.5052	0.19
3.5	0.50	−0.2415	0.786	0.3930	0.15

$$\sum l_x = 2.98 \qquad \sum l_x e^{-rx} = 2.661$$

Scaling l_x as a proportion, we obtain,

$$\begin{vmatrix} 0.33 \\ 0.30 \\ 0.20 \\ 0.17 \end{vmatrix}$$

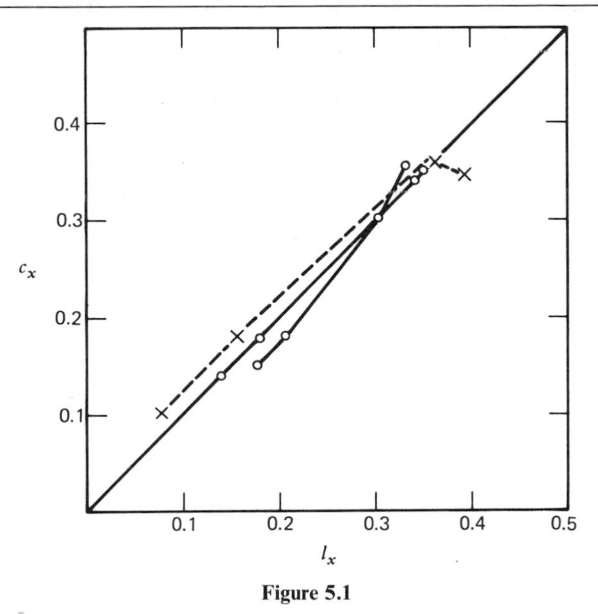

Figure 5.1

It is worth noting that this graph represents what seems to be an empirical generalization; that is, growing populations ($r > 0$) cross the 45° line one way and declining populations ($r < 0$), another way, as shown in the following diagram:

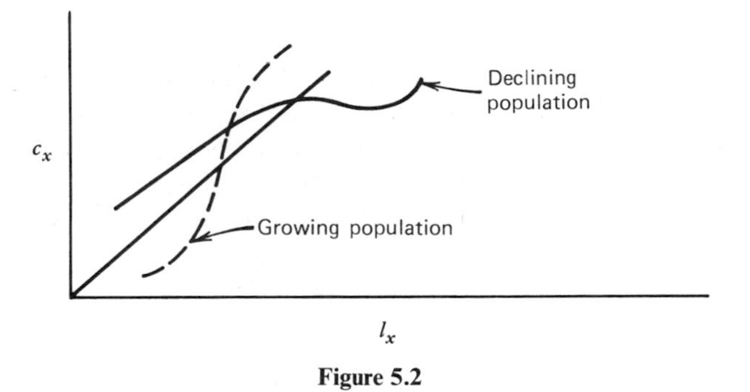

Figure 5.2

6 l_x for the five populations is

	Population Number				
x	1	2	3	4	5
0	1.0	1	1	1	1
1	0.90	0.90	0.90	0.90	0.90
2	0.75	0.75	0.75	0.75	0.75
3	0.50	0.50	0.50	0.50	0.50
4	0.10	0.10	0.10	0.10	0.10

7

x	l_x
0	1.000
1	0.675
2	0.300
3	1.500
4	0.500
5	0

Clearly, these values are totally unreasonable. First, by definition, l_x must be ≤ 1.0 for any value of x (as long as $l_0 = 1.0$ by convention). Second, the numbers bear no resemblance to those computed directly in exercise 6.

8 Recall that in Chapter 2 the transition probability is given as

$$P_{x+1,x} = \frac{N(x+1, t+1)}{N(x, t)}$$

Now recall that $N(0, 4)$ is the number of individuals between ages zero and one ($0 \leq x < 1$) that are alive at time 4. Because the exact age is known for each individual, we also know that the number of individuals in the first age category ($0 \leq x < 1$) at time 4 is exactly 200; thus $N(0, 4) = 200$. Following similar reasoning for the rest of the probabilities, we have

$$P_{10} = \frac{N(1, 5)}{N(0, 4)} = \frac{180}{200} = 0.9$$

$$P_{21} = \frac{N(2, 5)}{N(1, 4)} = \frac{112}{135} = 0.83$$

$$P_{32} = \frac{N(3, 5)}{N(2, 4)} = \frac{40}{60} = 0.67$$

$$P_{43} = \frac{N(4, 5)}{N(3, 4)} = \frac{60}{300} = 0.20$$

9 Recall that

$$P_{x+1,x} = \frac{L_{x+1}}{L_x} = \frac{\int_{x+1}^{x+2} l_x \, dx}{\int_x^{x+1} l_x \, dx}$$

In the present example all the individuals in a given age category are totally concentrated at the beginning of that category; for example, if we graph the number of individuals against age, we obtain for time 5

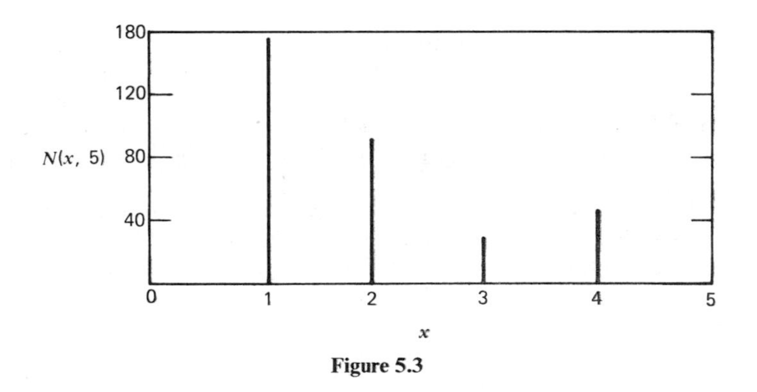

Figure 5.3

that is, we do not have a continuous distribution of individuals among ages. Thus

$$l_x = \int_x^{x+1} l_x \, dx$$

(the individuals at the $x + 1$ point belong to the next age category). So, for this example we have

$$P_{x+1,x} = \frac{L_{x+1}}{L_x} = \frac{l_{x+1}}{l_x}$$

and

$$P_{10} = \frac{l_1}{l_0} = \frac{0.9}{1}$$

$$\therefore l_1 = 0.9$$

$$P_{21} = \frac{l_2}{l_1}$$

$$l_2 = P_{21} l_1 = (0.83)(0.9) = 0.75$$

$$P_{32} = \frac{l_3}{l_2}$$

$$l_3 = P_{32} l_2 = (0.67)(0.75) = 0.50$$

$$P_{43} = \frac{l_4}{l_3}$$

$$l_4 = P_{43} l_3 = (0.20)(0.50) = 0.1$$

All the survivorship values correspond to the original l_x estimates.

10 (a)

	1976 Jan.	March	May	↓ July	Sept.	Nov.	1977 Jan.	March	↓ May
t	1	1	1	1	1	1	1	1	1
$N(t)$*	100	90	80	70	60	50	40	30	20
x	0	0.2	0.4	0.6	0.8	1.0	1.2	1.4	1.6

* $N(t)$ refers to the total number in the cohort, all of which are the same age. On July 1, 1976, there will be 70 individuals between 0 and 10. On May 1, 1977, there will be 20 alive, all between 10 and 20. Therefore $P_{1,0} = 20/70 = 0.286$.

(b)

	May	↓ July	Sept.	Nov.	Jan.	March	↓ May
t	1	1	1	1	1	1	1
$N(t)$	100	90	80	70	60	50	40
x							

On July 1, 1976, there will be 90 individuals between 0 and 10. On May 1, 1977, there will be 40 alive, all between 10 and 20. Therefore $P_{1,0} = \frac{40}{90} = 0.444$.

(c) The distribution of individuals in categories is extremely important when computing survivorships from a horizontal life table. To be perfectly correct the distribution should be equal to (proportional to) the stable age distribution in the two categories for which the transition probabilities are being estimated.

11

$$\begin{vmatrix} 0 & 2 & 11 & 16 \\ 0.2 & 0 & 0 & 0 \\ 0 & 0.2 & 0 & 0 \\ 0 & 0 & 0.2 & 0 \end{vmatrix} \begin{vmatrix} 110 \\ 17 \\ 3 \\ 2 \end{vmatrix} = \begin{vmatrix} 99 \\ 22 \\ 3 \\ 1 \end{vmatrix}$$

Of the 110 newborns $(110 - 22) = 88$ died. Of the 17 one-year-olds $(17 - 3) = 14$ died. Of the 3 two-year-olds $(3 - 1) = 2$ died. Of the 2 three-year-olds 2 died. Thus the total number that died is $88 + 14 + 2 + 2 = 106$. The total number in the population dropped from 132 to 125, which is an average of $(132 + 125)/2 = 128.5$. Thus the crude death rate is $106/128.5 = 0.825$. There was a total of 99 births during the time interval. Thus the crude birth rate is $99/128.5 = 0.770$.

12

$$\begin{vmatrix} 0 & 2 & 11 & 16 \\ 0.2 & 0 & 0 & 0 \\ 0 & 0.2 & 0 & 0 \\ 0 & 0 & 0.2 & 0 \end{vmatrix} \begin{vmatrix} 100 \\ 20 \\ 4 \\ 1 \end{vmatrix} = \begin{vmatrix} 100 \\ 20 \\ 4 \\ 1 \end{vmatrix}$$

By the same calculations used in exercise 11 the total number of deaths is $80 + 16 + 3 + 1 = 100$. The average number of individuals is 125. Thus the crude death rate is $\frac{100}{125} = 0.800$. The total number of births is 100; the crude birth rate is $\frac{100}{125} = 0.800$.

13 Approximate the integral with a sum (remember $r = 0$):

x	l_x
0	1.00
1	0.20
2	0.04
3	0.008
4	0.002

$$\sum l_x = 1.25$$

$$b = \frac{1}{\sum l_x} = \frac{1}{1.25} = 0.80$$

14 The intrinsic rate of natural increase is the intrinsic birth rate minus the intrinsic death rate; that is,

$$r = b - d$$

Because $r = 0$ and $b = 0.80$,

$$0 = 0.80 - d$$

$$d = 0.80$$

15

x	l_x	m_x	$l_x m_x$
0.5	0.600	0	0
1.5	0.120	4	0.480
2.5	0.024	25	0.600
3.5	0.004	40	0.160

$$\sum l_x m_x = 1.240$$

$R_0 = 1.240$ and the stable population will be increasing.
Try $r = 0.1$:

x	l_x	m_x	$-rx$	e^{-rx}	$l_x m_x e^{-rx}$
1.5	0.120	4	−0.15	0.861	0.4133
2.5	0.024	25	−0.25	0.779	0.4674
3.5	0.004	40	−0.35	0.705	0.1128

$$\sum l_x m_x e^{-rx} = 0.9934$$

We need a smaller value, try $r = 0.05$:

x	l_x	m_x	$-rx$	e^{-rx}	$l_x m_x e^{-rx}$
1.5	0.120	4	−0.075	0.928	0.4454
2.5	0.024	25	−0.125	0.882	0.5292
3.5	0.004	40	−0.175	0.839	0.1342

$$\sum l_x m_x e^{-rx} = 1.1088$$

Try $r = 0.08$:

x	l_x	m_x	$-rx$	e^{-rx}	$l_x m_x e^{-rx}$
1.5	0.120	4	−0.120	0.887	0.4258
2.5	0.024	25	−0.200	0.819	0.4914
3.5	0.004	40	−0.280	0.756	0.1210

$$\sum l_x m_x e^{-rx} = 1.038$$

By the usual graphical technique $r = 0.095$. Check

x	l_x	m_x	$-rx$	e^{-rx}	$l_x m_x e^{-rx}$
1.5	0.120	4	−0.1425	0.8670	0.4162
2.5	0.024	25	−0.2375	0.7885	0.4731
3.5	0.004	40	−0.3325	0.7170	0.1147

$$\sum l_x m_x e^{-rx} = 1.004$$

$R_0 = 1.240$, $r = 0.095$:

$$T = \frac{\ln R_0}{r} = \frac{0.215}{0.095} = 2.263$$

16

$$C_x = \frac{l_x e^{-rx}}{\int_0^\infty l_x e^{-rx}\,dx}$$

x	l_x	$-rx$	e^{-rx}	$l_x e^{-rx}$	C_x
0.5	0.6	−0.0475	0.9540	0.5724	0.8193
1.5	0.12	−0.1425	0.8670	0.1040	0.1494
2.5	0.024	−0.2375	0.7885	0.0189	0.0270
3.5	0.004	−0.3325	0.7170	0.0029	0.0042

$$\sum l_x e^{-rx} = 0.6986$$

17

$$\begin{vmatrix} 0 & 2.04 & 12.73 & 20.37 \\ 0.2 & 0 & 0 & 0 \\ 0 & 0.2 & 0 & 0 \\ 0 & 0 & 0.2 & 0 \end{vmatrix} \begin{vmatrix} 8194 \\ 1494 \\ 270 \\ 42 \end{vmatrix} = \begin{vmatrix} 7340 \\ 1639 \\ 299 \\ 54 \end{vmatrix} = \begin{vmatrix} 8250 \\ 1468 \\ 328 \\ 60 \end{vmatrix} = \begin{vmatrix} 8392 \\ 1650 \\ 293 \\ 66 \end{vmatrix}$$

$$= \begin{vmatrix} 8440 \\ 1678 \\ 330 \\ 59 \end{vmatrix} = \begin{vmatrix} 8826 \\ 1688 \\ 336 \\ 66 \end{vmatrix} = \begin{vmatrix} 9065 \\ 1765 \\ 338 \\ 67 \end{vmatrix} = \begin{vmatrix} 9268 \\ 1813 \\ 353 \\ 68 \end{vmatrix}$$

Divide the total number by 100 to facilitate looking up logs. $N(t) = 100.00$, 93.32, 101.06, 104.01, 105.07, 109.16, 112.35, 115.02; $\ln N(t) = 4.605$, 4.533, 4.615, 4.644, 4.654, 4.691, 4.718, 4.745.

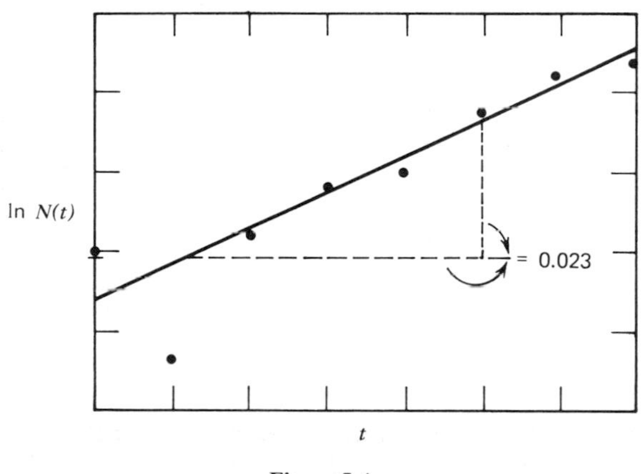

Figure 5.4

Thus the estimate of r by this technique is $r = 0.023$ and the stable age distribution (average of the last two projections) is

$$\begin{vmatrix} 0.8065 \\ 0.1575 \\ 0.0305 \\ 0.0060 \end{vmatrix}$$

which is not exactly the same as that computed in exercise 16. As we have noted before, these two techniques give slightly different answers. Clearly, in this case the l_x used was $(l_x + l_{x+1})/2$ rather than $\int_x^{x+1} l_x \, dx$. This approximation, of course, will inflate the value of r for a concave survivorship curve and deflate r for a convex survivorship curve.

18 $R_0 = 1.24$, $\ln R_0 = 0.215$.

Note that $N(t + T) = R_0 N(t)$ and $\ln N(t + T) - \ln N(t) = \ln R_0$.

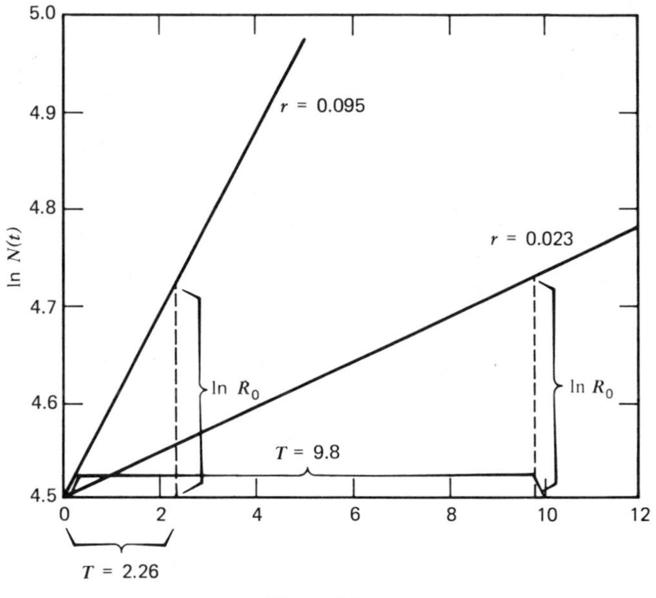

Figure 5.5

Thus, from $r = 0.023$ and $R_0 = 1.24$, $T = 9.8$. Obviously this is a ridiculous estimate of mean generation time (the other estimate was 2.263); but, if we take $r = 0.095$ and $R_0 = 1.24$, $T = 2.26$.

19

x	l_x	m_x	$l_x m_x$
0.5	0.45	0	0
1.5	0.09	4	0.36
2.5	0.018	25	0.45
3.5	0.004	40	0.16

$$R_0 = 0.97$$

that is, the new estimate of r will be less than zero.

Thus for the same data, depending on how various parameters are approximated, we estimate r from less than zero to 0.023 to 0.095. It is obviously important to be as accurate as possible in approximating integrals. It might be added that these computational difficulties virtually disappear if we are dealing with a population with a large number of age categories. Also, such errors are most severe when r is near zero.

20

x	l_x	m_x	$l_x m_x$
1.5	0.09	40	3.60
2.5	0.018	250	4.50
3.5	0.004	400	1.60

$$R_0 = 9.700$$

Try $r = 1.0$:

x	l_x	m_x	$-rx$	e^{-rx}	$l_x m_x e^{-rx}$
1.5	0.09	40	−1.5	0.224	0.8064
2.5	0.018	250	−2.5	0.082	0.3690
3.6	0.004	400	−3.5	0.030	0.0480

$$\sum l_x m_x e^{-rx} - 1.2234$$

We need a larger r. Try $r = 1.5$:

x	l_x	m_x	$-rx$	e^{-rx}	$l_x m_x e^{-rx}$
1.5	0.09	40	−2.25	0.1050	0.3780
2.5	0.018	250	−3.75	0.0235	0.1058
3.5	0.004	400	−5.25	0.0050	0.0080

$$\sum l_x m_x e^{-rx} = 0.4918$$

We need a smaller r. Try $r = 1.2$:

x	l_x	m_x	$-rx$	e^{-rx}	$l_x m_x e^{-rx}$
1.5	0.09	40	−1.80	0.165	0.594
2.5	0.018	250	−3.00	0.050	0.225
3.5	0.004	400	−4.20	0.015	0.024

$$\sum l_x m_x e^{-rx} = 0.843$$

We need a smaller r. Try $r = 1.1$:

x	l_x	m_x	$-rx$	e^{-rx}	$l_x m_x e^{-rx}$
1.5	0.09	40	−1.65	0.192	0.6912
2.5	0.018	250	−2.75	0.064	0.2880
3.5	0.004	400	−3.85	0.058	0.0928

$$\sum l_x m_x e^{-rx} = 1.072$$

Clearly, $1.1 < r < 1.2$. Try $r = 1.15$:

x	l_x	m_x	$-rx$	e^{-rx}	$l_x m_x e^{-rx}$
1.5	0.09	40	−1.725	0.178	0.6408
2.5	0.018	250	−2.875	0.057	0.2565
3.5	0.004	400	−4.025	0.018	0.0288

$$\sum l_x m_x e^{-rx} = 0.9261$$

Graphically we obtain $r = 1.112$.
Check

x	l_x	m_x	$-rx$	e^{-rx}	$l_x m_x e^{-rx}$
1.5	0.09	40	−1.683	0.186	0.6696
2.5	0.018	250	−2.805	0.061	0.2745
3.5	0.004	400	−3.927	0.020	0.0320

$$\sum l_x m_x e^{-rx} = 0.9761$$

Try 1.11:

x	l_x	m_x	$-rx$	e^{-rx}	$l_x m_x e^{-rx}$
1.5	0.09	40	−1.665	0.1890	0.68040
2.5	0.018	250	−2.775	0.0625	0.28125
3.5	0.004	400	−3.885	0.0205	0.03280

$$\sum l_x m_x e^{-rx} = 0.9944$$

Try 1.105:

x	l_x	m_x	$-rx$	e^{-rx}	$l_x m_x e^{-rx}$
1.5	0.09	40	−1.6575	0.191	0.6876
2.5	0.018	250	−2.7625	0.063	0.2835
3.5	0.004	400	−3.8675	0.021	0.0336

$$\sum l_x m_x e^{-rx} = 1.0047$$

Thus, rounded to the second decimal, $r = 1.11$.

$$T = \frac{\ln R_0}{r} = \frac{\ln 9.7}{1.11} = \frac{2.272}{1.11} = 2.047$$

Stable age distribution ($r = 1.11$):

x	l_x	$-rx$	e^{-rx}	$l_x e^{-rx}$	c_x
0.5	0.45	−0.555	0.574	0.2583	0.934
1.5	0.09	−1.665	0.189	0.0170	0.0615
2.5	0.018	−2.775	0.0625	0.0011	0.0040
3.5	0.004	−3.885	0.0205	0.00008	0.0003

$$\sum l_x e^{-rx} = 0.2765$$

21

$$\begin{vmatrix} 0 & 20.4 & 127.3 & 203.7 \\ 0.2 & 0 & 0 & 0 \\ 0 & 0.2 & 0 & 0 \\ 0 & 0 & 0.2 & 0 \end{vmatrix} \begin{vmatrix} 930 \\ 62 \\ 4 \\ 0 \end{vmatrix} \begin{vmatrix} 1774 \\ 186 \\ 12 \\ 1 \end{vmatrix} \begin{vmatrix} 5526 \\ 355 \\ 37 \\ 2 \end{vmatrix} \begin{vmatrix} 12360 \\ 1105 \\ 71 \\ 7 \end{vmatrix} \begin{vmatrix} 33006 \\ 2472 \\ 221 \\ 14 \end{vmatrix}$$

996 1973 5920 13543 35713

$\ln (996/100) = 2.299 \qquad \ln 1973/100 = 2.981 \qquad \ln 5920/100 = 4.081$

$\ln 13543/100 = 4.905 \qquad \ln 35713/100 = 5.878$

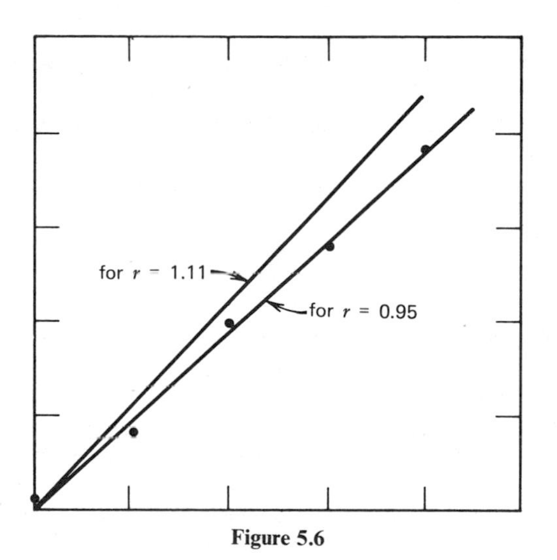

Figure 5.6

REFERENCES

Beddington, J. R. 1974. Age distribution and the stability of simple discrete time population models. *J. Theor. Biol.* **47**:65–74.

Birch, L. C. 1948. The intrinsic rate of natural increase of an insect population. *J. Ecol.* **17**:15–26.

Birch, L. C. 1953. Experimental background to the study of the distribution and abundance of insects. II. The relation between innate capacity for increase in numbers and the abundance of three grain beetles in experimental populations. *Ecology* **34**:712–726.

Brass, W. 1958. The distribution of births in human populations. *Popul. Stud. London* **12**:51–72.

Burnett, T. 1951. Effects of temperature and host density on the rate of increase of an insect parasite. *Am. Nat.* **85**:337–352.

Chapman, D. G. and G. I. Murphy. 1965. Estimates of mortality and population from survey-removal records. *Biometrics* **21**:921–935.

Chiang, C. L. 1960a. A stochastic study of the life table and its applications: I. Probability distributions of the biometric functions, *Biometrics* **16**:618–635.

Chiang, C. L. 1960b. A stochastic study of the life table and its applications: II. Sample variance of the observed expectation of life and other biometric functions. *Hum. Biol.* **32**:275–285.

Chiang, C. L. 1961. A stochastic study of the life table and its applications: III. The follow-up study with consideration of competing risks. *Biometrics* **17**:57–78.

Coale, Ansley J. 1957a. How the age distribution of a human population is determined. *Cold Spring Harbor Symp. Quant. Biol.* **22**:83–89.

Coale, A. J. 1957b. A new method for calculating Lotka's r—the intrinsic rate of growth in a stable population. *Popul. Stud. London.* **11**:92–94.

Coale, A. J. 1967. Convergence of a human population to a stable form. Annual meeting of Population Association of America. Annual Meeting of the Population Association of America.

Deevey, E. S. 1947. Life tables for natural populations of animals. *Q. Rev. Biol.* **22**:283–314.

Dublin, L. I. and A. J. Lotka. 1925. On the true rate of natural increase. *J. Am. Stat. Assoc.* **20**:395–339.

Elton, C. 1942. Voles, mice and lemmings: problems in population dynamics. Oxford: Oxford University Press.

Fenchel, T. 1974. Intrinsic rate of natural increase; the relationship with bodysize. *Oecologia, Berlin* **14**:317–326.

Hairston, N. G. 1965a. On the mathematical analysis of schistosome populations. *Bull. Wld. Hlth. Org.* **33**:45–62.

Hairston, N. G. 1965b. An analysis of age-prevalence data by catalytic models. *Bull. Wld. Hlth. Org.* **33**:163–175.

Keyfitz, N. 1970. Finding probabilities from observed rates, or how to make a life table. *Am. Stat.* **24**:28–33.

Lack, D. 1973. The numbers and species of hummingbirds in the West Indies, **27**:326–337.

May, R. M. 1976. Estimating *r*: a pedagogical note. *Am. Nat.* **110**:496–499.

Newsome, A. E. 1969. A population study of house-mice temporarily inhabiting a South Australian wheatfield. *J. Anim. Ecol.* **38**:341–359.

Neyman, J., T. Park, and E. L. Scott. 1956. Struggle for existence. The *Tribolium* model: biological and statistical aspects. In Proc. 3rd Berkeley Symp. Mathematical Statistics and Probability. Vol. IV, pp. 41–79. Berkeley: University California Press.

Southwood, T. R. E. 1970. The natural and manipulated control of animal populations. In L. R. Taylor, Ed. The optimum population for Britain. *Inst. Biol. Symp.* 19. London: Academic, pp. 87–102.

Waugh, W. A. O'N. 1955. An age-dependent birth and death process. *Biometrika* **42**:291–306.

6. The Analysis of Spatial Pattern

Ecologists have always been interested in pattern. Most pattern studies are concerned with plants, but more recently animal ecologists have been investigating the distribution of organisms in space. Many contemporary problems in population and community ecology require an understanding of the measurement of pattern and what it means once it is made.

Although much recent work has been done on the spatial patterns of two or more populations, in this chapter we shall discuss the distribution of individuals in only one biological population. For an excellent coverage of patterns of two or more species the reader is referred to Pielou (1969).

Central to an understanding of most of this chapter are the notions of random, and superdispersed. Suppose you fired a shotgun at a 6-in. target from a distance of 30 ft. What sort of pattern might the pellets of shot make on that 6-in. target? Clearly, after traveling 30 ft the shot will be dispersed in a pattern much greater than 6 in. Under the presumption that the individual pellets of shot will not interact with one another, they might make a pattern (see Figure 6.1):

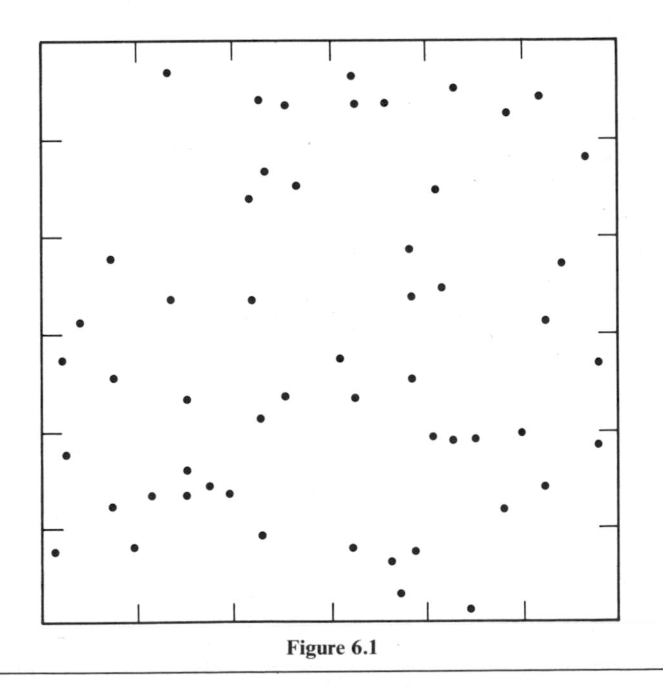

Figure 6.1

This pattern is random. Technically, each point on the target is just as likely to be hit by a pellet as every other point on the target.

Now suppose that you fire the same shotgun with the end of the barrel 6 in. from the target. You might obtain a pattern like this (Figure 6.2):

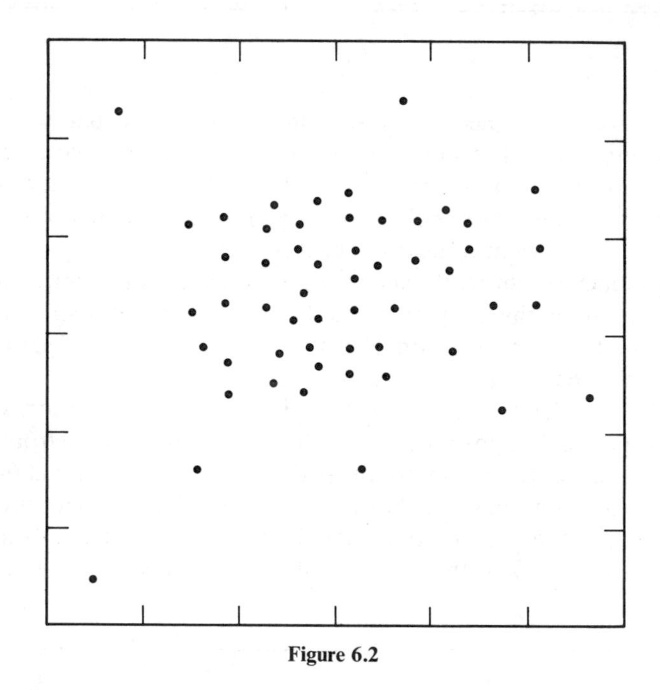

Figure 6.2

Most of the pellets are clumped near the center of the target. This is called a *clumped* or *contagious* distribution.

Now suppose that all the pellets in the shell contain some sort of repelling force that tends to separate them. As the shotgun is fired, the pellets, much like repelling magnets, repel one another as they fly through the air toward the target. You might then obtain a pattern something like this (Figure 6.3):

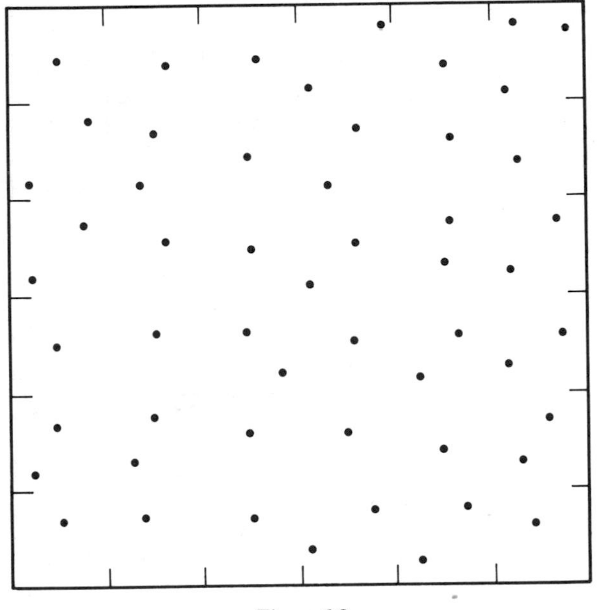

Figure 6.3

This pattern is called hyperdispersed or superdispersed.

Contagious and hyperdispersed patterns are frequently referred to collectively as nonrandom. It will be clear in what follows that a nonrandom distribution is much more difficult to describe than a random distribution. You might say that there is only one way to be random but an infinite number of ways to be nonrandom.

THE POISSON DISTRIBUTION. Suppose that the arrangement of individual points on a plane surface is random. Let us concern ourselves, in particular, with the pattern shown in Figure 6.1. This, in fact, is a random pattern. If it is strictly random, we should be able to describe it in a statistical sense. Suppose we superimpose quadrat lines (Figure 6.4):

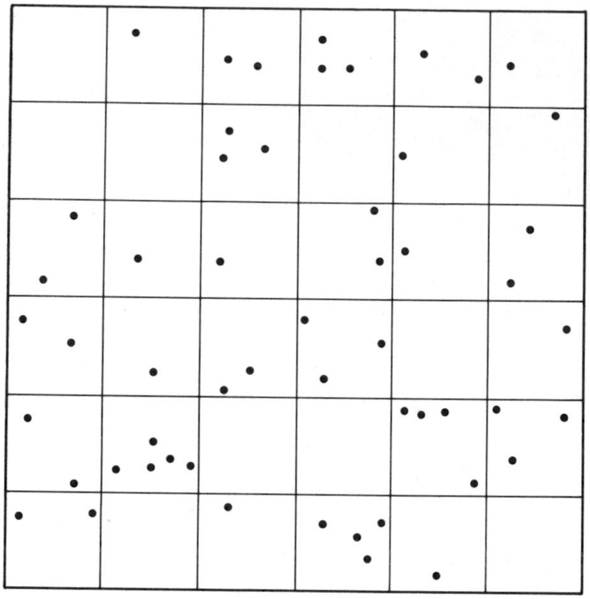

Figure 6.4

How many of the quadrats will contain 0 points, how many will contain 1 point, how many 2 points, and so on? Can such a question be answered in a rigorous mathematical way?

To answer this question let us suppose that each quadrat is subdivided into a large number of subquadrats. Suppose the subquadrats are so small that no more than one individual (particle or point) can occupy each of them.

The probability that a given subquadrat is occupied is p. It is the same for each subquadrat. Note that we have now set things up such that each subquadrat can be occupied or not occupied. There are no other possibilities. What is the probability that exactly r positions will be occupied?

Take an example. Suppose that $n = 3$, $r = 2$, and $p = 0.4$. What is the probability that exactly two of the three spaces will be occupied? We must enumerate all possibilities. The possibilities are listed in Figure 6.5 with associated probabilities:

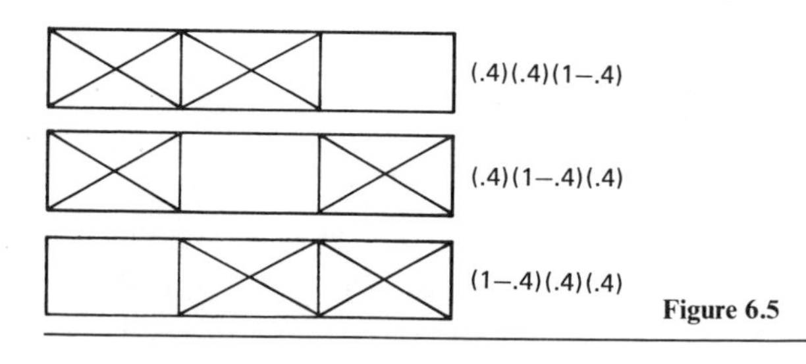

Figure 6.5

In other words, the probability that two of the three spaces will be occupied is $(3)(0.4)^2(1 - 0.4)$.

In general, the probability that r of n spaces will be occupied if p is the probability of being occupied is

$$Cp^r(1 - p)^{n-r}$$

where C is a constant equal to the number of ways r things can fit into the n locations. In general, the constant C is given as

$$\frac{n!}{(n - r)!r!}.$$

In the preceding case $r = 2$ and $n = 3$; therefore

$$C = \frac{(3)(2)(1)}{[(1)][(2)(1)]} = \tfrac{6}{2} = 3$$

Thus, in general, the probability that exactly r locations will be occupied out of a possible n for which the probability of being occupied is p is

$$P_r = \frac{n!}{(n - r)!r!} P^r(1 - p)^{n-r}$$

which is the binomial probability (consult any elementary statistics text if you do not know what a binomial probability is).

Now we presume that each of the quadrats in Figure 6.4 is made up of a large number of subquadrats. The probability that exactly r of those subquadrats will be occupied is the same as the probability that each quadrat will contain r points. We already know that this probability is given by the binomial, but on our development here we have specified only that n is large. Therefore exactly what value of n should be used to compute the probability? Obviously, that is an impossible question because n has to be large enough so that no more than one point will be likely to land in any one subquadrat.

It is by presuming that n is large and p is small that we can derive an exact equation for such a random distribution. Assume that r is negligibly small compared with n. Then

$$\frac{n!}{(n - r)!r!} P^r(1 - p)^{n-r} = \frac{n(n - 1)(n - 2) \cdots (n - r + 1)}{r!} P^r(1 - p)^{n-r}$$

is approximately equal to

$$\frac{n(n - 1)(n - 2) \cdots (n - r + 1)}{r!} P^r(1 - p)^n$$

The $n(n - 1)(n - 2) \cdots (n - r + 1)$ term is essentially the same as n multiplied by itself r times (since n is so large in comparison to r); thus we have,

$$\frac{n^r p^r (1-p)^n}{r!}$$

Symbolizing np with λ, we have

$$\frac{\lambda^r (1-\lambda/n)^n}{r!}$$

Recall from calculus (or, if you don't recall, take it on faith) that in general $\lim_{n\to\infty}: (1 - x/n)^n = e^{-x}$. Thus, if we allow n to become very large,

$$P_r = \frac{\lambda^r e^{-\lambda}}{r!} \tag{1}$$

This is the equation for the Poisson series. The parameter λ is the mean number of points per quadrat (recall $\lambda = np$, which was the mean of the original binomial distribution).

☐ EXERCISES

1 For the diagram in Figure 6.4 how many quadrats contain 0 points, how many 1 point, and so on?
2 What is the mean number of points per quadrat?
3 Compute $p_0, p_1, p_2, \ldots$ for these data, using equation 1. Multiply the probabilities by the total number of quadrats and compare with the data tabulated in Exercise 1.
4 Draw quadrats on Figure 6.2 (use the same size quadrats as in Figure 6.4; follow the guide lines in the figure if you want your answer to be *exactly* the same as in this text). How many quadrats contain no points? How many one point? And so on.
5 Draw quadrats on Figure 6.3 (use the same size quadrats as in Figure 6.4). How many quadrats contain 0, 1, 2, ... points?
6 Graph "observed number of quadrats" [call this $f_0(x)$] against the number of points (call this x) for the data in exercises 1, 4, and 5.
7 For the following pattern compute $f_0(x)$ (the observed number of quadrats with 0, 1, 2, ... points). Compute the Poisson expectation [call this $f_e(x)$ and compare with $f_0(x)$]. (Superimpose the same size quadrat grid as before.)

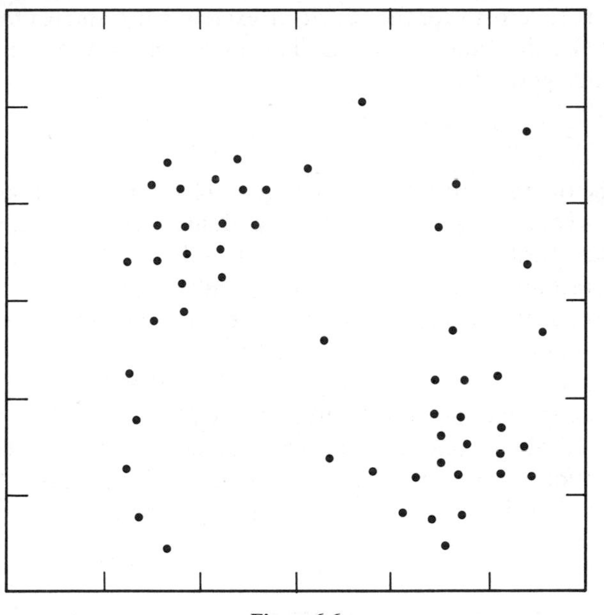

Figure 6.6

8 For the following pattern compute $f_0(x)$ (the Poisson expectation) and compare it with $f_e(x)$. (Superimpose the same size quadrat grid as before.)

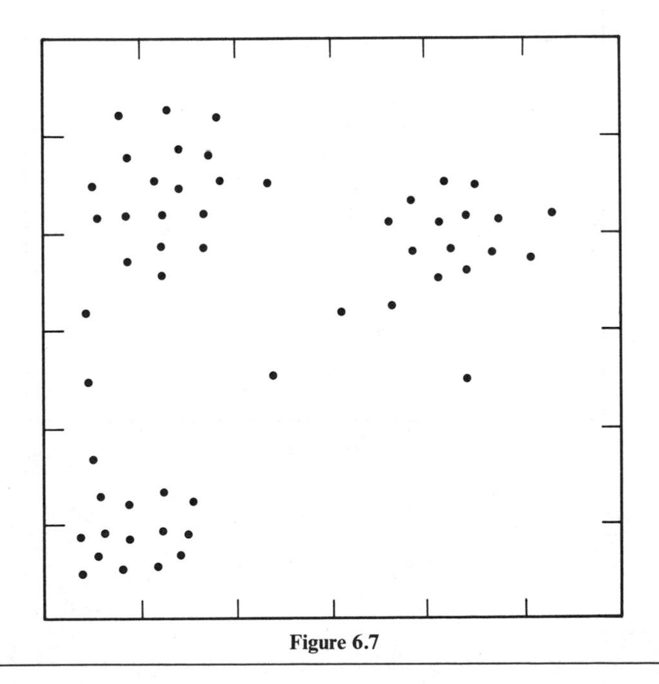

Figure 6.7

9 Plot $f_0(x)$ against x for the data from exercises 1, 7, and 8. On the same graph plot $f_e(x)$, the Poisson expectation for all three. What do you conclude from this graph? □

It should be obvious from the preceding exercises that it is reasonably easy to determine whether a given pattern is random. A simple χ^2 test will provide rigorous statistical proof. If the observed data do not differ significantly from a Poisson distribution, the pattern will not differ significantly from a random pattern. Frequently, however, we are interested in more than a qualitative determination of randomness or nonrandomness.

In particular, consider exercises 7 and 8. Both deviate from randomness, yet obviously each is qualitatively different. In fact, the material in exercises 7 and 8 is difficult to deal with and will be covered in the next section. In this section we wish to develop an intuition and measurement of the relative degree of nonrandomness. Consider the six diagrams in Figure 6.8:

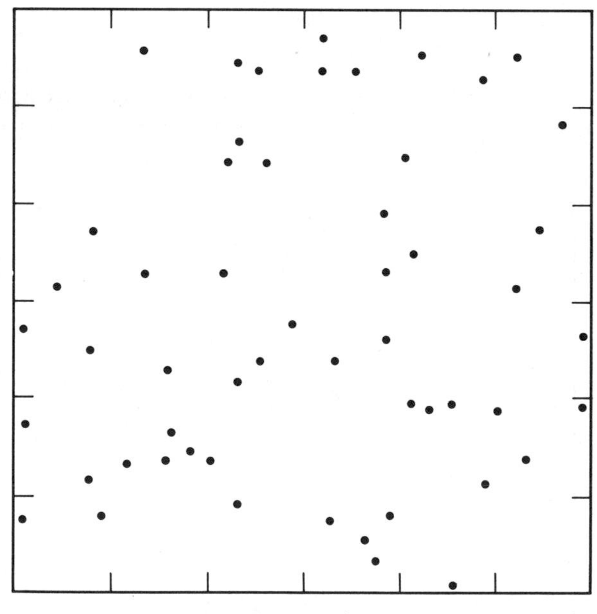

Figure 6.8 (*a*)

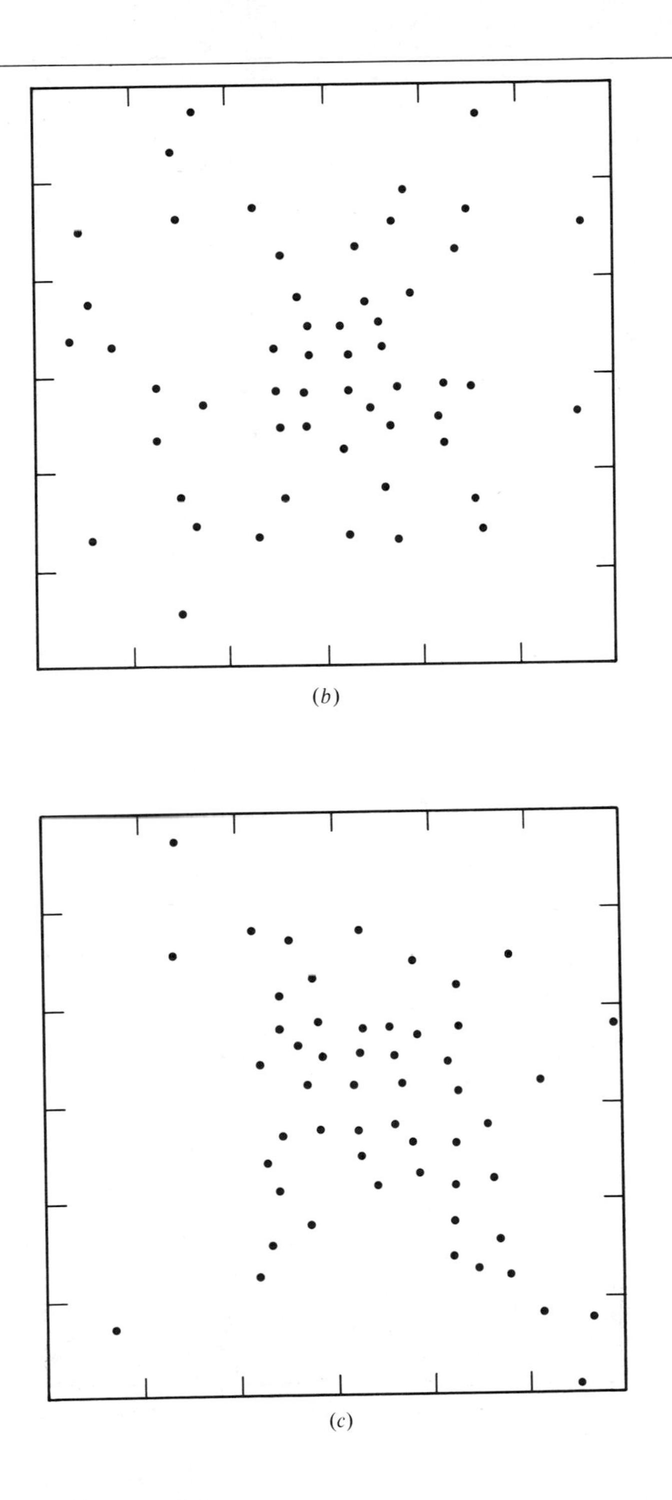

(b)

(c)

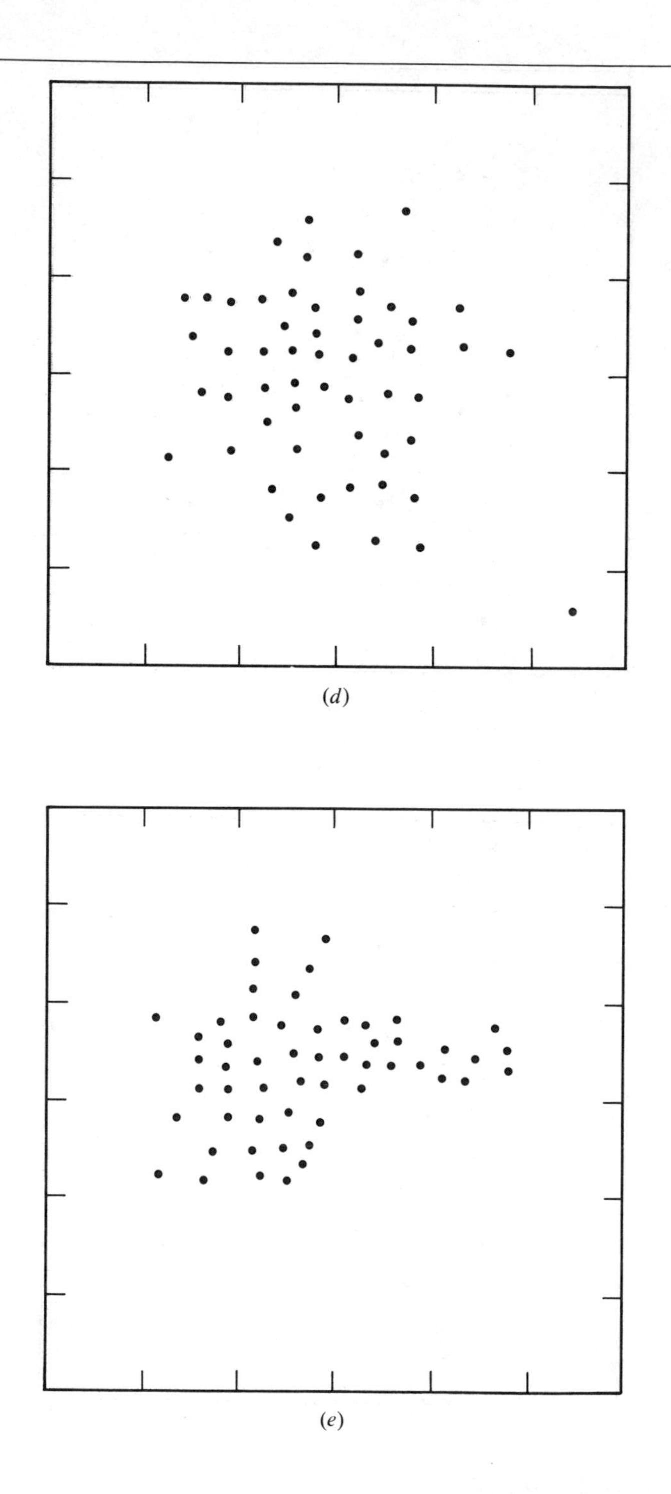
(d)

(e)

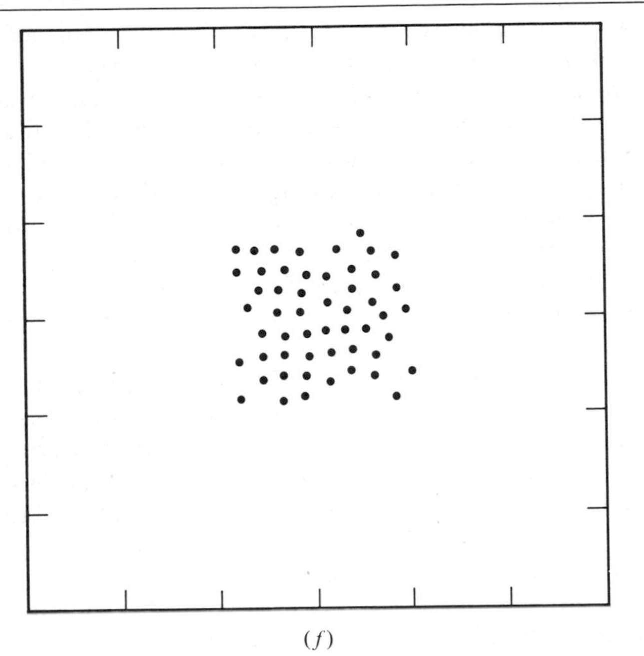

(f)

As we go from Figure 6.8a to Figure 6.8f, we pass from a totally random pattern (Figure 6.8a) to a highly clumped pattern (Figure 6.8f). The transition is relatively smooth and obviously could be made even smoother by adding intermediate cases. All cases, except Figure 6.8a are nonrandom, yet we can easily recognize a "natural" series that passes from a to b to c to d to e to f.

☐ EXERCISES

10 Compute $f_0(x)$ for all the cases in Figure 6.8 (use the guidelines to make your quadrats). Plot $f_0(x)$ against x for all six cases. (You already have done Figure 6.8a—a perfectly random pattern of Figure 6.1.) ☐

What seems fairly obvious is that as clumping increases variability increases; for example, in Figure 6.8a the maximum number of dots per quadrat is 5, the minimum is 0, whereas for Figure 6.8f the maximum number is 14, the minimum 0. If all we really desire is a number that reflects "degree of clumping" in the sense it is displayed in Figure 6.8, perhaps something related to the variability in number of individuals per quadrat would suffice.

At this point it is worth diverging for a moment to consider what it means to derive or invent a measurement. Frequently in ecology we attempt to provide measurement techniques for common sense phenomena. Examples include species diversity, complexity, predictability, niche breadth, and degree of clumping. Usually, with no theory to rely on, we are forced to base a measurement solely on a correspondence with our intuitive feeling about the phenomenon. Whether this is a good idea is at least debatable. To take the present example we wish to define a procedure in which the pattern in Figure 6.8 *f* will be automatically classified as clumped and that in Figure 6.8*a* will be automatically classified as random. Furthermore, we wish to have a procedure that will give us a number that is small for the situation in Figure 6.8*a*, larger for Figure 6.8*b*, yet larger for Figure 6.8*c*, and so on. (Equivalently, we might wish to have a small number for Figure 6.8*f*, a larger number for Figure 6.8*e*, and so on). The point is that we frequently wish to quantify a common-sense phenomenon. It makes no difference how we do it as long as the quantification reflects our intuitive notion of that phenomenon. We might question, philosophically, the wisdom of such an approach by pointing out that not everyone's common sense is in agreement. I would agree. Nevertheless, for numerous concepts in ecology, lacking any sort of theoretical framework, we are forced to conceptualize in this mindless operational fashion.

In the present case we wish to devise a measurement technique that will give us a small number for Figure 6.8*a* and a large number for Figure 6.8*f*; the intermediate patterns will have intermediate values. At this point the exact procedure does not matter. All that does matter is that the behavior of the measurement correspond to our intuition.

As hinted in exercise 10, the spread of numbers of points per quadrat might be a measurement of the degree of clumping. Indeed, such a measure corresponds perfectly to our intuition for the cases in Figure 6.8. The variance of the distribution, actually an arbitrary choice, is the measure often used. The important point is that the variance is generally large in situations like Figure 6.8*f* (some quadrats with large numbers of points, others with none) and generally small for Figure 6.8*a*. Furthermore, in hyperdispersed patterns the variance is even smaller (almost all the quadrats have the same number of points).

The equation for variance is

$$S^2 = \frac{\sum (x_1 - \lambda)^2}{Q - 1}$$

where X_1 is the number of points in the *i*th quadrat, λ is the mean number of points per quadrat, and Q is the number of quadrats.

☐ EXERCISES

11 Compute the variances for all six examples in Figure 6.8 and for the example in Figure 6.3. Do they correspond to your intuition of degree of clumping? ☐

It is a well-known property of the Poisson distribution that the variance is equal to the mean (recall the mean and variance for Figure 6.8*a*, $\lambda = 1.5$, $s^2 = 1.68$). Because the condition of randomness requires the mean to be equal to the variance and the variance seems to satisfy our intuitive notion of a measurement of degree of clumping, the ratio of the variance to the mean appears to be an ideal measure. If the ratio is equal to one, the pattern is random. If the ratio is greater than one, the pattern is clumped. If the ratio is less than one, the pattern is superdispersed. Furthermore, in an elementary way the greater the clumping, the larger the variance to mean ratio. This is more fully discussed in the next section.

☐ EXERCISES

12 Compute the variance-to-mean ratio for *b*, *d*, and *f* in Figure 6.8 and in Figure 6.3.

13 Compute the variance-to-mean ratio for Figure 6.2 and the pattern in exercises 7 and 8.

14 Using the guidelines provided in Figure 6.9, compute the variance-to-mean ratio. Is the pattern random?

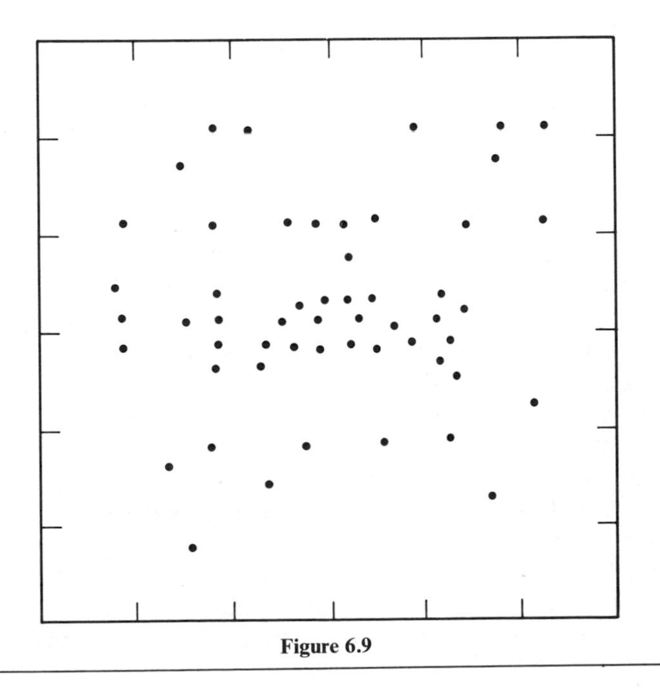

Figure 6.9

15 For this pattern construct quadrats four times the size of the guidelines (use every other guideline as new guidelines). What is the variance-to-mean ratio? What do you conclude?

16 In the diagram in Figure 6.10 one-half of the points were removed from Figure 6.2; 27 points were removed at random. Thus in a sense only the density should have changed, not the "pattern." Compute the variance and mean and their ratio. Compare them with similar statistics for the pattern in Figure 6.2. What do you conclude? □

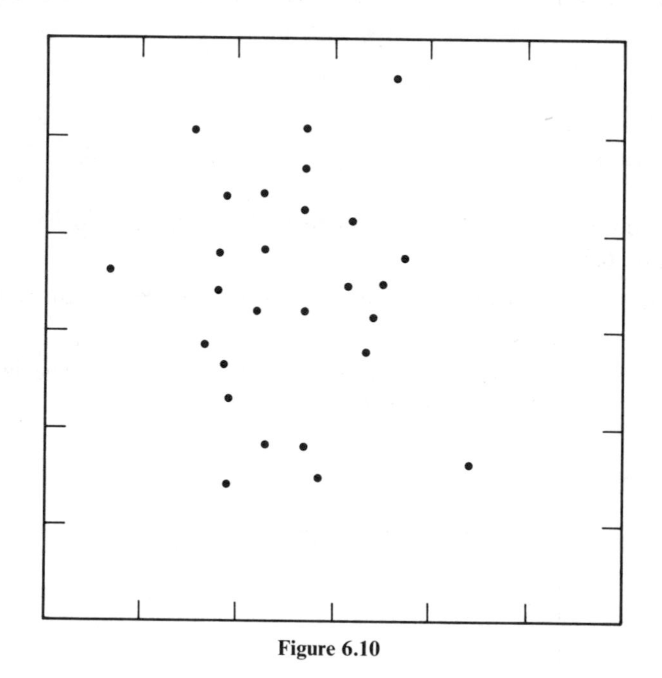

Figure 6.10

THE METHODS OF LLOYD AND MORISITA. We have noticed three disturbing facts about measuring degrees of clumpness. First, the measured pattern seems to be highly dependent on the quadrat size used (see exercises 14 and 15). Second, different patterns can give exactly the same index of clumping (see exercise 13). Third, two patterns that should be, in some sense, the same can easily be judged to be different, perhaps wrongly. We now seek to resolve these three problems.

Compare the patterns of Figures 6.2 and 6.3. Suppose each dot is a plant or animal. The individuals in Figure 6.3 seem to have plenty of elbow room, whereas those in Figure 6.2 seem to be crowded and irritable. How crowded are they? In either situation we might represent the amount of crowding quantitatively as the average crowding felt by an individual; that is, after dividing the area into quadrats we might ask, for each quadrat, how many individuals will share the quadrat with each individual in that quadrat. For example, if a quadrat contained 10 individuals, each of those individuals will share that quadrat with nine others. If an adjacent quadrat contained five individuals, each of those individuals will share the quadrat with four others. Considering both quadrats together, the average amount of crowding (the average number of individuals that will share a quadrat) is (10(9) + 5(4))/15. In general, if X_i is the number of individuals in quadrat i, the "mean crowding" is defined by Lloyd (1967) as

$$\overset{*}{m} = \frac{1}{N} \sum_{i=1}^{Q} X_i(X_i - 1)$$

where Q is the number of quadrats and N is the total number of individuals.

☐ EXERCISES

17 Compute mean crowding ($\overset{*}{m}$) for the examples in Figures 6.1, 6.2, and 6.3 (use the guidelines for establishing the quadrats).

18 The following pattern is nearly random (like the pattern in Figure 6.1). Compute the mean and variance and their ratio. Compute $\overset{*}{m}$ and compare with $\overset{*}{m}$ as the pattern in Figure 6.1. What do you conclude? ☐

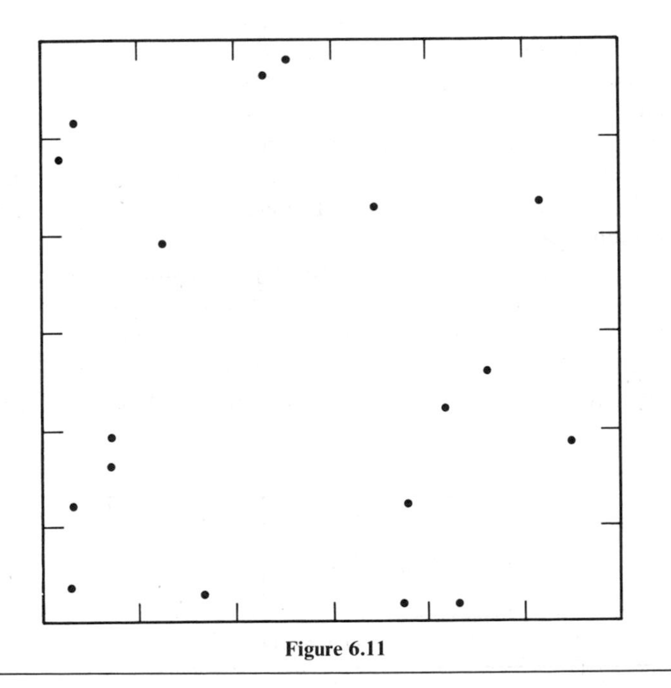

Figure 6.11

Thus the degree of crowding (as measured by $\overset{*}{m}$) is dependent on the degree of clumping and the population density. We can easily scale the population density effect out of the measurement. This is exactly what Lloyd did when he derived the patchiness index. Patchiness is defined as

$$C = \overset{*}{m}/\lambda$$

☐ EXERCISES

19 Compute patchiness for the examples in Figures 6.1, 6.2, and 6.3 and the pattern in exercise 18.

20 Compute patchiness for the example in exercise 16. Compare with the patchiness in Figure 6.2. What do you conclude?

21 Express mean crowding and patchiness in terms of the mean and variance in the distribution. (The variance is computed by $\{\sum X_i^2 - [(\sum X)]^2/Q\}/(Q-1)$ as well as $[\sum (X_i - \lambda)^2]/(Q-1)$. For this derivation be careful also to distinguish N, the total number of individuals, from Q, the total number of quadrats. Let Q be large such that $Q - 1 = Q$.

22 Another simple measure of clumping is $I = (s^2/\lambda) - 1$, which is nothing more than the mean variance ratio scaled such that $I = 0$ for a random pattern, $I > 0$ for a clumped pattern, and $I < 0$ for a hyperdispersed pattern. Express $\overset{*}{m}$ and C in terms of I. ☐

In summary, the index of mean crowding reflects the degree of clumping and the population density. The index of patchiness reflects only the degree of clumping in the sense that if the population density is altered at random the index of patchiness will remain constant (as in exercise 20). But clearly, both the mean crowding measure and the patchiness measure are dependent in general on quadrat size. Next, then, we consider quadrat size, the second of the three problems mentioned earlier.

The following method was developed by Morisita (1959). If we draw two individuals randomly from any of the patterns we have seen so far, what is the probability that both individuals will have come from the same quadrat? Consider a simple example to begin with. Suppose we have only the following two quadrats, in which the individuals have been numbered for convenience.

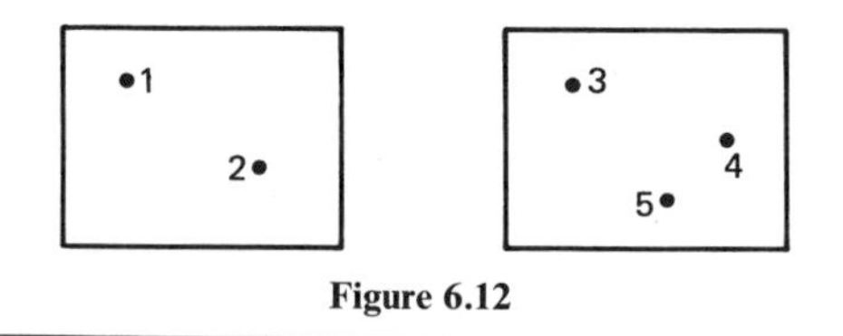

Figure 6.12

The possible combinations from the same quadrat are (1, 2), (3, 4), (4, 5), (3, 5); that is, there are four possibilities. The possible combinations that do not come from the same quadrat are (1, 3), (1, 4), (1, 5), (2, 3), (2, 4), (2, 5); that is, six possibilities. Thus the probability that two randomly drawn individuals will come from the same quadrat is $4/(4 + 6) = 0.4$. In general, the probability is equal to the number of possible pairs that will come from the same quadrat, divided by the total number of possible pairs. Of any collection of objects, the total number of possible pairs is $\frac{1}{2}N(N - 1)$, where N is the number of objects. Then, if X_i is the number of individuals in the ith quadrat, $\frac{1}{2}X_i(X_i - 1)$ is the number of possible pairs in that quadrat. Thus the total number of pairs that will come from the same quadrats is $\frac{1}{2}\sum_{i=1}^{Q} X_i(X_i - 1)$. The probability that the individuals of a randomly drawn pair will come from the same quadrat is

$$\delta = \frac{\sum_{i=i}^{Q} X_i(X_i - 1)}{N(N - 1)}$$

If the individuals are dispersed among the quadrats at random, $\delta = 1/Q$. Morisita takes the ratio between δ and the δ expected in a random population as his measurement of dispersion; that is,

$$I_\delta = \frac{Q\sum_{i=i}^{Q} X_i(X_i - 1)}{N(N - 1)}$$

is the measure of dispersion (note that this bears no relation to I in exercise 22). A moment's reflection will convince the reader that this measure, as well as any other we have proposed, satisfies our intuition about clumpedness (low for Figure 6.8*a* and high for Figure 6.8*f*).

□ EXERCISES

23 Show how I_δ is related to mean crowding and patchiness. How are they related as N becomes large.

24 Compute I_δ for the patterns in Figures 6.1, 6.2, and 6.3. □

In most practical situations Lloyd's index of patchiness and Morisita's I_δ are nearly the same. In the following analysis we use I_δ rather than C only because Morisita used it. The reader should be aware that the following could just as easily be developed with C.

If the space in which the pattern occurs is broken up into small quadrats, we can compute a value of I_δ. The same space can be broken up into larger quadrats (usually twice the size of the original quadrats) and a new value of I_δ, computed. The same space can then be broken up into still larger quadrats and a new value of I_δ, computed. In this way we can generate a graph of I_δ against quadrat size. neither I_δ nor C will change with quadrat size if the distribution on the whole is random. If the distribution is nonrandom, I_δ will change as a function of quadrat size. These changes can tell us a lot about the pattern.

☐ EXERCISES

25 Beginning with a quadrat size one-half that indicated in the diagram in Figure 6.1, compute I_δ for successively larger quadrat sizes (i.e., from a quadrat size of 1 × 1 to 2 × 2 to 3 × 3 to 4 × 4 to 6 × 6). Plot I_δ against quadrat size. The pattern in Figure 6.1 is reproduced in Figure 6.13 with the new smaller quadrats:

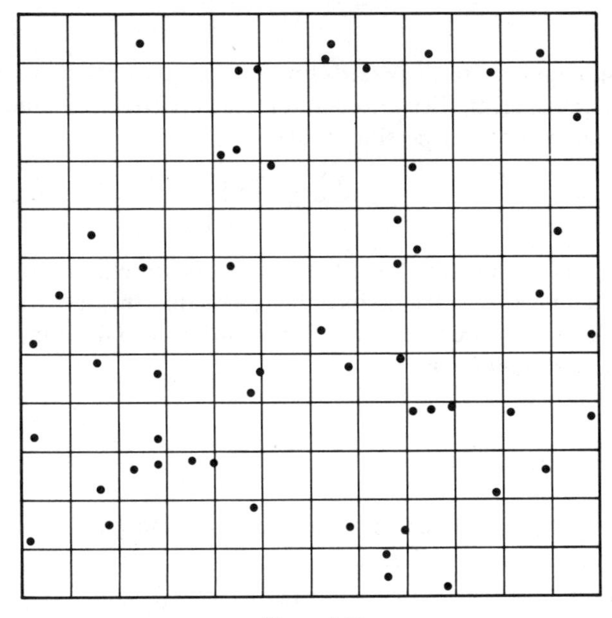

Figure 6.13

26 The following pattern is the same as that in exercise 14 with finer quadrats superimposed:

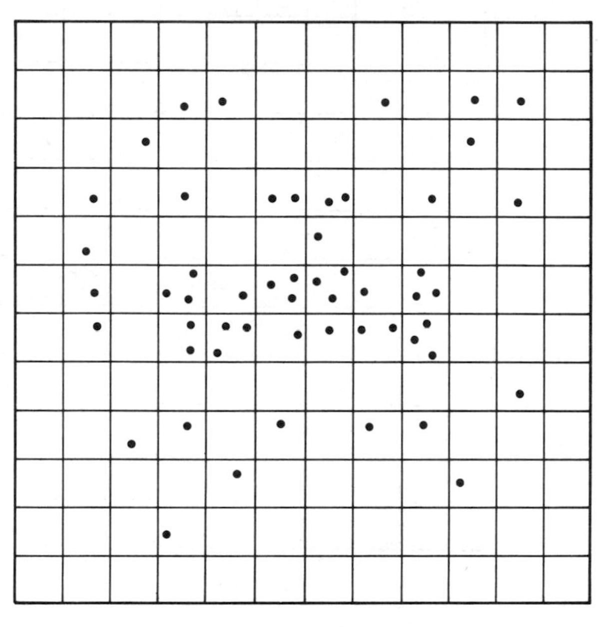

Figure 6.14

Compute I_δ for the various quadrat sizes, as in exercise 25. Plot I_δ against quadrat size. What do you conclude?

27 Construct the graph of I_δ against quadrat size for the pattern in Figure 6.15:

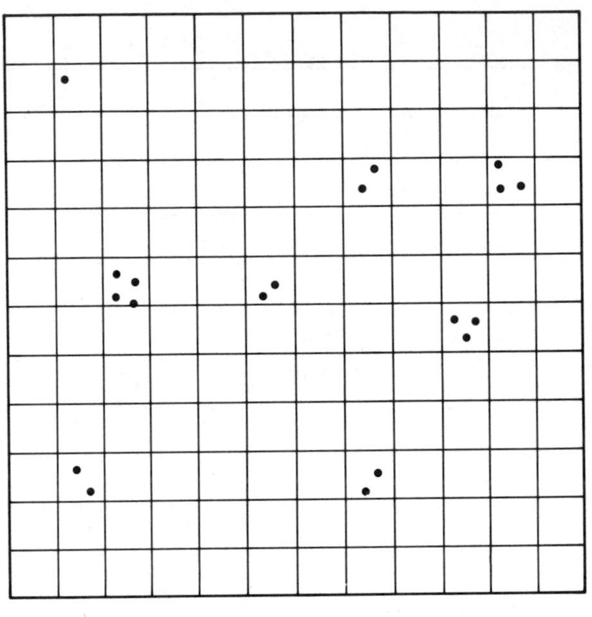

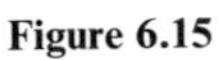

Figure 6.15

28 Construct the graph of I_δ against quadrat size for the pattern in Figure 6.16:

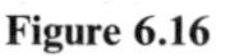

Figure 6.16

29 Construct the graph of I_δ against quadrat size for the pattern in Figure 6.17: □

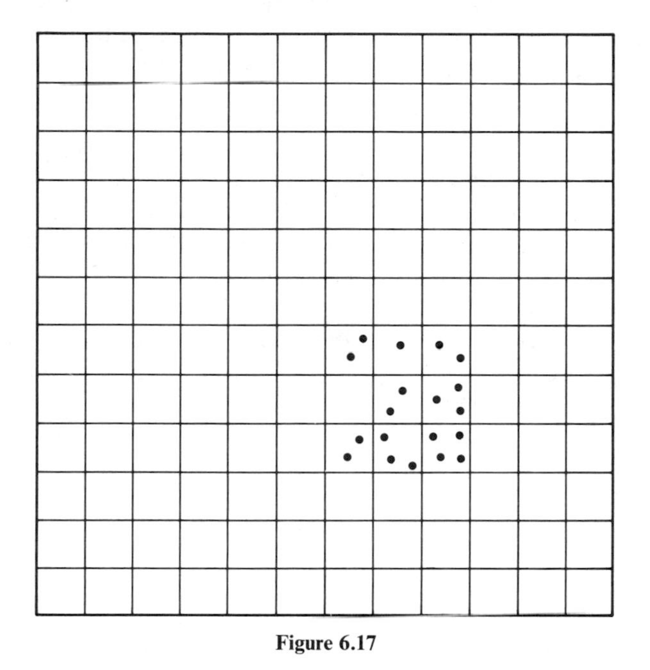

Figure 6.17

From the examples in exercises 25, 26, 27, 28, and 29 we see that the way I_δ changes with respect to quadrat size will vary, depending on the nature of the pattern. On the one extreme I_δ may be greater than one for all quadrat sizes, in which case the pattern can generally be referred to as clumped. On the other extreme I_δ may be less than one for all quadrat sizes, in which case the pattern can be generally referred to as hyperdispersed. But look at the I_δ versus quadrat-size graph for the patterns in exercises 28 and 29. Both are clumped but the graphs reflect the obviously different patterns. In general, a rapid drop in I_δ indicates a clump size that corresponds approximately to the quadrat size at which the rapid drop occurs; for example, in exercise 27 a rapid drop at a quadrat size between 1 and 2 reflects a clump size of approximately 1 to 2 units (the units are qual to one side of the smallest quadrat). An examination of the pattern in exercise 27 reveals these clumps (more or less). In exercise 28 the most rapid drop occurs between quadrat sizes 2 and 3. An examination of the pattern reveals these clumps. In exercise 29 the most rapid drop occurs between quadrat sizes 3 and 4. The pattern shows this clump.

Morisita's method [and a similar one invented by Greig Smith (1952)] is probably the most commonly applied method of pattern analysis in ecology. Unfortunately its shortcomings are many. In practice the number of different sized quadrats obtainable is usually restricted, making the pattern of I_δ against quadrat size frequently difficult to interpret. Furthermore, natural patterns of plants and animals, although usually clumped, do not commonly consist of nice discrete clumps all of the same size. The rapid fall in the value of I_δ does not necessarily occur at the "average" clump size. In practice the analysis of pattern in ecology seems to remain in a primitive stage.

NEAREST NEIGHBOR TECHNIQUES. The methods described are useful in a variety of situations, especially in those in which the exact location of each organism is not known, but samples are taken at particular localities. We then ask how many individuals are in each sample and apply one of the foregoing techniques to the data.

An alternative technique exists if the exact location of each of the individuals is known. The technique is known as nearest neighbor analysis and has been widely applied in plant ecology.

□ EXERCISES

30 In the following diagram measure the distance from point 1 to point 2, from point 2 to point 3, from point 3 to point 4, and from point 4 to point 3. What is the average of the four measurements?

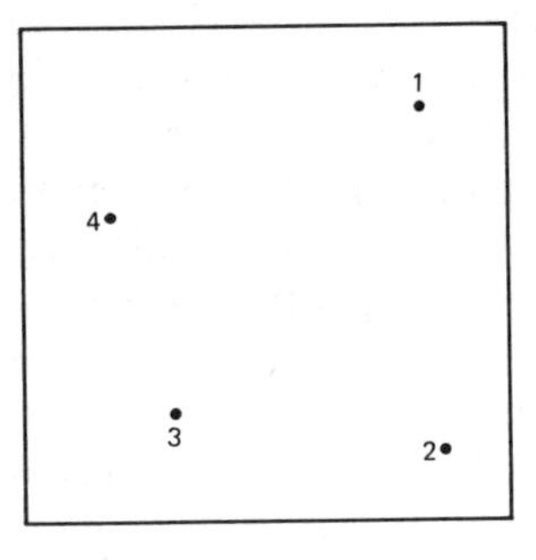

Figure 6.18

31 For each point in the diagram in Figure 6.19 measure the distance to the next *nearest* point and compute the average of the distances. □

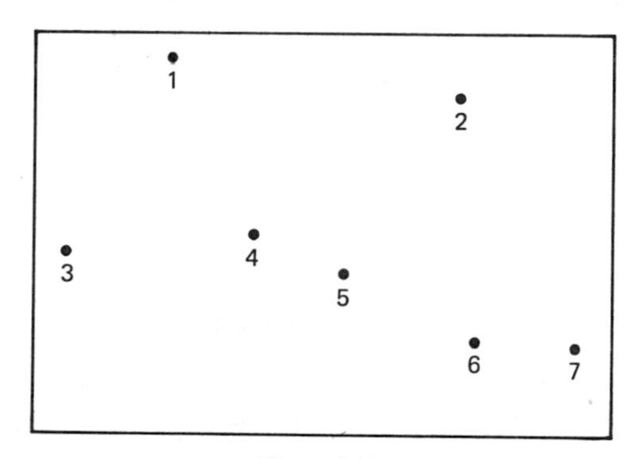

Figure 6.19

Given a distribution of individuals over a plane, the average of the nearest neighbor distances is

$$\bar{r} = \frac{\sum_{i=i}^{N} r_i}{N} \tag{2}$$

If the points are distributed at random, the expected value of r is

$$\bar{r}_{\text{rand}} = \frac{1}{2(N/A)^{1/2}}$$

where A is the total area (measured in the same units as the distances). If the actual mean is computed with a random sample of individuals from the population, the expected value must still use the true density; that is, it is necessary to know the density, independent of the nearest neighbor measurements.

According to the intuitive reasoning used earlier to derive measurements of aggregation, it is obvious that

$$R = \frac{\bar{r}}{\bar{r}_{\text{rand}}}$$

is a measure of aggregation.

□ EXERCISES

32 Compute R for the pattern in exercise 31.

33 For the pattern in Figure 6.20 compute the mean nearest neighbor distance and the measure of aggregation R.

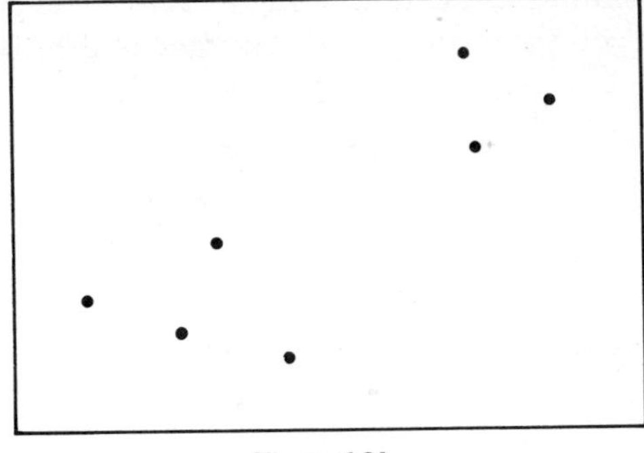

Figure 6.20

34 For the pattern in Figure 6.21 compute the mean nearest neighbor distance and R. □

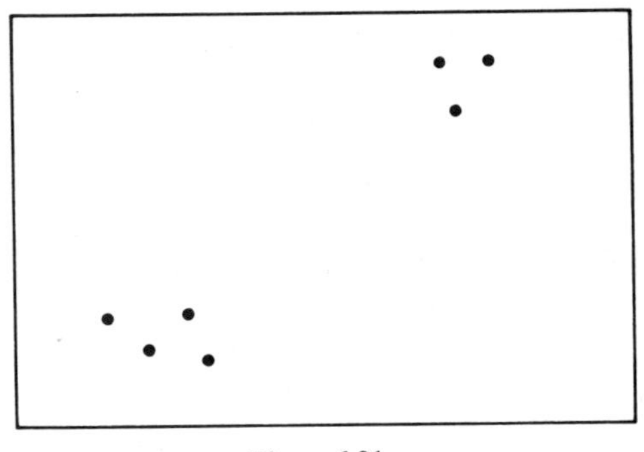

Figure 6.21

ANSWERS TO EXERCISES

1

Number of points	Number of quadrats
0	9
1	11
2	9
3	4
4	2
5	1

2 Total of 54 points and 36 quadrats. Mean = λ = 1.5.

3 For computational purposes the logarithmic form of equation 1 is most convenient:

$$l_n p_r = r \ln \lambda - \lambda - \ln (r!)$$

for this example $\lambda = 1.53$ and $\ln \lambda = 0.405$.

r	$r \ln\lambda$	$r!$	$\ln r!$	$\ln P_r$	P_r
0	0	1	0	−1.5	0.223
1	0.405	1	0	−1.125	0.335
2	0.810	2	0.693	−1.413	0.251
3	1.215	6	1.792	−2.107	0.126
4	1.620	24	3.178	−3.088	0.047
5	2.025	120	4.787	−4.292	0.014
6	2.430	720	6.579	−5.679	0.004

r	Expected number of quadrats (P_r) (36)	Observed number of quadrats
0	8.03	9
1	12.06	11
2	9.04	9
3	4.54	4
4	1.69	2
5	0.50	1
6	0.14	0

4

Number of points	Number of quadrats
0	16
1	7
2	4
3	3
4	3
5	1
6	1
7	1
8	0

5

Number of points	Number of quadrats
0	0
1	19
2	16
3	1
4	0

6

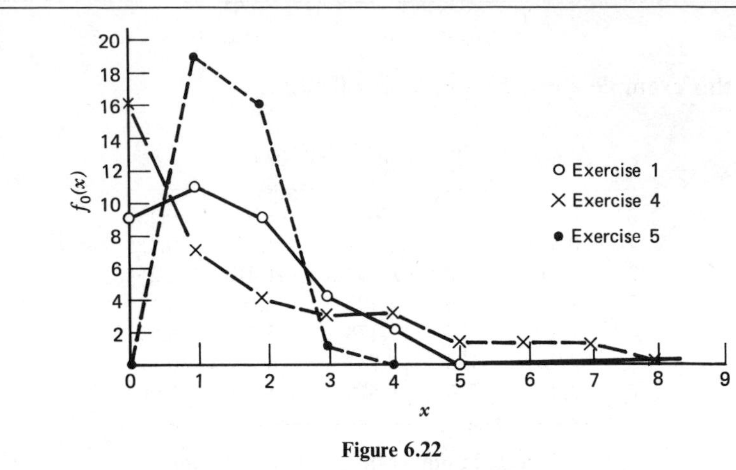

Figure 6.22

7 The expected number has already been computed in exercise 3. (Again we have 36 quadrats and 54 points $\lambda = 54/36 = 1.5$.)

x	$f_0(x)$	$f_e(x)$
0	16	8.03
1	7	12.06
2	4	9.04
3	3	4.54
4	3	1.69
5	1	0.50
6	1	0.14
7	1	0
8	0	0

8 Same exact answer as in 7.

9

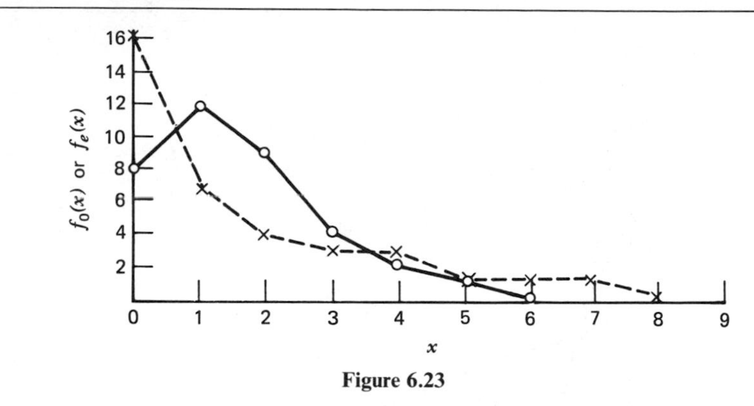

Figure 6.23

Obviously different patterns can show the same deviation from randomness.

10

x	$f_0(x)$ for 5a	$f_0(x)$ for 5b	$f_0(x)$ for 5c	$f_0(x)$ for 5d	$f_0(x)$ for 5e	$f_0(x)$ for 5f
0	8	14	20	24	29	32
1	12	7	3	1	0	0
2	9	6	3	1	0	0
3	4	4	3	2	0	0
4	2	3	3	2	0	0
5	1	1	1	2	1	0
6		1	2	2	1	0
7			1	1	1	0
8				1	1	0
9					2	0
10					1	0
11						0
12						0
13						2
14						2

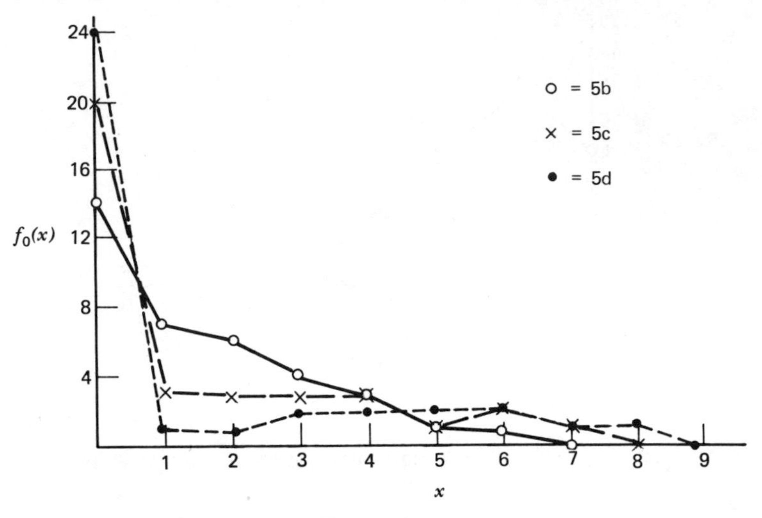

Figure 6.24

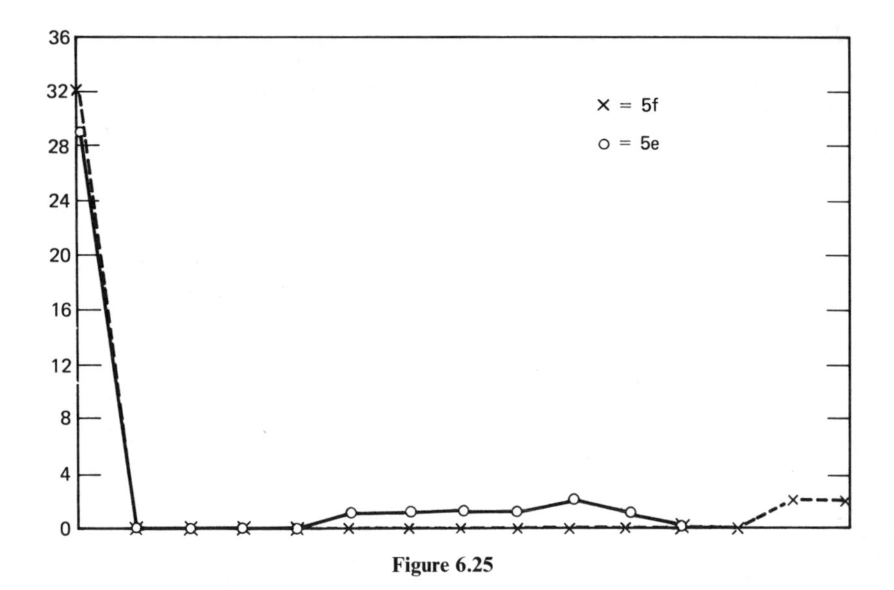

Figure 6.25

11 For Figure 6.5*a* $s^2 = [9(1.5)^2 + 11(0.5)^2 + 9(0.5)^2 + 4(1.5)^2 + 2(2.5)^2 + (3.5)^2]/35$
$= 1.68$
For Figure 6.5*b* $s^2 = 2.714$
For Figure 6.5*c* $s^2 = 4.429$
For Figure 6.5*d* $s^2 = 5.971$
For Figure 6.5*e* $s^2 = 10.143$
For Figure 6.5*f* $s^2 = 18.54$
For Figure 3 $s^2 = 0.314$

The variance behaves in accordance with what our intuitive notion of degree of clumping would be.

12 We have already computed the variances in exercise 11. All the means are 1.5 (a total of 54 points in 36 quadrats). Thus for Figure 6.5*a* $1.68/1.5 = 1.12$; for Figure 6.5*b*, $2.714/1.5 = 1.81$, for Figure 6.5*c*, $4.429/1.5 = 2.95$; for Figure 6.5*d*, $5.971/1.5 = 3.98$; for Figure 6.5*e*, $10.143/1.5 = 6.76$; for Figure 6.5*f*, $18.54/1.5 = 12.36$, and for Figure 6.3, $1.34/1.5 = 0.893$.

13 For Figure 6.2 $s^2 = [16(1.5)^2 + 7(0.5)^2 + 4(0.5)^2 + 3(1.5)^2 + 3(2.5)^2 + (4.5)^2 + (5.5)^2]/35 = 127/35 = 3.63$. The mean is still 1.5, and s^2/mean $= 3.63/1.5 = 2.42$. For 7 and 8 s^2/mean $= 3.63/1.5 = 2.42$. Thus for three patterns that appear to be totally different from one another the variance-to-mean ratios are identical.

14 The mean is again 1.5, the variance is 1.68, and the ratio is 1.12. Apparently the pattern is nearly random even though it obviously looks clumped.

15 The number of points per quadrat are 4, 6, 5, 7, 16, 8, 3, 3, and 2. The mean number per quadrat is $54/9 = 6$; the variance is 18.00; s^2/mean $= 3.00$. Now by this measure the same pattern is highly clumped. Clearly, quadrat size makes a difference.

16 Mean is $27/36 = 0.75$. Variance is 1.279. For Figure 6.2 the mean was 1.5 and the variance, 3.63. Variance/mean $= 1.279/0.75 = 1.71$, whereas it was 2.42 for Figure 6.2. The variance, or mean, or variance/mean ratio is not measuring purely pattern, at least in the sense that Figure 6.2 and this figure have the same pattern.

17 For Figure 6.1 the x's are 1, 1, 1, 1, 1, 1, 1, 1, 1, 1, 1, 2, 2, 2, 2, 2, 2, 2, 2, 2, 3, 3, 3, 3, 4, 4, 5. $N\overset{*}{m} = 2(1) + 2(1) + 2(1) + 2(1) + 2(1) + 2(1) + 2(1) + 2(1) + 2(1) + 3(2) + 3(2) + 3(2) + 3(2) + 4(3) + 4(3) + 5(4)$ $\overset{*}{m} = 86/54 = 1.592$. For Figure 6.2 $\overset{*}{m} = 154/54 = 2.852$. For Figure 6.3 $\overset{*}{m} = 38/54 = 0.704$.

18 The mean is $18/36 = 0.5$; variance is 0.54. The variance/mean ratio is 1.08; $\overset{*}{m} = 0.56$. Thus for this random pattern mean crowding is 0.56, whereas for the random pattern in Figure 6.1 mean crowding is 1.592. It must be concluded that the measurement mean crowding ($\overset{*}{m}$) is dependent on the number of points in the sampling area (biologically, the population density of the organisms in question).

19 For Figure 6.1 $\overset{*}{m} = 1.592$, $\lambda = 1.5$, $\overset{*}{m}/\lambda = 1.061$. For Figure 6.2 $\overset{*}{m} = 2.852$, $\lambda = 1.5$, $\overset{*}{m}/\lambda = 1.901$. For Figure 6.3, $\overset{*}{m} = 0.704$, $\lambda = 1.5$, $\overset{*}{m}/\lambda = 0.469$. For exercise 17 $\overset{*}{m} = 0.56$, $\lambda = 0.5$, $\overset{*}{m}/\lambda = 1.12$.

20 $\overset{*}{m} = 38/27 = 1.41$ (for Figure 6.2 $\overset{*}{m} = 2.852$). Patchiness $= 1.41/0.75 = 1.88$ (for Figure 6.2 patchiness $= 1.90$). At least in this case, randomly removing one-half of the population did not change patchiness, although it changed every other measure of clumpness we have considered.

21 Remember $\sum X_i = N$

$$\overset{*}{m} = \frac{1}{N} \sum X_i(X_i - 1)$$

$$= \frac{1}{\sum X_i} \sum X_i^2 - \frac{1}{\sum X_i} \sum X_i \tag{1}$$

$$= \frac{\sum X_i^2}{\sum X_i} - 1$$

The equation for the variance is

$$s^2 = \frac{\sum X_i^2 - [(\sum X_i)^2]/Q}{Q}$$

Remember, Q is very large; therefore $Q \approx Q - 1$ whence

$$\sum X_i^2 = s^2(Q) + \frac{(\sum X)^2}{Q} \tag{2}$$

The equation for the mean is $\lambda = \sum X_i/Q$, whence

$$\sum X_i = \lambda Q \tag{3}$$

Substituting equations 2 and 3 into 1, we obtain

$$\overset{*}{m} = \frac{s^2(Q) + [(\sum X)^2]/Q}{\lambda Q} - 1$$

$$= \frac{s^2(Q)}{\lambda Q} + \frac{[(\sum X)^2]/Q}{\lambda Q} - 1$$

$$= \frac{s^2 Q}{\lambda Q} + \frac{\lambda^2 Q^2}{\lambda Q^2} - 1$$

$$= \frac{s^2}{\lambda} + \lambda - 1$$

Thus mean crowding is one less than the sum of the variance-to-mean ratio and the mean itself. Patchiness, then, is

$$C = \frac{\overset{*}{m}}{\lambda} = \frac{s^2}{\lambda^2} + 1 - \frac{1}{\lambda}$$

22

$$\overset{*}{m} = \frac{(s^2 - 1)}{\lambda} + \lambda = I + \lambda$$

$$C = \frac{\overset{*}{m}}{\lambda} = 1 + \frac{I}{\lambda}$$

23

$$I_\delta = \frac{Q \sum X_i(X_i - 1)}{N(N-1)} = \frac{Q}{N-1} \frac{[\sum X(X-1)]}{N}$$

and because the term in parentheses is m

$$I_\delta = \frac{Q}{N-1} \overset{*}{m}$$

patchiness is $\overset{*}{m}/\lambda = \overset{*}{m}Q/N = C$:

$$I_\delta = \frac{Q}{N-1} \overset{*}{m} = \frac{N}{N} \frac{Q}{N-1} \overset{*}{m} = \frac{\overset{*}{m}Q}{N} \frac{N}{N-1} = \frac{N}{N-1} C$$

where C is the symbol for patchiness. As $N \to \infty$, the difference between $N - 1$ and N is negligible and $I \to C = \overset{*}{m}/\lambda$.

24 From exercise 19 $C = 1.061$ for Figure 6.1; $C = 1.901$ for Figure 6.2; $C = 0.469$ for Figure 6.3. $N = 54$ for all three figures; therefore $I_\delta = (54/53)1.061 = 1.081$ for Figure 6.1, $I_\delta = 1.937$ for Figure 6.2, $I_\delta = 0.478$ for Figure 6.3.

25 For the smallest quadrat size (1 unit on a side)

x	$f_0(x)$
0	99
1	37
2	7
3	1

$$I_\delta = \frac{[7(2)(1) + 3(2)]144}{54(53)}$$

$$= 1.006$$

For the next largest quadrat (2 × 2)

x	$f_0(x)$
0	9
1	11
2	9
3	4
4	2
5	1

$$I_\delta = \frac{[9(2)(1) + 4(3)(2) + 2(4)(3) + 5(4)]36}{54(53)}$$

$$= 1.08$$

For the next largest quadrat size (3 units on a side)

x	$f_0(x)$
0	0
1	1
2	3
3	4
4	3
5	4
6	1

$$I_\delta = \frac{[3(2)(1) + 4(3)(2) + 3(4)(3) + 4(5)(4) + 6(5)]16}{54(53)}$$

$$= 0.98$$

For the next largest quadrat (4 × 4)

x	$f_0(x)$
0	0
1	1
2	0
3	0
4	1
5	2
6	1
7	0
8	3
9	1

$$I_\delta = \frac{[4(3) + 2(5)(4) + 6(5) + 3(8)(7) + 9(8)]9}{54(53)}$$

$$= 1.012$$

For the next largest quadrat (6 units on a side) there are only four quadrats. The number of points in each of the quadrats is 13, 15, 16, 10.

$$I_\delta = \frac{[13(12) + 15(14) + 16(15) + 10(9)]4}{54(53)} = 0.972$$

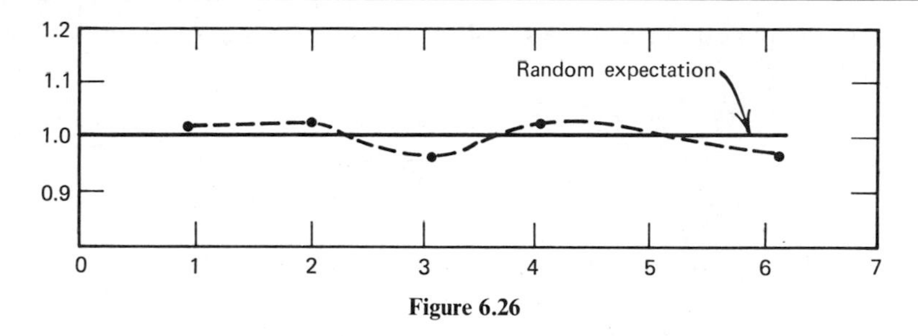

Figure 6.26

26 For the smallest quadrat size (1 × 1)

x	$f_0(x)$
0	106
1	28
2	4
3	6
4	0

$$I_\delta = \frac{[4(2) + 6(3)(2)]144}{54(53)}$$

$$= 2.213$$

For the next size quadrat (2 × 2)

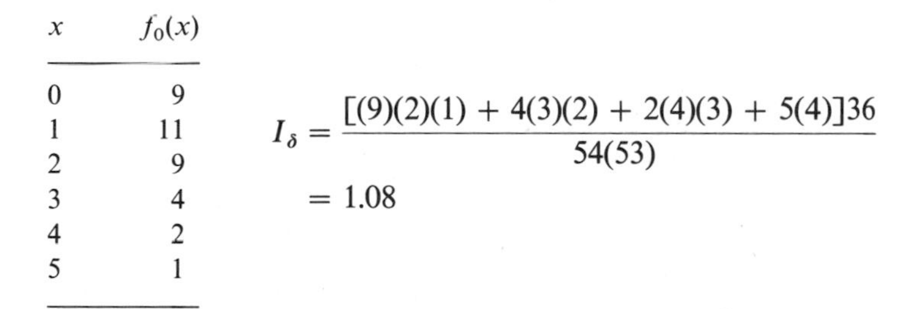

x	$f_0(x)$
0	9
1	11
2	9
3	4
4	2
5	1

$$I_\delta = \frac{[(9)(2)(1) + 4(3)(2) + 2(4)(3) + 5(4)]36}{54(53)}$$

$$= 1.08$$

For the next quadrat size (3 × 3)

x	$f_0(x)$
0	2
1	5
2	3
3	2
8	2
10	1
11	1

$$I_\delta = \frac{[(3)(2) + 2(3)(2) + 2(8)(7) + 10(9) + 11(10)]16}{54(53)}$$

$$= 1.85$$

For the next quadrat size (4 × 4) the individuals per quadrat are 4, 6, 5, 8, 16, 7, 2, 3, 3.

$$I_\delta = \frac{[2(1) + 2(3)(2) + 4(3) + 5(4) + 6(5) + 7(6) + 8(7) + 16(15)]9}{54(53)}$$

$$= 1.30$$

For the next quadrat size (6 × 6) the numbers are 16, 16, 12, 10.

$$I_\delta = \frac{[2(16)(15) + 12(11) + 10(9)]4}{54(53)} = 0.981$$

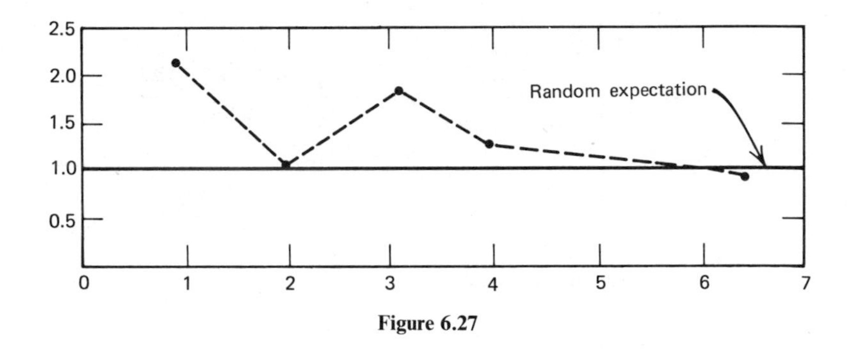

Figure 6.27

27 For the smallest quadrant size (1 × 1)

x	$f_0(x)$
0	136
1	1
2	4
3	2
4	1

$$I_\delta = \frac{[4(2) + 2(3)(2) + 4(3)]144}{19(18)}$$

$$= 13.47$$

For the next largest (2 × 2)

x	$f_0(x)$
0	28
1	1
2	4
3	2
4	1

$$I_\delta = \frac{[4(2) + 2(3)(2) + 4(3)]36}{19(18)}$$

$$= 3.36$$

For the next largest (3 × 3)

x	$f_0(x)$
0	8
1	1
2	4
3	2
4	1

$$I_\delta = \frac{[4(2) + 2(3)(2) + 4(3)]16}{19(18)}$$
$$= 1.50$$

For the next largest (4 × 4)

x	$f_0(x)$
0	1
1	1
2	4
3	2
4	1

$$I_\delta = \frac{[4(2) + 2(3)(2) + 4(3)]9}{19(18)}$$
$$= 0.84$$

For the next largest (6 × 6) the number in each quadrat is 5, 7, 5, 2.

$$I_\delta = \frac{[2(5)(4) + 7(6) + 2]4}{19(18)} = 0.98$$

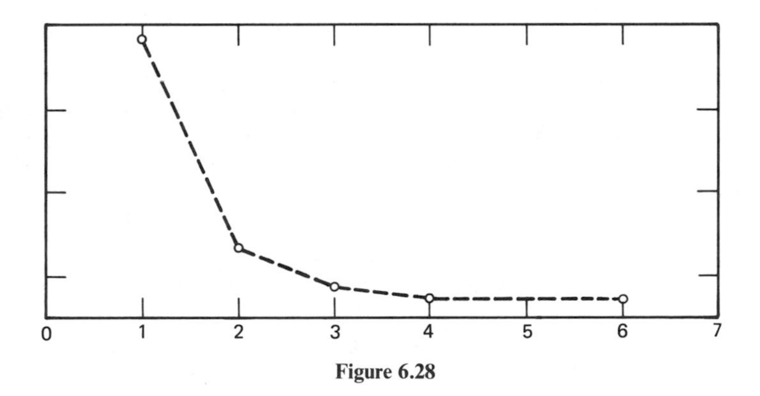

Figure 6.28

28 For the smallest quadrat (1 × 1)

x	$f_0(x)$
0	136
1	1
2	4
3	2
4	1

$$I_\delta = 13.47$$

For the next largest (2 × 2)

x	$f_0(x)$
0	33
5	1
7	2

$$I_\delta = \frac{[5(4) + 2(7)(16)]36}{19(18)}$$
$$= 10.95$$

For the next largest (3 × 3)

x	$f_0(x)$
0	10
2	2
3	2
4	1
5	1

$$I_\delta = \frac{[2(2) + 2(3)(2) + 4(3) + 5(4)]16}{19(18)}$$
$$= 2.25$$

For the next largest (4 × 4)

x	$f_0(x)$
0	6
5	1
7	2

$$I_\delta = \frac{[5(4) + 2(7)(6)]9}{19(18)}$$
$$= 2.74$$

For the next largest (6 × 6)

x	$f_0(x)$
0	1
5	1
7	2

$$I_\delta = \frac{[5(4) + 2(7)(6)]4}{19(18)}$$
$$= 1.22$$

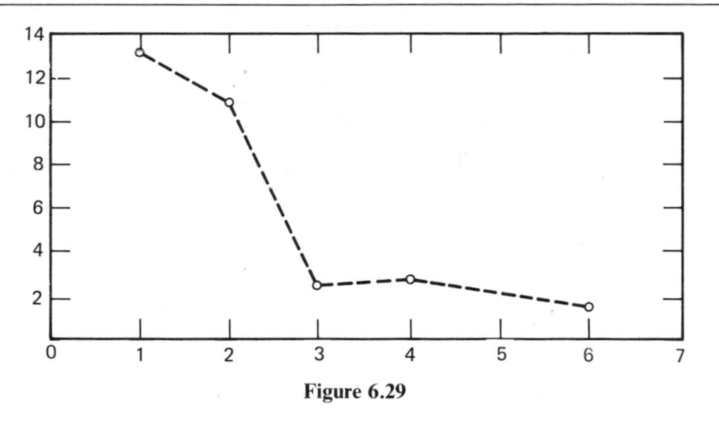

Figure 6.29

29 For the smallest quadrat (1 × 1) $I_\delta = 13.47$.
For the next size (2 × 2)

x	$f_0(x)$
0	12
4	1
5	3

$$I_\delta = \frac{[4(3) + 3(5)(4)]36}{19(18)} = 7.58$$

For the next largest (3 × 3)

x	$f_0(x)$
0	14
9	1
10	1

$$I_\delta = \frac{[19(18)]16}{19(18)} = 16.0$$

For the next largest (4 × 4) the number of points per quadrat is 5, 5, 4, 5.

$$I_\delta = \frac{[4(3) + 3(5)(4)]9}{19(18)} = 1.89$$

For the next largest (6 × 6) the number of points per quadrat is 19.

$$I_\delta = \frac{[(19)(18)]4}{(19)(18)} = 4.00$$

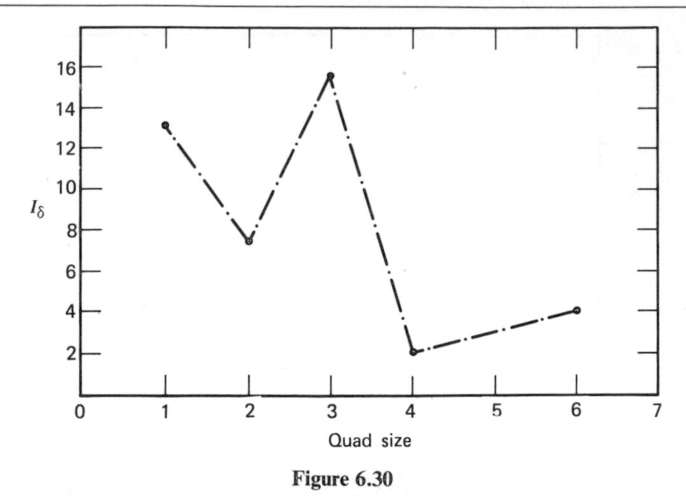

Figure 6.30

30 From 1 to 2 = 2.5 cm, from 2 to 3 = 2.0 cm, from 3 to 4 = 1.5 cm, from 4 to 3 = 1.5 cm. The average is 2.5 + 2.0 + 1.5 + 1.5 = 7.5/4 = 1.875.

31 From 1 to 4 = 2 cm, from 2 to 5 = 2 cm, from 3 to 4 = 2 cm, from 4 to 5 = 1 cm, from 5 to 4 = 1 cm, from 6 to 7 = 1 cm, from 7 to 6 = 1 cm. The average is 2 + 2 + 2 + 1 + 1 + 1 + 1 = 10/7 = 1.43.

32 From the answer to exercise 31 we know that $r = 1.43$. There are seven points in an area 6 × 4 cm = 24 cm^2. Thus the density is 7/24 = 0.29. The square root = 0.54 and $R = 1.43/2(0.54) = (1.43)(1.08) = 1.32$. The pattern is close to random but slightly on the overdispersed side.

33 The mean nearest neighbor distance is 1.0. There are seven individuals in an area of 24 cm^2. From exercise 32 we know that the density is 0.29. Thus $R = 1.0/1.08 = 0.92$. Again the pattern is close to random, but this time it is slightly on the clumped side.

34 Again density is 0.29. The mean nearest neighbor distance is 0.5 cm and $R = 0.5/1.08 = 0.55$. The pattern clearly deviates from random and is clumped.

REFERENCES

Austin, M. P., P. S. Asgon, and P. Greig-Smith. 1972. The application of quantitative methods to vegetation survey. III. A re-examination of rain forest data from Brunei. *J. Ecol.* **60**:305–324.

Burnett, T. 1958. Dispersal of an insect parasite over a small plot. *Can. Entomol.* **90**:279–283.

Clark, P. J. and F. E. Evans. 1954. Distance to nearest neighbor as a measure of spatial relationships in populations. *Ecology* **35**:445–453.

Evans, D. A. 1953. Experimental evidence concerning contagious distributions in ecology. *Biometrika* **40**:186–211.

Grieg-Smith, P. 1952. The use of random and contiguous quadrats in the study of the structure of plant communities. *Ann. Bot. Lond. N.S.* **16**:293–316.

Knight, D. H. 1975. A phytosociological analysis of species-rich tropical forest on Barro Colorado Island, Panama. *Ecol. Monogr.* **45**:259–284.

Lloyd, M. 1967. Mean Crowding. *J. Anim. Ecol.* **36**:1–30.

Morisita, M. 1959. Measuring the dispersion of individuals and analysis of the distributional patterns. *Mem. Fac. Sci. Kyushu Univ. Ser.* **2**:215–235.

Pielou, E. C. 1960. A single mechanism to account for regular, random, and aggregated populations. *J. Ecol.* **48**:575–584.

Pielou, E. C. 1961. Segregation and symmetry in two-species populations as studied by nearest neighbor relations. *J. Ecol.* **49**:255–269.

Pielou, E. C. 1964. The spatial pattern of two phase patchworks of vegetation. *Biometrics* **20**:156–167.

Pimentel, D. 1961. The influence of plant spatial patterns on insect populations. *Ann. Entomol. Soc. Am.* **54**:61–69.

Rohlf, F. J. 1969. The effect of clumped distributions in sparse populations, *Ecology* **50**:716–721.

Skellam, J. G. 1952. Studies in statistical ecology. I. Spatial patterns. *Biometrika* **39**:346–362.

Watt, A. S. 1947. Pattern and process in the plant community. *J. Ecol.*, **35**:1–22.

Webb, L. J., J. G. Tracey, and W. T. Williams. 1972. Regeneration and pattern in the subtropical rain forest. *J. Ecol.* **60**:675–695.

7. Interspecific Competition

In this chapter we return to the dynamic approach of Chapter 1. We are concerned fundamentally with a modification of the logistic equation. One would be well advised to keep in mind the limitations of that equation, for the material in this chapter is even more riddled with restrictive assumptions.

THE LOTKA–VOLTERRA EQUATIONS. Recall the most general form of density dependence

$$\frac{dN}{N\,dt} = f(N)$$

where N is the population density and $f(N)$ is some function of population density such that $\partial f(N)/\partial N < 0$. We now wish to extend the concept of density dependence from intra (within) to inter (between), from intraspecific (or intrapopulational) to interspecific (interpopulational). Instead of considering the number of individuals in the population, we take the number of individuals in one population (N_1) and the number in another (N_2) where the two populations interact with one another in a negative fashion, that is, the two populations compete. Although in the logistic equation the individuals in a single population interacted negatively (in competition), we now have individuals in two populations and each individual interacts negatively with two types of individual, those in population 1 (of which there are N_1) and those in population 2 (of which there are N_2). Obviously the most general representation is

$$\frac{dN_1}{N_1\,dt} = f_1(N_1, N_2)$$
$$\frac{dN_2}{N_2\,dt} = f_2(N_1, N_2) \qquad (1)$$

where f_1 and f_2 are functions of both population densities and $\partial f_1(N_1, N_2)/\partial N_1 < 0$, $\partial f_1(N_1, N_2)/\partial N_2 < 0$, $\partial f_2(N_1, N_2)/\partial N_1 < 0$, and $\partial f_2(N_1, N_2)/\partial N_2 < 0$.

As in Chapter 1, we make the simplest mathematical assumption possible and let f_1 and f_2 take on a simple linear form; that is, suppose that when N_1 and N_2 are very small ($N_1 \to 0$, $N_2 \to 0$) both populations will grow like exponential equations, and for every individual added to the population the per capita growth rate will be decreased by a particular amount. To be as general as possible, we do not require that the effect of an individual of population 1 will be the same as the effect of an individual of population 2, nor do we require that the effect of one individual on the growth rate of population 1 be the same as its effect on the growth rate of population 2.

□ EXERCISES

1. Consider two populations whose densities are given by N_1 and N_2 and whose dynamics are described by equations 1. Suppose the intrinsic rate of increase for population 1 is 1.0 and for population 2, 1.5. Suppose that for each individual of population 1 added to the environment the per capita rate of increase of population 1 is decreased by 0.05 and that of population 2, by 0.03. For each individual of population 2 added to the environment the per capita rate of population 1 is decreased by 0.001 and that of population 2, by 0.07. Assuming that f_1 and f_2 are linear, write the specific form of equations 1 for these figures.
2. Write the general form of equations 1 if f_1 and f_2 are linear.
3. If $N_1 = 0$, what will be the equilibrium population density of N_2 in terms of the equations derived in exercise 2? If $N_2 = 0$, what will be the equilibrium population density of N_1? □

Classically, the competitive effect of one population on another is quantified as the degree to which an individual of the other population affects the per capita growth rate of the population in question, divided by the degree to which an individual of the same population affects the per capita rate of the population in question; that is, the interspecific (interpopulational) competitive effect is expressed in terms of intraspecific (intrapopulational) units. Thus, if, for example, the per capita growth rate of population 1 is decreased by 0.01 for each individual of population 2 added to the environment and by 0.02 for each individual of population 1 added to the environment (individuals of its own population), the competitive effect of population 2 on population 1 will be $0.01/0.02 = 0.5$. The coefficient of competition is then given as $\alpha_{ij} = a_{ij}/a_{ii}$, where a_{ij} is the reduction in the per capita growth rate of population i due to the introduction of a single individual of population j and a_{ii} is the reduction in the r of population i due to the introduction of a single individual of population i. Loosely, α_{ij} is the number of individuals of population i that it would take to equal one individual of population j in terms of the effect on the per capita rate of increase in population i.

□ EXERCISES

4. What are α_{12} and α_{21} for the data in exercise 1?
5. Write the equations you derived in exercise 2 in terms of a_{ii} and α_{ij}.

6 Write the equations you derived in exercise 5 in terms of α_{ij}, r, and K (the carrying capacity derived in exercise 3.)

7 Define β_{ij} as the effect on the per capita rate of increase of population i and of the introduction of one individual of population j in relation to the effect of the introduction of one individual of population j (the same population) on its own per capita rate; that is,

$$\beta_{ij} = \frac{a_{ij}}{a_{jj}}$$

Rewrite the equations from exercise 6 in terms of β_{ij}, r, and K. □

The usual way of modeling interspecific competition is with the equations derived in exercise 6;

$$\frac{dN_1}{N_1 dt} = \frac{r_1(K_1 - N_1 - \alpha_{12}N_2)}{K_1}$$
$$\frac{dN_2}{N_2 dt} = \frac{r_2(K_2 - N_2 - \alpha_{21}N_1)}{K_2} \tag{2}$$

These equations are frequently referred to as the Guase–Volterra equations or more commonly as the Lotka–Volterra equations. It is important to recognize the distinction between equations 1 and 2. Equations 1 are much more general because they merely state that interspecific competition is occurring. Equations 2 assume that both intra- and interspecific density dependence have a particular linear form. As with the logistic equation, an infinite number of other particular models might be derived to represent the process of interspecific competition. Equations 2 are not uniquely true; sometimes they are thought to be totally "wrong" but they are the most commonly used model of competition.

□ EXERCISES

8 Suppose that the per capita rate of increase of population 1 is decreased by a_{12} as the result of adding an individual of population 2 to the environment, and by a_{11} as the result of adding an individual of population 1. Suppose, further, that a_{12} and a_{11} are decreased by 0.0001 with the introduction of each individual of population 1 to the environment. Assume that $a_{12} = A_1$ and $a_{11} = A_2$ when $N_1 = 0$. Express this particular form of equations 1 in terms of A_1, A_2, and r (only the equation for N_1).

9 Suppose that α_{12} and α_{21} vary with the population density of both populations; that is, the competition coefficients themselves are density-dependent. Furthermore, suppose that α_{12} and α_{21} are linearly related to N_1 and N_2. Rewrite equations 2 under these assumptions.

10 Suppose that a_{11}, a_{22}, a_{12}, and a_{21} from the equations in exercise 2 all vary linearly with the population density of N_1 and N_2. Rewrite the equations in exercise 2. □

The constants g_1, g_2, h_1, and h_2 in exercises 9 and 10 are frequently called "higher order interaction coefficients." Much debate has centered on whether higher order interactions exist in natural competitive systems. Such debate is largely misdirected. In fact, if equations 1 are reasonable descriptors of a particular system, equations 2 almost certainly will be accurate descriptors of that system near equilibrium even if there are very large higher order interaction coefficients (this is made clear later). In terms of the population dynamics far from equilibrium, equations 2 are most likely far too simple to be accurate descriptors of any but the simplest systems. Like the logistic equation, their use is mainly as a heuristic tool to further our understanding of the qualitative nature of interspecific competition.

To this end let us examine the behavior of equations 2 in detail. Under what conditions will both populations be in equilibrium? Assuming that $N_1 > 0$ and $N_2 > 0$, we set both derivatives equal to zero to obtain

$$K_1 = N_1 + \alpha_{12}N_2 \tag{3}$$

$$K_2 = N_2 + \alpha_{21}N_1 \tag{4}$$

where equation 3 applies when $dN_1/dt = 0$ and equation 4 applies when $dN_2/dt = 0$.

□ EXERCISES

11 What are the conditions under which $dN_1/dt > 0, dN_2/dt < 0, dN_2/dt > 0, dN_1/dt < 0$? Express those conditions on a graph of N_1 against N_2.

12 Suppose that $K_1 = 200$, $K_2 = 300$, $\alpha_{12} = 0.8$, $\alpha_{21} = 2.0$. Plot equations 3 and 4 on one graph.

13 Suppose that $K_1 = 200$, $K_2 = 300$, $\alpha_{12} = 0.8$, $\alpha_{21} = 0.8$. Plot equations 3 and 4 on one graph.

14 On the graph from exercise 13 shade in the area for which $dN_1/dt > 0$ *and* $dN_2/dt > 0$.

15 On the graph from exercise 12 shade in the area for which $dN_1/dt < 0$ *and* $dN_2/dt > 0$.

16 On the graph from exercise 12 shade in the area for which $dN_1/dt > 0$ *and* $dN_2/dt < 0$.

17 On the graph from exercise 12 shade in the area for which $dN_1/dt > 0$ *and* $dN_2/dt > 0$.

18 On the graph from exercise 12 shade in the area for which $dN_1/dt < 0$ *and* $dN_2/dt < 0$.

19 Suppose that $K_1 = 100$, $K_2 = 100$, $\alpha_{12} = 0.5$, $\alpha_{21} = 0.5$. Graph equations 3 and 4.

20 On the graph from exercise 19 shade in the area for which $dN_1/dt < 0$ *and* $dN_2/dt > 0$. Compare with the answer in exercise 15.

21 On the graph from exercise 19 shade in the area for which $dN_1/dt > 0$ *and* $dN_2/dt < 0$. □

If a particular combination of population 1 and population 2 is in a portion of the graph for which dN_1/dt and dN_2/dt are greater than zero, we expect both populations to increase; that is, if we pick a particular value of N_1 and N_2 and graph that point on a graph of N_1 against N_2 and if the point falls in an area for which both derivatives are greater than zero, in the next instant of time we expect both populations to be larger. If the point falls in an area for which $dN_1/dt > 0$ and $dN_2/dt < 0$, in the next instant of time N_1 will be larger and N_2 will be smaller; for example, in the graph of exercise 19 we might indicate the behavior of the system as a whole (Figure 7.1):

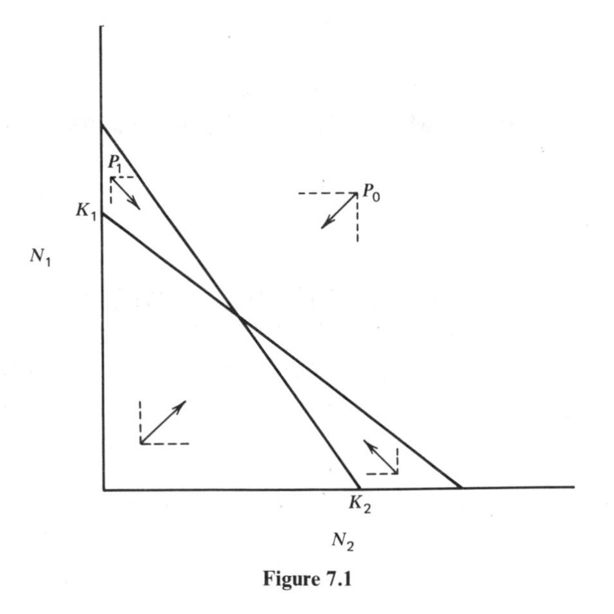

Figure 7.1

Thus, if N_1 and N_2 are plotted as the point P_0, they will fall in an area of the graph for which both derivatives must be negative (see exercise 18). Thus the direction of movement, indicated by the arrow in Figure 7.1, must be such that both populations will decrease. If we indicate the expected direction of change with a small arrow, that arrow must fall between the two dotted lines, each of which represents no population change for one of the populations (i.e., the dotted lines refer to $dN_1/dt = 0$ *or* $dN_2/dt = 0$). If N_1 and N_2 are plotted as point P_1, they will fall in an area of the graph for which $dN_1/dt < 0$ and $dN_2/dt > 0$ (see exercise 20). Thus we expect N_1 to decrease and N_2 to increase. Again the arrow from point P_1 in Figure 7.1 indicates the general direction in which the system must go.

□ EXERCISES

22 Let $K_1 = 500$, $K_2 = 300$, $\alpha_{12} = 0.8$, $\alpha_{21} = 1.0$. Graph equations 2 and 4. Indicate the general direction of change expected for all points on the graph.

23 Let $K_1 = 500$, $K_2 = 400$, $\alpha_{12} = 0.8$, $\alpha_{21} = 0.6$. Graph equations 3 and 4. Indicate the general direction of change expected for all points on the graph.

24 Let $K_1 = 200$; K_2, α_{21}, α_{12} are as in exercise 23. Graph equations 3 and 4. Indicate the general direction of change expected for all points on the graph.

25 Let $K_1 = 500$, $K_2 = 600$, $\alpha_{12} = 2.0$, $\alpha_{21} = 1.8$. Graph equations 3 and 4. Indicate the general direction of change expected for all points on the graph.

26 What are the N_1 and N_2 intercepts for equations 3 and 4? Graph both equations and label the intercepts. □

The answers to exercises 22, 23, 24, and 25 represent the four qualitatively different ways in which interspecific competition can occur (i.e., if we represent the process with equations 2). Using the intercept labels derived in exercise 26, we represent the four cases in Figure 7.2:

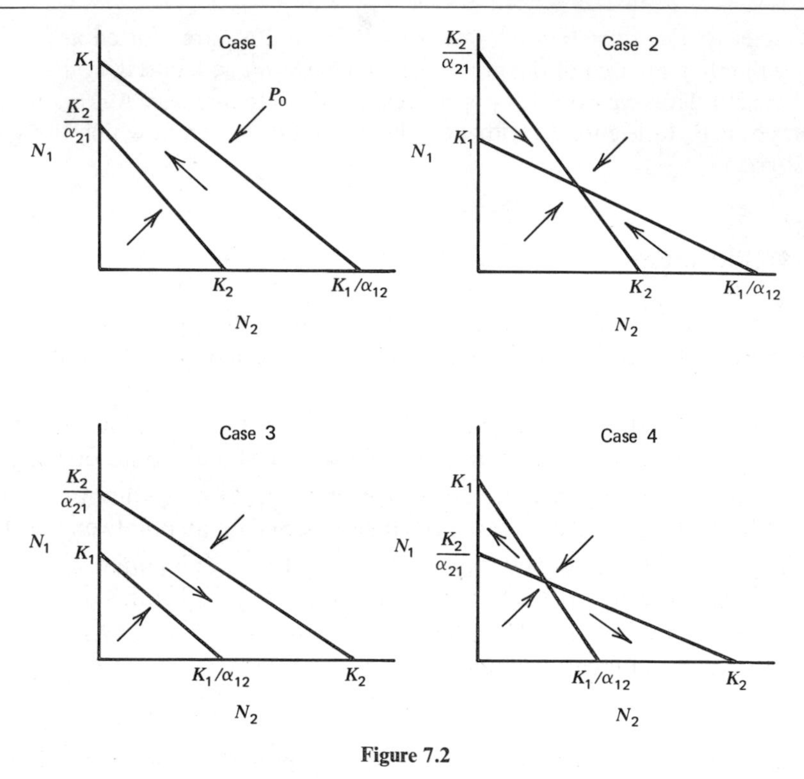

Figure 7.2

In Figure 7.2 the general dynamics of the system has been represented by the arrows, as in exercises 22–25. Consider case 1. If we begin with a particular combination of N_1 and N_2 (a point), say P_0 in the figure, both populations will decrease in numbers as they reproduce and die. Thus a new point will soon represent the system. That new point will be closer to the line K_1, K_1/α_{12}. Soon still another point will represent the system. Again the new point will be closer to the line K_1, K_1/α_{12}. This process will repeat itself until the system is represented by a point in between the lines K_1, K_1/α_{12} and K_2, K_2/α_{21}. This process is illustrated in Figure 7.3 for several starting points:

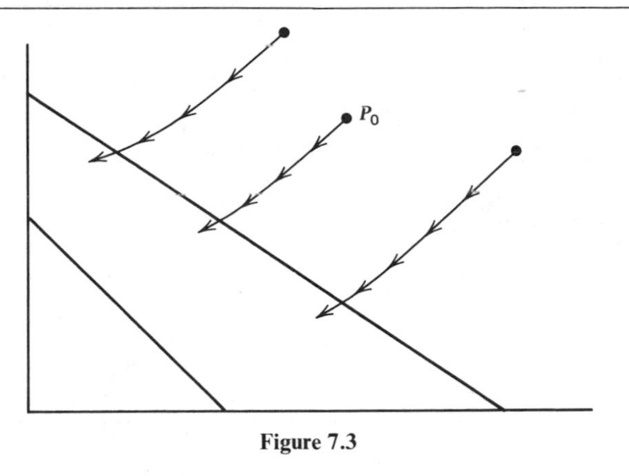

Figure 7.3

If the point is located in between the two lines, the system must travel upward and to the left (because in between the two lines is the area for which $dN_1/dt > 0$ and $dN_2/dt < 0$). The process is illustrated in Figure 7.4:

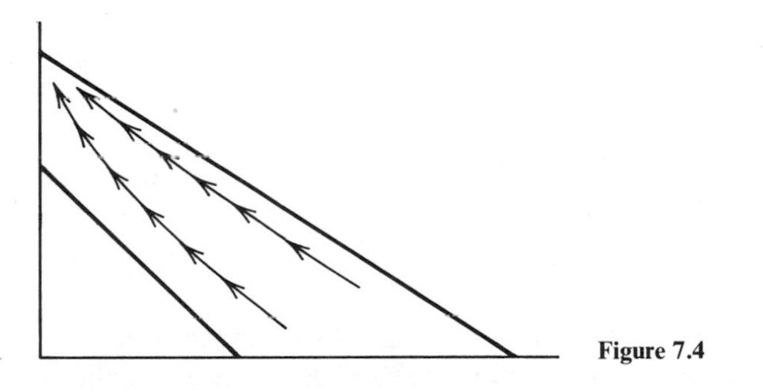

Figure 7.4

Similar arguments apply to points below the line $K_2, K_2/\alpha_{21}$; that is, eventually they will wind up in between the two lines. If a point should be exactly on one of the axes (i.e., N_1 or $N_2 = 0$), the whole system will revert to the logistic equation and travel along the axis to K_1 or K_2. Thus for the case we have been considering here (case 1) we can illustrate the dynamic behavior of the entire system in Figure 7.5:

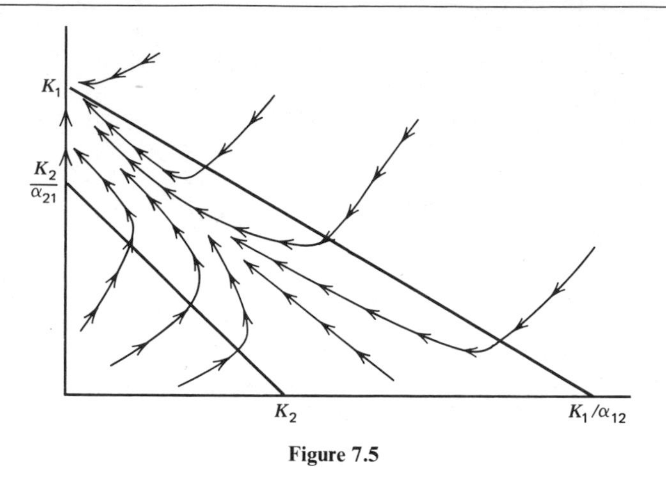

Figure 7.5

Thus it is intuitively obvious that this case must result in the system coming to rest at $N_1 = K_1$, $N_2 = 0$. The qualitative result is that population 1 "wins" in competition in the sense that population 2 is driven to extinction as population 1 attains its carrying capacity (K).

□ EXERCISES

27 Illustrate the qualitative dynamics of cases 2, 3, and 4, as was done for case 1. Describe the qualitative outcome for each case.

28 Distinguish between cases 1 through 4 on the basis of inequalities. (Use the intercept values to get two inequalities for each case.)

29 Show how you can distinguish between cases 2 and 4 by using only α_{12} and α_{21} (i.e., presume that cases 1 and 3 cannot occur and derive the inequalities that mark the two remaining cases).

30 Can you distinguish between cases 1 and 3 by using only α_{12} and α_{21}?

31 Express the equilibrium condition of the Lotka–Volterra equations in matrix form (i.e., express equations 3 and 4 in matrix form).

32 Use the equilibrium state of the Lotka–Volterra equations to derive equations that will give the equilibrium value of N_1 and N_2 in terms of K_1, K_2, α_{12}, and α_{21}.

33 The "determinant" of a 2×2 matrix is easy to calculate. In later chapters we shall use determinants of larger matrices, but for now you need only to compute determinants of 2×2 matrices. For the 2×2 matrix

$$B = \begin{vmatrix} a & b \\ c & d \end{vmatrix}$$

the determinant is defined as

$$\text{Det } B = ad - bc$$

What is the determinant of the community matrix?

34 Express the equilibrium values of N_1 and N_2, as derived in exercise 32, in the form of the ratio between two determinants.

35 Express the difference between cases 2 and 4 as conditions on the determinant of the community matrix. (Remember the result of exercise 29).

From the foregoing discussion we see that there are four qualitatively different outcomes of interspecific competition. This result is somewhat misleading. Recall from exercises 29 and 35 that it is possible to distinguish between cases 2 and 4 by examining the competition coefficients. Thus, if Det $A > 0$, we will have a stable equilibrium (case 2) and if Det $A < 0$ we will have an unstable equilibrium (case 4). This is presuming, of course, that cases 1 and 3 are impossible. What if we have a condition of case 1 (species 1 wins)? If Det $A > 0$, is the situation no different than if Det $A < 0$? In fact, the two situations are quite different in very important ways.

Consider the case 1 situation in Figure 7.6:

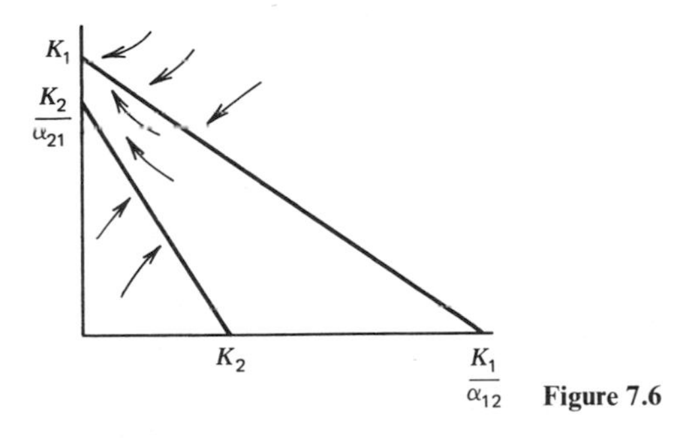

Figure 7.6

Remember from the basic equilibrium form of the Lotka–Volterra equations that the angle of the lines is given by α_{12} and α_{21}. Suppose that we hold everything except K_1 constant. Suppose that we modify the environment in such a way that K_1 decreases (remember everything else remains the same). If we decreased K_1 just a little, we might obtain the following (Figure 7.7):

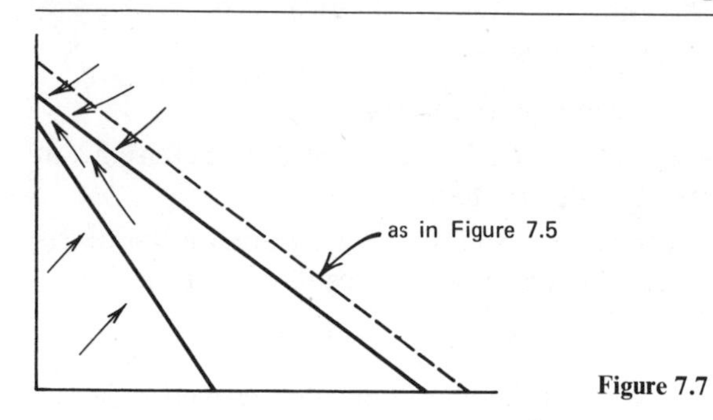

Figure 7.7

Qualitatively nothing happens. Species 1 still wins, but suppose we decrease K_1 still further. We might obtain Figure 7.8:

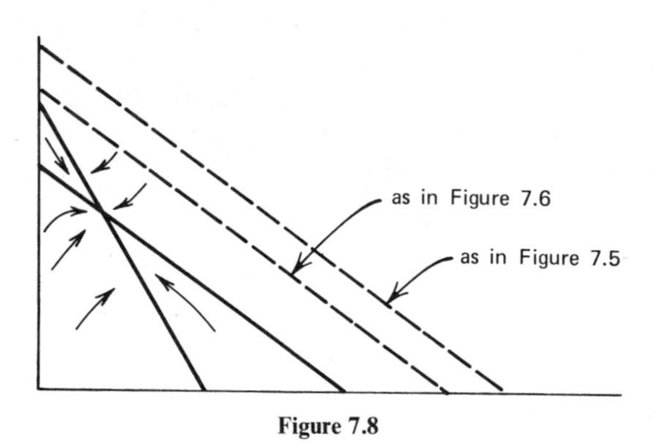

Figure 7.8

We have converted case 1 into a stable equilibrium (case 2). Alternatively, consider the case 1 situation in Figure 7.9:

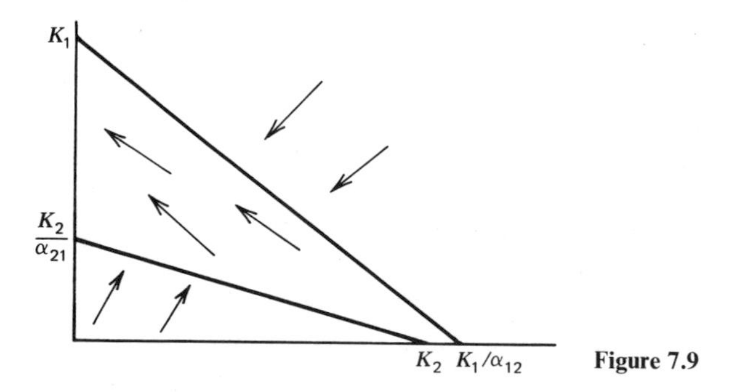

Figure 7.9

Again let us decrease K_1 and hold everything else constant. We obtain Figure 7.10:

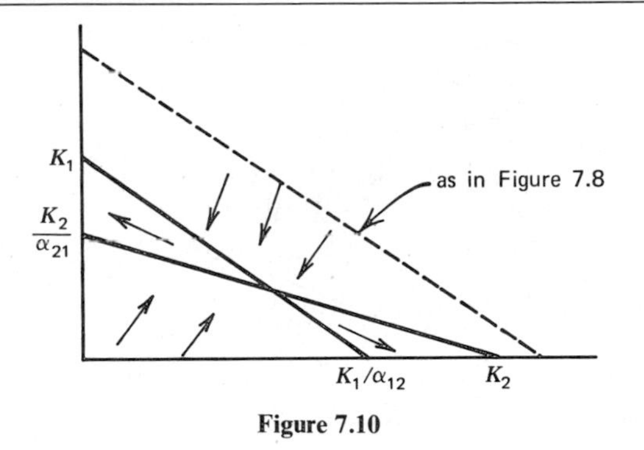

Figure 7.10

We see the remarkable result that by the same "biological" modification (lowering K_1) that changed a case 1 situation into a stable equilibrium we have now changed a case 1 situation into an unstable equilibrium.

If a case 1 or case 3 situation can be changed into a stable equilibrium, it is true that Det $A > 0$. If a case 1 or case 3 situation can be changed into an unstable equilibrium, it is true that Det $A < 0$. Thus, in general, we really have six qualitatively different competitive outcomes: (1) stable equilibrium, (2) case 1 *and* Det $A > 0$, (3) case 3 *and* Det $A > 0$, (4) unstable equilibrium, (5) case 1 *and* Det $A < 0$, and (6) case 3 *and* Det $A < 0$.

These six cases can most easily be understood if we present them graphically in K_1, K_2 space. Suppose first that Det $A > 0$. We have case 1 if

$$K_1 > \frac{K_2}{\alpha_{21}}$$

$$K_1 > K_2\alpha_{12}$$

case 3 if

$$K_1 < \frac{K_2}{\alpha_{21}}$$

$$K_1 < K_2\alpha_{12}$$

and case 2 if

$$K_1 < \frac{K_2}{\alpha_{21}}$$

$$K_1 > K_2\alpha_{12}$$

Graphically

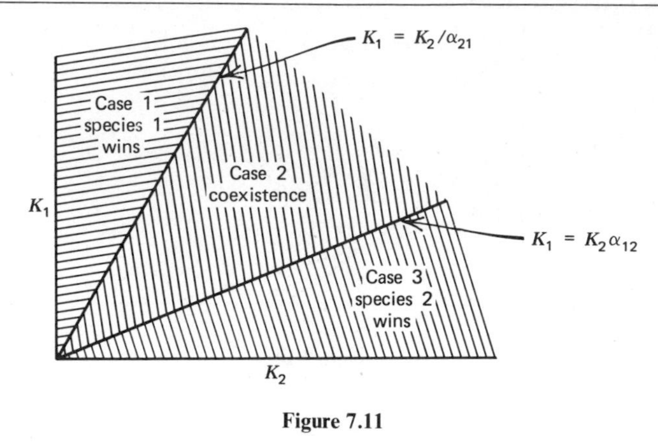

Figure 7.11

Thus, if α_{12} and α_{21} are fixed so that Det $A > 0$, we can quickly see the outcome of competition in Figure 7.11 for any pair of values of K_1 and K_2.

□ EXERCISES

36 Construct the graph analogous to Figure 7.11 for Det $A < 0$. □

With the aid of the graphs in Figure 7.11 and exercise 36, we can easily visualize all six cases of competition. Furthermore, it is possible to see how they grade into one another simply by rotating the lines, which is the same as changing the competition coefficients. Nevertheless, it remains true that the classical method of presentation, with four possible competitive outcomes, is heuristic. The six-way classification provides a deeper understanding of the underlying dynamics that go into producing the four outcomes.

□ EXERCISES

37 Express the inequalities for each of the four classical cases of competition in terms of β (see exercise 7) and r.

38 Distinguish between cases 2 and 4 by using β_{12} and β_{21}.

39 Construct graphs similar to Figure 7.11 and the graph in exercise 36 to show how the six cases of competition can be expressed in terms of r and β (in r_1, r_2 space). □

This analysis becomes especially important when we try to apply competition theory to a field situation. Frequently an ecologist working in the field is not fully cognizant of his or her own working definition of competition. What is really meant when we say, "*Nessiestrus* is a good competitor" or "*Ptherus pubis* is a poor competitor"? If the field worker has the impression that species A is a "good" competitor, does that mean that its α is high, its β is high, its a_{ij} is high, or something different? Obviously we could construct cases in which α_{12} is high and β_{12} is low.

All too often ramifications of simple competition theory are applied in a loose qualitative sense to natural situations. If this is to be done at all, we must be certain that the working definition of "relative intensity of competition" is objective and that different field workers are truly applying the same, or at least similar, notions of competition.

Recently a plethora of field and laboratory studies has purported to estimate intensity of competition (usually α) from experimental or observational data. We now turn our attention to such work.

ESTIMATING COMPETITION COEFFICIENTS IN THE REAL WORLD. Basically there are two situations to be faced when we try to estimate competition coefficients. First, data are frequently available in which two populations have been isolated from one another such that the carrying capacity (K) and the intrinsic rate (r) are known for both of them. Then in the same environment data are available for the two populations living together. In such a situation it is a relatively simple task to compute meaningful competition coefficients.

Second, various sorts of observational or quasi-experimental data are available in which one can examine distributions or feeding habits or behavior and gain an impression of competitive relationships and/or degree of intensity of competition. Although these data are also used to calculate competition coefficients, the validity of such use remains to be demonstrated.

☐ EXERCISES

40 In a laboratory situation two species of flour beetle have been cultured separately. Both populations approach an apparent carrying capacity asymptotically: that of the first is 200; that of the second is 400. Additionally, both have been cultured together. The population densities they reach in mixed culture are 150 for the first and 120 for the second. Assuming that this competitive process is reasonably modeled by the classic Lotka–Volterra competition coefficients, what are the values of α_{12} and α_{21}?

41 In the example in exercise 40 r_1 is estimated as 0.2, r_2 as 0.1. Compute β_{12} and β_{21}.

42 Of the two populations in exercise 40 which is the better competitor?

43 Two species of ant live in old termite nests. They do not live anywhere else. Each termite nest can contain one and only one ant colony and both ant species are equally capable of occupying a termite colony if undisturbed by a competitor. After censusing all the dead termite nests, 150 colonies of species 1 and 100 colonies of species 2 were found in 250 old termite nests. No unoccupied nests were found. Assume that the two populations are close to equilibrium and compute α_{12} and α_{21}. Could these be valid estimates of α?

44 Suppose that there are three types of termite host: A, B, and C. Ant species 1 can live in types A and B. Ant species 2 can live in types B and C. The environment consists of 200 type A nests, 50 type B, and 100 type C. All type A nests are inhabited by species 1 and all type C nests, by species 2. Among 50 type B nests 10 are inhabited by species 1 and 40 by species 2. Assuming equilibrium, compute α_{12} and α_{21} and check to see whether the equilibrium is consistent with the α's.

45 Modify the conditions in exercise 44 to increase the resource base of type B termite nests. In particular, suppose that 300 type B nests are occupied in the same proportions by ant species 1 and ant species 2; that is, 60 nests are occupied by species 1 and 240, by species 2. Assume a stable equilibrium and compute α_{12} and α_{21}. Compare with the estimate in exercise 44.

46 Graph the results of exercises 44 and 45 on a graph of N_1 against N_2.

47 Using the graph in exercise 46, determine how many colonies of species 1 would have to occupy the type B nests in exercise 45 (300 of them) to keep α_{12} and α_{21} constant from exercises 44 to 45? □

These exercises are illustrative of the conceptual difficulties we encounter when we try to estimate competition coefficients from field observations. Indeed, rarely if ever is information so complete as in the exercises (44 and 45) available from nature. Yet even under such relatively ideal conditions grave conceptual errors can occur in the estimation of competition coefficients. In those examples we had estimates of K_1 and K_2 from obvious habitat considerations, an exact knowledge of N_1 and N_2, and, by assumption, N_1 and N_2 were exactly at their stable equilibrium values. Even under these conditions we were unable to compute alphas with certainty.

In spite of these considerations a bulk of current ecological research is devoted to the estimation of alphas with drastically less information. Frequently all that is available, using the old termite nests and ants as a metaphor, is the distribution of ant colonies in the various termite nest types, without knowledge of which ant species *could* occupy a given nest type. This approach is based on certain ideas associated with the "ecological niche." Briefly, consider (1) the distribution of individuals in a population over an environmental gradient, such that N_{ij} is the number of individuals of species i at position j along the environmental gradient, or (2) the distribution of individuals in discrete habitat units, such that N_{ij} is the number of individuals in habitat j, or (3) the use of different resource types, such that N_{ij} is the amount of time (or energy) an average individual of species i expends on resource type j. In any of these cases "niche overlap" is a concept that refers to the degree to which two species have similar values of N_{ij} over a spectrum of j's; that is, if two species occupy an environmental gradient equally, their niche overlap is great, whereas if one species occupies one end of the environmental continuum and the second species occupies the other their niche overlap is small. Calculating niche overlap values and assuming they have anything whatsoever to do with competition is a highly questionable technique.

ANSWERS TO EXERCISES

1 Both functions must have the general form $f = a + bN_1 + cN_2$ because both functions are linear. If both N_1 and N_2 are zero, $f = a$; therefore a must be equal to r, the intrinsic rate of natural increase. Consider f_1 first. The effect of one individual of population 1 on the r of population 1 is 0.05. The effect of one individual of population 2 on the r of population 1 is 0.001. Thus $f_1(N_1, N_2) = r - 0.05N_1 - 0.001N_2$. Similar reasoning applies to the growth rate of the second population. Thus

$$\frac{dN_1}{N_1 dt} = 1.0 - 0.05N_1 - 0.001N_2$$

$$\frac{dN_2}{N_2 dt} = 1.5 - 0.03N_1 - 0.07N_2$$

2 Subscript all constants so that each can be distinguished from all the others.

$$\frac{dN_1}{N_1 dt} = r_1 - a_{11}N_1 - a_{12}N_2$$

$$\frac{dN_2}{N_2 dt} = r_2 - a_{22}N_2 - a_{21}N_1$$

where a_{ij} is the decrease in r of the ith population due to the addition of one individual of the jth population.

3 If $N_2 = 0$,

$$\frac{dN_1}{N_1 dt} = r_1 - a_{11}N_1$$

$$0 = r_1 - a_{11}N_1$$

$$N_1 = \frac{r_1}{a_{11}}$$

If $N_1 = 0$,

$$\frac{dN_2}{N_2\,dt} = r_2 - a_{22}N_2$$

$$0 = r_2 - a_{22}N_2$$

$$N_2 = \frac{r_2}{a_{22}}$$

4

$$\alpha_{12} = \frac{a_{12}}{a_{11}} = \frac{0.001}{0.05} = 0.02$$

$$\alpha_{21} = \frac{a_{21}}{a_{22}} = \frac{0.03}{0.07} = 0.43$$

5

$$\alpha_{ij} = \frac{a_{ij}}{a_{ii}}$$

$$\frac{dN_1}{N_1\,dt} = r_1 - a_{11}N_1 - a_{12}N_2$$

Divide both sides by a_{11}:

$$\frac{1}{a_{11}}\frac{dN_1}{N_1\,dt} = \frac{r_1}{a_{11}} - N_1 - \frac{a_{12}}{a_{11}}N_2$$

$$\frac{dN_1}{N_1\,dt} = a_{11}\left(\frac{r_1}{a_{11}} - N_1 - \alpha_{12}N_2\right)$$

Similar reasoning gives

$$\frac{dN_2}{N_2\,dt} = a_{22}\left(\frac{r_2}{a_{22}} - N_2 - \alpha_{21}N_1\right)$$

6 Because $K_1 = r_1/a_{11}$ (see exercise 3), we have from exercise 5,

$$\frac{dN_1}{N_1\,dt} = a_{11}(K_1 - N_1 - \alpha_{12}N_2)$$

but $a_{11} = r_1/K_1$; therefore

$$\frac{dN_1}{N_1\,dt} = \frac{r_1}{K_1}(K_1 - N_1 - \alpha_{12}N_2) = r_1\left(\frac{K_1 - N_1 - \alpha_{12}N_2}{K_1}\right)$$

Similar reasoning gives

$$\frac{dN_2}{N_2\,dt} = r_2\,\frac{(K_2 - N_2 - \alpha_{21}N_1)}{K_2}$$

7
$$\frac{dN_1}{N_1 dt} = r_1 - a_{11}N_1 - a_{12}N_2$$

Divide by a_{22}:

$$\frac{1}{a_{22}}\frac{dN_1}{N_1 dt} = \frac{r_1}{a_{22}} - \frac{a_{11}}{a_{22}}N_1 - \beta_{12}N_2$$

Note that $a_{11}/a_{22} = r_1K_2/r_2K_1$, $a_{22} = r_2/K_2$; thus

$$\frac{dN_1}{N_1 dt} = \frac{r_2}{K_2}\left(\frac{r_1K_2}{r_2} - \frac{r_1K_2}{r_2K_1}N_1 - \beta_{12}N_2\right)$$

$$\frac{dN_1}{N_1 dt} = r_1\frac{(K_1 - N_1)}{K_1} - \frac{r_2}{K_2}\beta_{12}N_2$$

Similarly,

$$\frac{dN_2}{N_2 dt} = r_2\frac{(K_2 - N_2)}{K_2} - \frac{r_1}{K_1}\beta_{21}N_1$$

8 From exercise 2 recall that

$$\frac{dN_1}{N_1 dt} = r_1 - a_{11}N_1 - a_{12}N_2$$

but both a_{11} and a_{12} arc lincarly related to N_1; that is,

$$a_{12} = A_1 - 0.0001N_1$$
$$a_{11} = A_2 - 0.0001N_1$$

Thus

$$\frac{dN_1}{N_1 dt} = r_1 - (A_2 - 0.0001N_1)N_1 - (A_1 - 0.0001N_1)N_2$$
$$= r_1 - A_2N_1 + 0.0001N_1^2 - A_1N_2 + 0.0001N_1N_2$$

9
$$\alpha_{12} = g_0 + g_1N_1 + g_2N_2$$
$$\alpha_{21} = h_0 + h_1N_1 + h_2N_2$$

where g_i and h_i are arbitrary constants. Thus

$$\frac{dN_1}{N_1 dt} = r_1\left[\frac{K_1 - N_1 - (g_0 + g_1N_1 + g_2N_2)N_2}{K_1}\right]$$
$$\frac{dN_2}{N_2 dt} = r_2\left[\frac{K_2 - N_2 - (h_0 + h_1N_1 + h_2N_2)N_1}{K_2}\right]$$

which simplifies to

$$\frac{dN_1}{N_1 dt} = r_1\left(\frac{K_1 - N_1 - g_0 N_2 - g_1 N_1 N_2 - g_2 N_2^2}{K_1}\right)$$

$$\frac{dN_2}{N_2 dt} = r_2\left(\frac{K_2 - N_2 - h_0 N_1 - h_1 N_1^2 - h_2 N_1 N_2}{K_2}\right)$$

10 For the first equation the assumptions are

$$a_{11} = g_0 + g_1 N_1 + g_2 N_2$$
$$a_{12} = h_0 + h_1 N_1 + h_2 N_2$$

where g_i and h_i are constants. The first equation is,

$$\frac{dN_1}{N_1 dt} = r_1 - a_{11} N_1 - a_{12} N_2$$

which becomes

$$\frac{dN_1}{N_1 dt} = r_1 - (g_0 + g_1 N_1 + g_2 N_2)N_1 - (h_0 + h_1 N_1 + h_2 N_2)N_2$$
$$= r_1 - g_0 N_1 - g_1 N_1^2 - g_2 N_1 N_2 - h_0 N_2 - h_1 N_1 N_2 - h_2 N_2^2$$
$$= r_1 - g_0 N_1 - g_1 N_1^2 - (g_2 + h_1) N_1 N_2 - h_0 N_2 - h_2 N_2^2$$

11 $dN_1/dt > 0$ can be written as

$$0 < r_1 \frac{(K_1 - N_1 - \alpha_{12} N_2)}{K_1}$$
$$0 < K_1 - N_1 - \alpha_{12} N_2$$
$$N_1 < K_1 - \alpha_{12} N_2$$

Thus in the shaded region in Figure 7.12 $dN_1/dt > 0$:

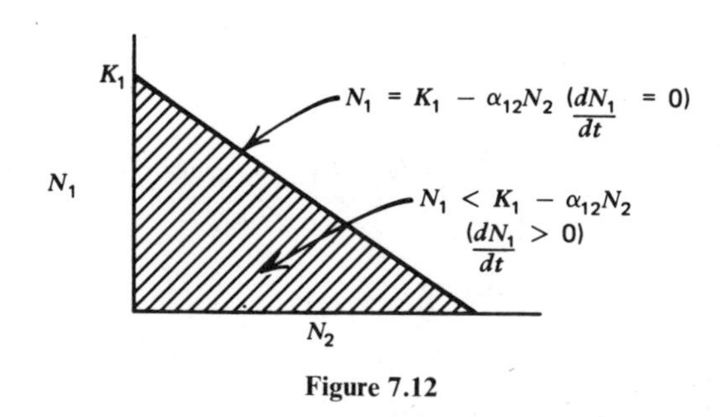

Figure 7.12

and $dN_1/dt < 0$ can be written as

$$0 > r_1 \frac{(K_1 - N_1 - \alpha_{12} N_2)}{K_1}$$

$$0 > K_1 - N_1 - \alpha_{12} N_2$$

$$N_1 > K_1 - \alpha_{12} N_2$$

In the shaded region in Figure 7.13 $dN_1/dt < 0$:

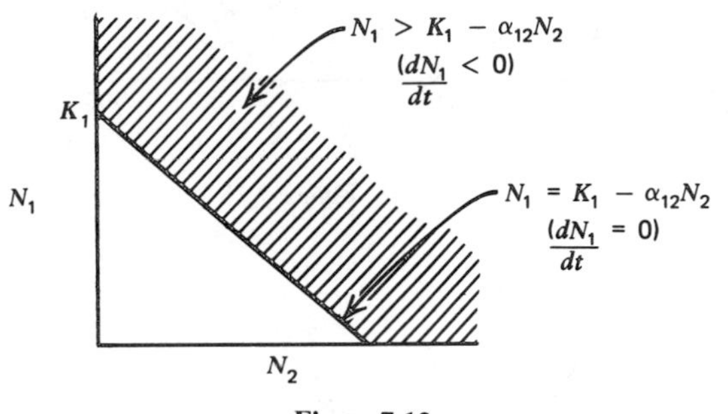

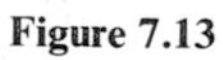

Figure 7.13

Similar reasoning for dN_2/dt yields the following:

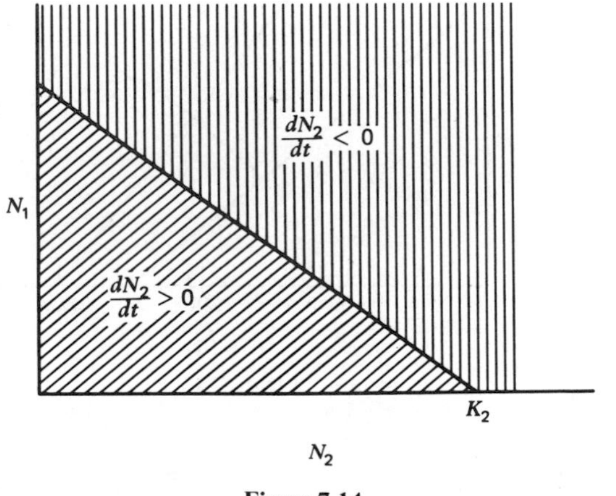

Figure 7.14

12

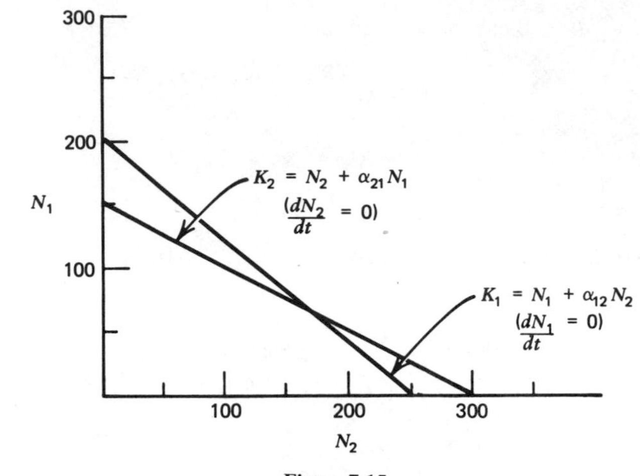

Figure 7.15

13

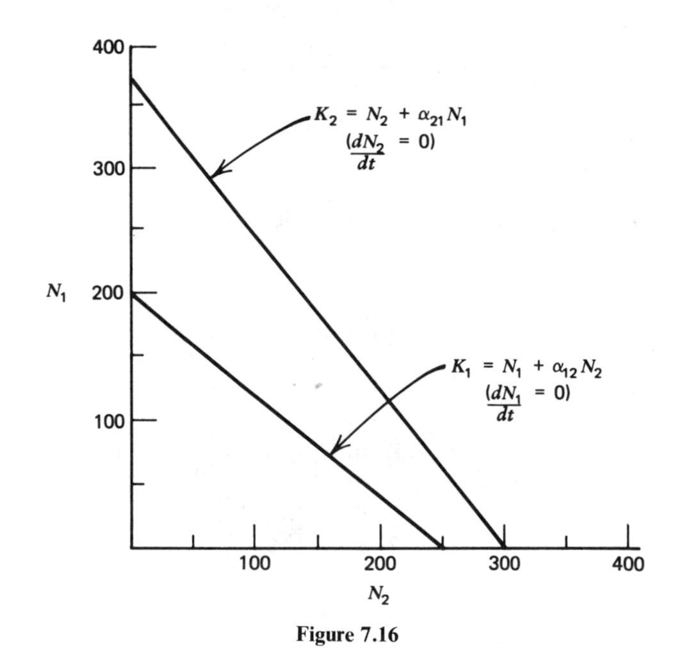

Figure 7.16

14

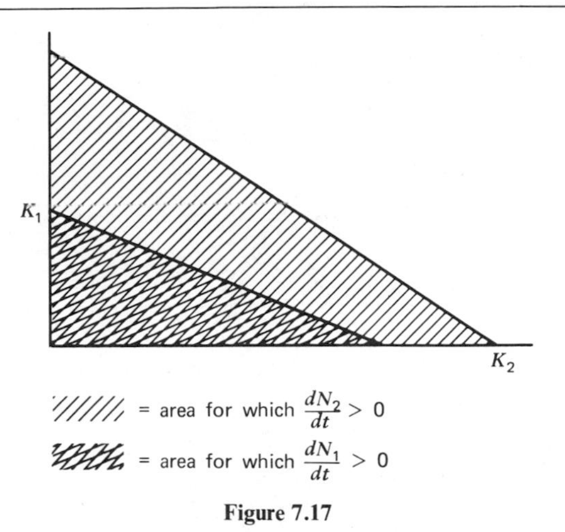

Figure 7.17

15

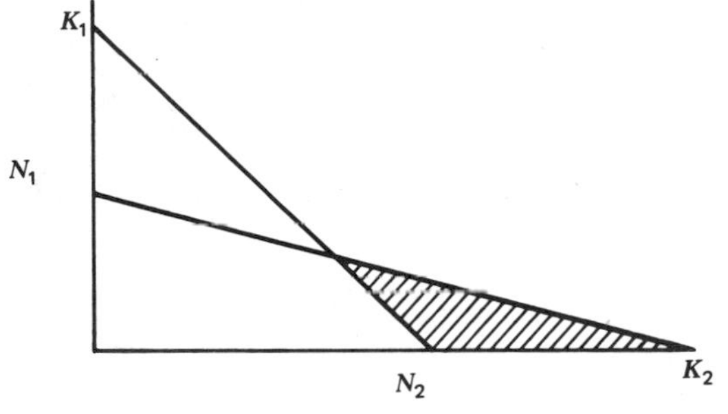

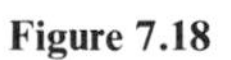

Figure 7.18

16

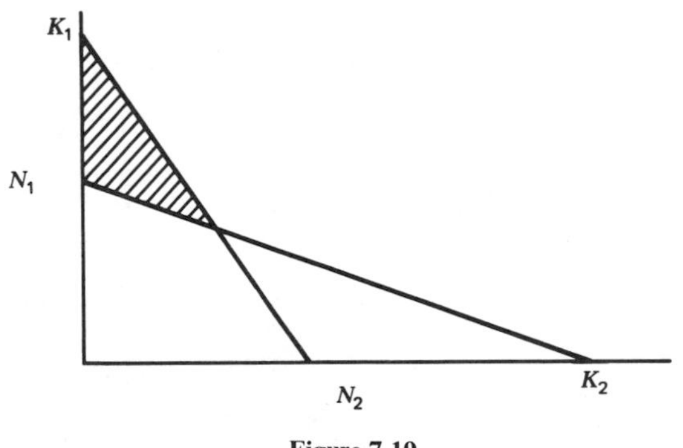

Figure 7.19

17

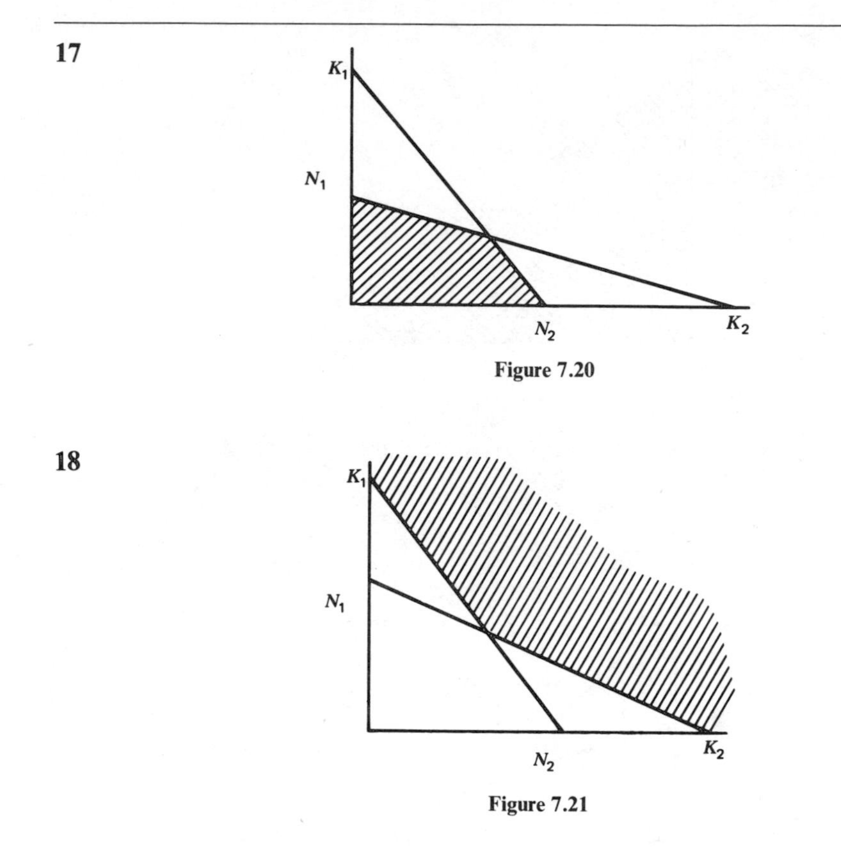

Figure 7.20

18

Figure 7.21

19

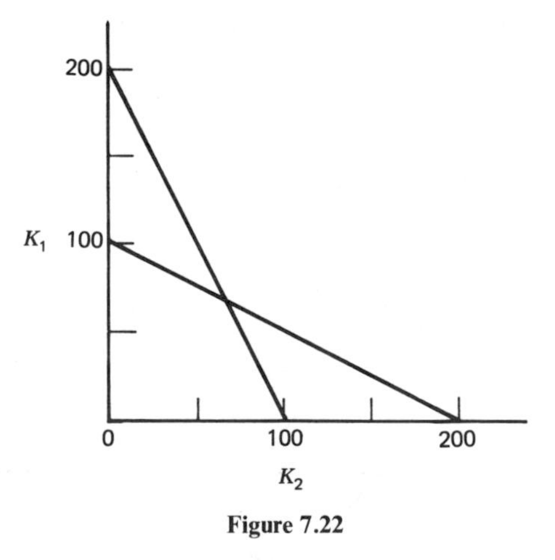

Figure 7.22

20

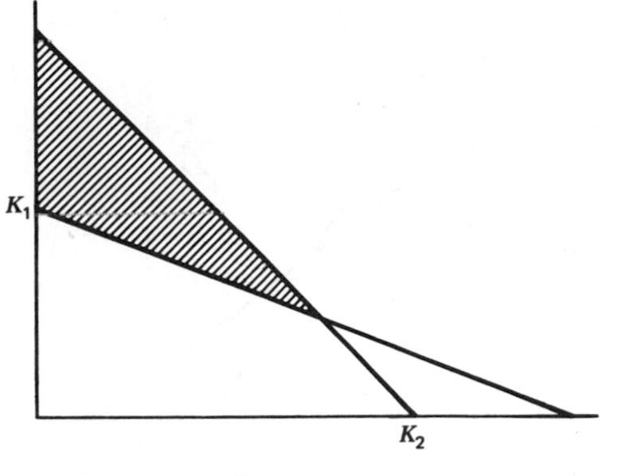

Figure 7.23

21

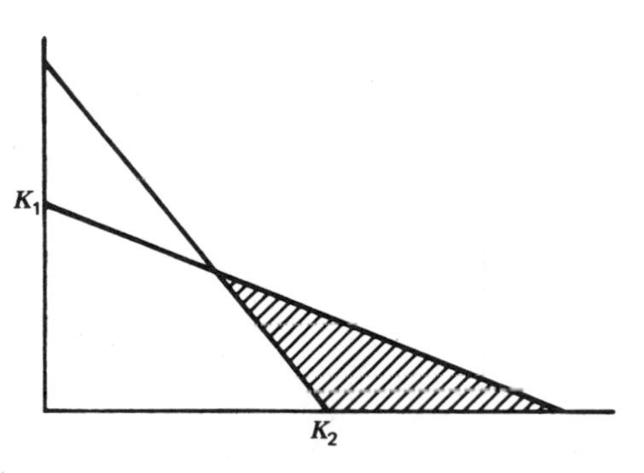

Figure 7.24

22

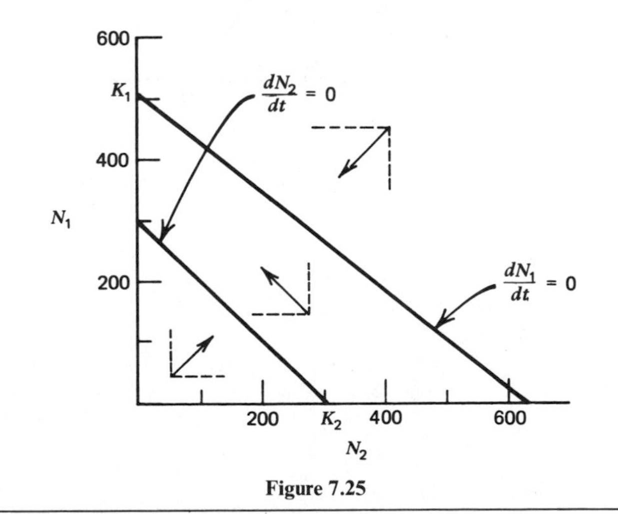

Figure 7.25

23

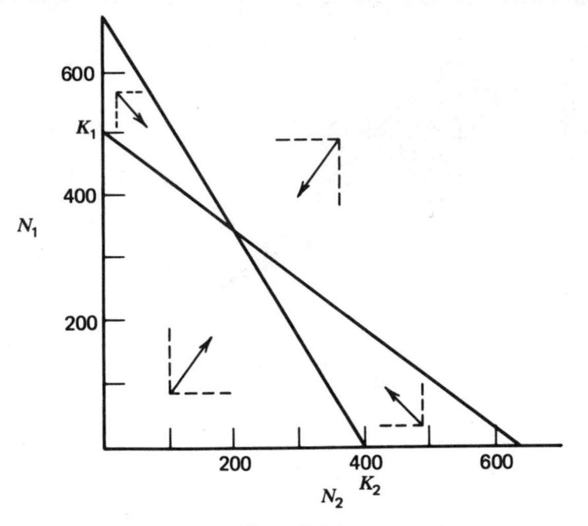

Figure 7.26

24

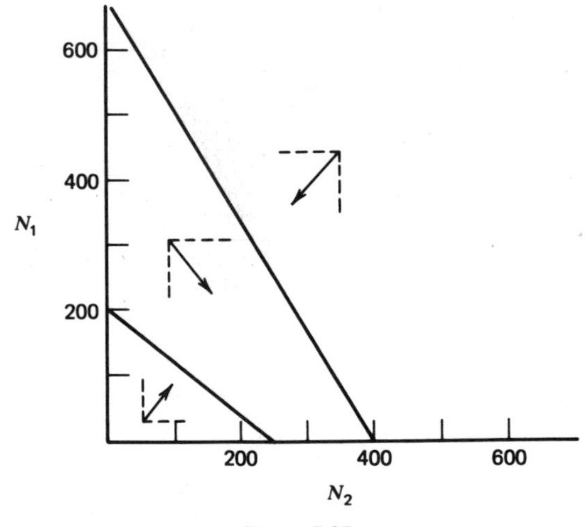

Figure 7.27

25

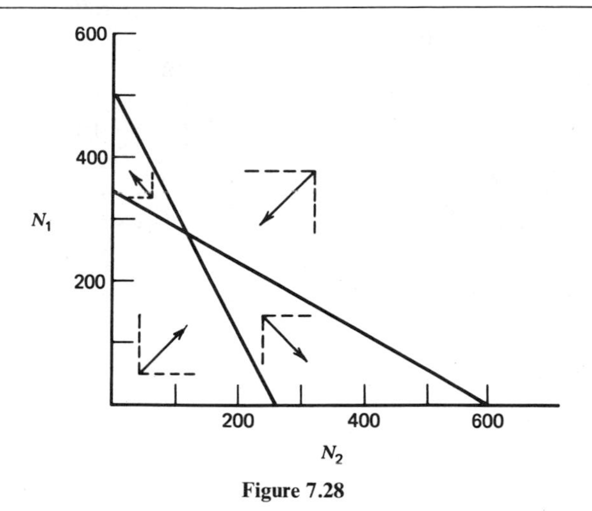

Figure 7.28

26

$$K_1 = N_1 + \alpha_{12} N_2$$

If $N_2 = 0$, $N_1 = K_1$. If $N_1 = 0$,

$$K_1 = \alpha_{12} N_2$$

$$N_2 = \frac{K_1}{\alpha_{12}}$$

Similarly, the N_2 intercept for equation 4 is K_2 and the N_1 intercept is K_2/α_{21}.

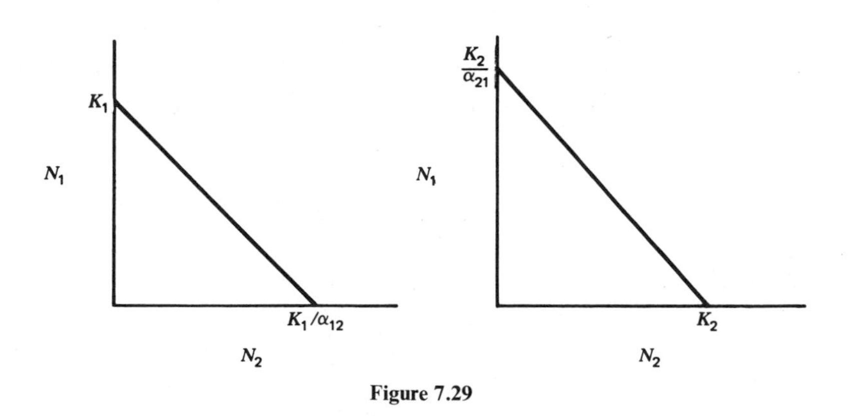

Figure 7.29

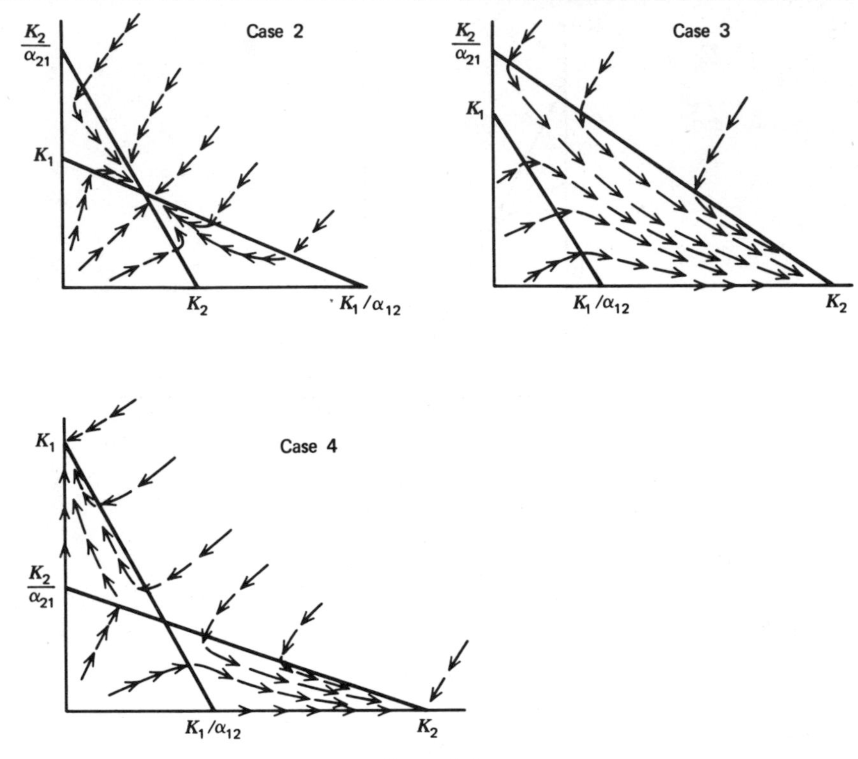

Figure 7.30

27 *Case* 2. Both species persist where the two lines cross: "stable equilibrium." *Case* 3. The opposite of case 1. Species 1 becomes extinct and species 2 reaches its carrying capacity. "Species 2 wins."
Case 4. An equilibrium is established where the two lines intersect. It is an "unstable" equilibrium (explained later); therefore the system eventually drifts to "species 1 wins" *or* "species 2 wins."

28 *Case* 1.

$$K_1 > \frac{K_2}{\alpha_{21}}$$

$$K_2 < \frac{K_1}{\alpha_{12}}$$

Case 2.

$$K_1 < \frac{K_2}{\alpha_{21}}$$

$$K_2 < \frac{K_1}{\alpha_{12}}$$

Case 3.

$$K_1 < \frac{K_2}{\alpha_{21}}$$

$$K_2 > \frac{K_1}{\alpha_{12}}$$

Case 4.

$$K_1 > \frac{K_2}{\alpha_{21}}$$

$$K_2 > \frac{K_1}{\alpha_{12}}$$

29 *Case* 2.

$$K_1 < \frac{K_2}{\alpha_{21}}$$

$$K_2 < \frac{K_1}{\alpha_{12}}$$

Case 4.

$$K_1 > \frac{K_2}{\alpha_{21}}$$

$$K_2 > \frac{K_1}{\alpha_{12}}$$

Case 2.

$$\alpha_{21}K_1 < K_2 \tag{1}$$

$$\alpha_{12}K_2 < K_1 \tag{2}$$

Because $\alpha_{12}K_2 < K_1$, we can substitute $\alpha_{12}K_2$ for K_1 in equation 1 and have the inequality remain true; that is,

$$\alpha_{21}(\alpha_{12}K_2) < K_2$$

$$\alpha_{12}\,\alpha_{21} < 1$$

$$0 < 1 - \alpha_{12}\alpha_{21}$$

Case 4.

$$K_1 > \frac{K_2}{\alpha_{21}}$$

$$K_2 > \frac{K_1}{\alpha_{12}}$$

$$\alpha_{21}K_1 > K_2$$

$$\alpha_{12}K_2 > K_1$$

Again, substituting $\alpha_{12}K_2$ for K_1,

$$\alpha_{21}(\alpha_{12}K_2) > K_2$$

$$\alpha_{21}\alpha_{12} > 1$$

$$0 > 1 - \alpha_{21}\alpha_{12}$$

30 *Case* 1.

$$K_1 > \frac{K_2}{\alpha_{21}} \qquad (1)$$

$$K_2 < \frac{K_1}{\alpha_{12}} \qquad (2)$$

Clearly the sort of substitution that could distinguish between cases 2 and 4 is not possible here.

31 The equilibrium condition is given as

$$K_1 = N_1 + \alpha_{12}N_2$$

$$K_2 = N_1\alpha_{21} + N_2$$

Thus in matrix form we have

$$K = AN$$

where

$$K = \begin{vmatrix} K_1 \\ K_2 \end{vmatrix} \qquad N = \begin{vmatrix} N_1 \\ N_2 \end{vmatrix} \qquad A = \begin{vmatrix} 1 & \alpha_{12} \\ \alpha_{21} & 1 \end{vmatrix}$$

where K is the "K-vector," N is the "N-vector," and A is the "community matrix."

32

$$K_1 = N_1 + \alpha_{12}N_2 \qquad (1)$$

$$K_2 = N_1\alpha_{21} + N_2 \qquad (2)$$

From equation 2

$$N_2 = K_2 - N_1\alpha_{21}$$

Substitute into equation 1

$$\begin{aligned} K_1 &= N_1 + \alpha_{12}(K_2 - N_1\alpha_{21}) \\ &= N_1 + K_2\alpha_{12} - N_1\alpha_{12}\alpha_{21} \\ &= N_1(1 - \alpha_{12}\alpha_{21}) + K_2\alpha_{12} \end{aligned}$$

and

$$N_1 = \frac{K_1 - K_2\alpha_{12}}{1 - \alpha_{12}\alpha_{21}}$$

Substitute into equation 2

$$K_2 = \frac{(K_1 - K_2\alpha_{12})}{1 - \alpha_{12}\alpha_{21}}\alpha_{21} + N_2$$

$$N_2 = K_2\left(\frac{1 - \alpha_{12}\alpha_{21}}{1 - \alpha_{12}\alpha_{21}}\right) - \left(\frac{K_1 - K_2\alpha_{12}}{1 - \alpha_{12}\alpha_{21}}\right)\alpha_{21}$$

$$= \frac{K_2 - \alpha_{12}\alpha_{21}K_2 - K_1\alpha_{21} + K_2\alpha_{12}\alpha_{21}}{1 - \alpha_{12}\alpha_{21}}$$

$$N_2 = \frac{K_2 - K_1\alpha_{21}}{1 - \alpha_{12}\alpha_{21}}$$

33 The community matrix is,

$$A = \begin{vmatrix} 1 & \alpha_{12} \\ \alpha_{21} & 1 \end{vmatrix}$$

Thus

$$\text{Det } A = 1 - \alpha_{12}\alpha_{21}$$

34 Recall from exercise 32

$$N_1 = \frac{K_1 - K_2\alpha_{12}}{1 - \alpha_{12}\alpha_{21}}$$

$$N_2 = \frac{K_2 - K_1\alpha_{21}}{1 - \alpha_{12}\alpha_{21}}$$

Clearly the denominator of both equations is Det A. We now seek matrices whose determinants are

$$K_1 - K_2\alpha_{12} \tag{1}$$

and

$$K_2 - K_1\alpha_{21}$$

Clearly, if we define the matrix

$$A_1 = \begin{vmatrix} K_1 & \alpha_{12} \\ K_2 & 1 \end{vmatrix}$$

that is, we replace the first column with the K vector, we have

$$\text{Det } A_1 = K_1 - K_2\alpha_{12}$$

Similarly, if we define the matrix

$$A_2 = \begin{vmatrix} 1 & K_1 \\ \alpha_{21} & K_2 \end{vmatrix}$$

that is, we replace the second column with the K vector, we have

$$\text{Det } A_2 = K_2 - K_1\alpha_{21}$$

Thus we have

$$N_1 = \frac{K_1 - \alpha_{12}K_2}{1 - \alpha_{12}\alpha_{21}} = \frac{\text{Det } A_1}{\text{Det } A}$$

$$N_2 = \frac{K_2 - \alpha_{21}K_1}{1 - \alpha_{12}\alpha_{21}} = \frac{\text{Det } A_2}{\text{Det } A}$$

35 From exercise 29 we have case 2 when

$$1 - \alpha_{12}\alpha_{21} > 0$$

and we have case 4 when

$$1 - \alpha_{12}\alpha_{21} < 0$$

From the definition of the determinant we can say that we have case 2 when

$$\text{Det } A > 0$$

and we have case 4 when

$$\text{Det } A < 0$$

36 For case 1

$$K_1 > \frac{K_2}{\alpha_{21}}$$

$$K_1 > K_2\alpha_{12}$$

For case 3

$$K_1 < K_2/\alpha_{21}$$

$$K_1 < K_2\alpha_{12}$$

For case 4

$$K_1 > K_2/\alpha_{21}$$
$$K_1 < K_2\alpha_{12}$$

Here, unlike Figure 7.11, we require $K_1 > K_2/\alpha_{21}$ and $K_1 < K_2\alpha_{12}$ to be true for some values of K_1 and K_2. Consider some particular value of K_1 (say K_1^*) and K_2 (say K_2^*) (Figure 7.31):

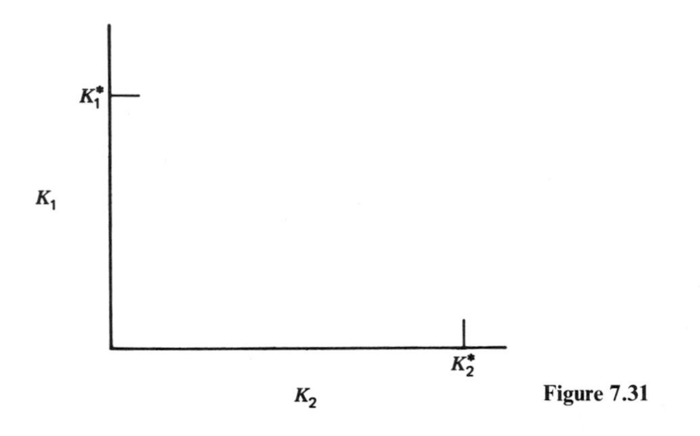

Figure 7.31

We know that the equations must look something like Figure 7.32:

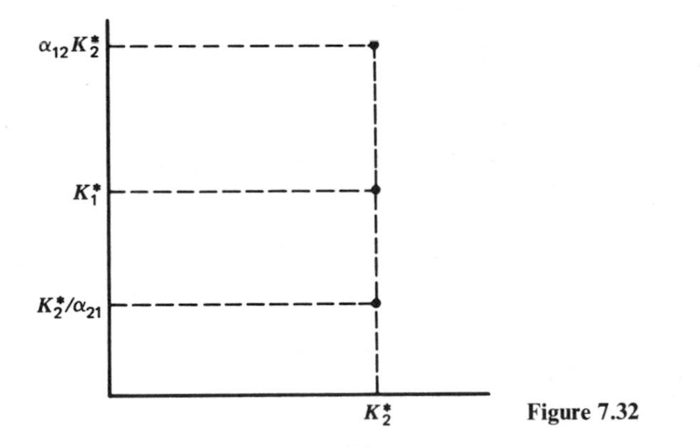

Figure 7.32

The lines that form the graphs of the equilibrium equations must pass through the two points made by the intersection of K_2^* with $K_2^*\alpha_{12}$ (K_1^* will be less than $K_2^*\alpha_{21}$) and by the intersection of K_2^* with K_2^*/α_{21} (K_1^* will be $> K_2^*/\alpha_{21}$). Thus we obtain Figure 7.33:

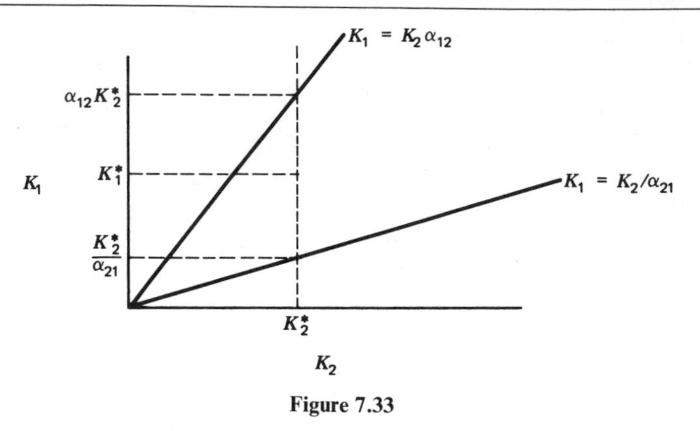

Figure 7.33

Shading in the appropriate part, we have Figure 7.34:

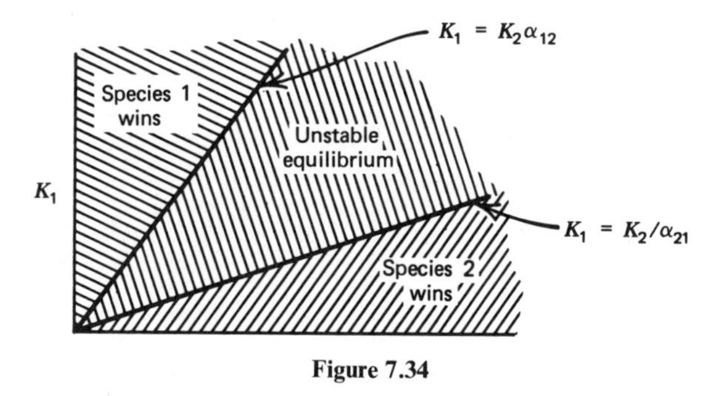

Figure 7.34

37 Recall the original definition of the Lotka–Volterra equations

$$\frac{dN_1}{N_1 dt} = r_1 - a_{11}N_1 - a_{12}N_2$$

$$\frac{dN_2}{N_2 dt} = r_2 - a_{22}N_2 - a_{21}N_1$$

and the definitions

$$K_i = \frac{r_i}{a_{ii}}$$

$$\alpha_{ij} = \frac{a_{ij}}{a_{ii}}$$

$$\beta_{ij} = \frac{a_{ij}}{a_{jj}}$$

Conditions for case 1 (see exercise 28):

$$\alpha_{21}K_1 > K_2$$
$$\alpha_{12}K_2 < K_1$$

Convert to the original notation:

$$\frac{a_{21}}{a_{22}}\frac{r_1}{a_{11}} > \frac{r_2}{a_{22}}$$
$$\frac{a_{12}}{a_{11}}\frac{r_2}{a_{22}} < \frac{r_1}{a_{11}}$$

Multiply the first equation by a_{22} and the second by a_{11}:

$$\frac{a_{21}}{a_{11}}r_1 > r_2$$
$$\frac{a_{12}}{a_{22}}r_2 < r_1$$

Thus

$$\beta_{21}r_1 > r_2$$
$$\beta_{12}r_2 < r_1$$

For case 2

$$\beta_{21}r_1 < r_2$$
$$\beta_{12}r_2 < r_1$$

For case 3

$$\beta_{21}r_1 < r_2$$
$$\beta_{12}r_2 > r_1$$

For case 4

$$\beta_{21}r_1 > r_2$$
$$\beta_{12}r_2 > r_1$$

38 For case 2

$$\beta_{21}r_1 < r_2 \qquad (1)$$
$$\beta_{12}r_2 < r_1 \qquad (2)$$

Substitute $\beta_{12}r_2$ for r_1 in equation 1:

$$\beta_{21}(\beta_{12}r_2) < r_2$$
$$\beta_{12}\beta_{21} < 1$$
$$1 - \beta_{12}\beta_{21} > 0$$

39 For $1 - \beta_{12}\beta_{21} > 0$ (Figure 7.35):

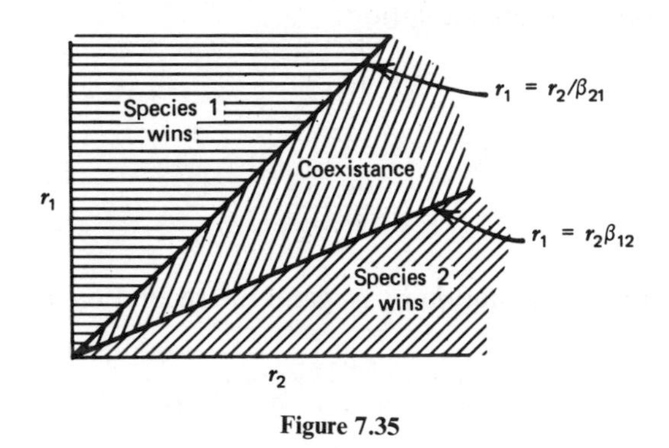

Figure 7.35

For $1 - \beta_{12}\beta_{21} < 0$ (Figure 7.36):

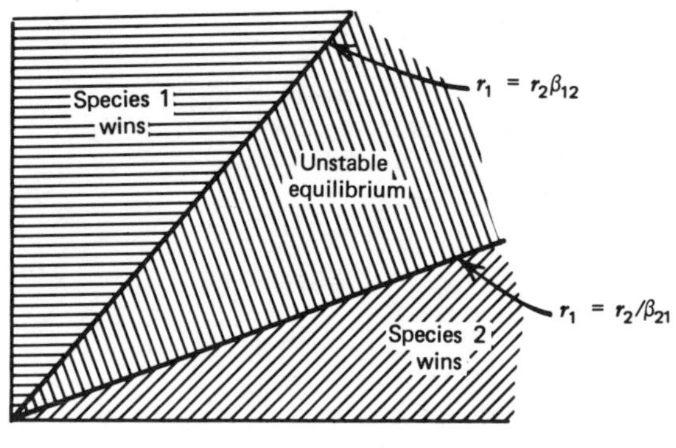

Figure 7.36

40 The equilibrium equations are

$$K_1 = N_1 + \alpha_{12}N_2$$
$$K_2 = N_2 + \alpha_{21}N_1$$

Substituting the experimental values

$$200 = 150 + \alpha_{12}120$$

$$400 = 120 + \alpha_{21}150$$

$$\alpha_{12} = \frac{200 - 150}{120} = 0.417$$

$$\alpha_{21} = \frac{400 - 120}{150} = 1.87$$

41

$$\beta_{12} = \frac{a_{12}}{a_{22}} = \frac{a_{12}}{a_{22}}\frac{a_{11}}{a_{11}} = \frac{a_{12}}{a_{11}}\frac{a_{11}}{a_{22}} = \alpha_{12}\frac{a_{11}}{a_{22}}$$

We know that $r_1/a_{11} = K_1$ and $r_2/a_{22} = K_2$. Thus $a_{11} = r_1/K_1$ and $a_{22} = r_2/K_2$. Therefore $a_{11} = 0.2/200 = 0.0001$ and $a_{22} = 0.1/400 = 0.00025$. Thus

$$\beta_{12} = \alpha_{12}\frac{a_{11}}{a_{22}} = 0.417\left(\frac{0.0001}{0.00025}\right) = 1.67$$

$$\beta_{21} = \frac{a_{21}}{a_{11}} = \frac{a_{21}}{a_{11}}\frac{a_{22}}{a_{22}} = \alpha_{21}\frac{a_{22}}{a_{11}}$$

$$= 1.87\left(\frac{0.00025}{0.0001}\right) = 0.47$$

42 This is an impossible question to answer as it stands because $\alpha_{12} < \alpha_{21}$ implies that population 1 is a better competitor and $\beta_{21} < \beta_{12}$ implies that population 2 is a better competitor. The question "which is a better competitor," at least in this case, is not valid.

43 $K_1 = K_2 = 250$ because the ants can live nowhere but in the old termite nests. Thus we have

$$\alpha_{12} = \frac{K_1 - N_1}{N_2} = \frac{250 - 150}{100} = 1.0$$

$$\alpha_{21} = \frac{K_2 - N_2}{N_1} = \frac{250 - 100}{150} = 1.0$$

These estimates of α could not be valid because Det $A = 1 - \alpha_{12}\alpha_{21} = 1 - 1 = 0$ and the two populations could never reach equilibrium (Det $A > 0$ is necessary for an equilibrium to be established). In general,

$$N_i = K_i - N_j$$

$$\alpha_{12} = \frac{K_1 - K_1 - N_2}{N_2} = \frac{N_2}{N_2} = 1$$

that is, no matter what the mix of the two species, $\alpha_{12} = \alpha_{21} = 1$.

44 $K_1 = 250, K_2 = 150$. N_1 is 210 and N_2 is 140, supposedly both equilibrium situations. Thus

$$\alpha_{12} = \frac{K_1 - N_1}{N_2} = \frac{250 - 210}{140} = 0.29$$

$$\alpha_{21} = \frac{K_2 - N_2}{N_1} = 0.05$$

Stable equilibrium requires

$$K_1 < \frac{K_2}{\alpha_{21}}$$

$$K_2 < \frac{K_1}{\alpha_{12}}$$

which, in this case, is

$$250 < \frac{150}{0.05}$$

$$150 < \frac{250}{0.29}$$

Conditions for stable equilibrium are satisfied.

45

$$K_1 = 200 + 300 = 500;\ K_2 = 100 + 300 = 400$$

$$N_1 = 200 + 60 = 260,\ N_2 = 100 + 240 = 340$$

$$\alpha_{12} = \frac{500 - 260}{340} = 0.70$$

$$\alpha_{21} = 0.23$$

In exercise 44 $\alpha_{12} = 0.29$, and $\alpha_{21} = 0.05$.

46

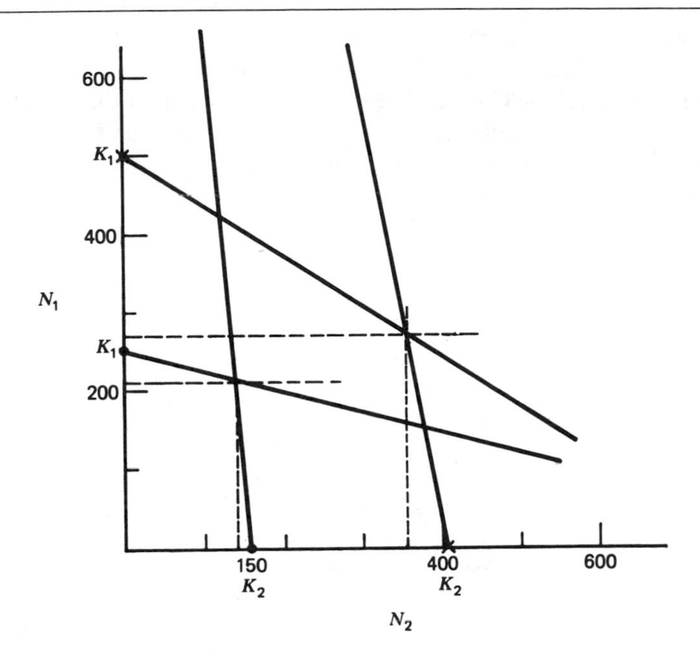

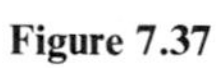

Figure 7.37

47

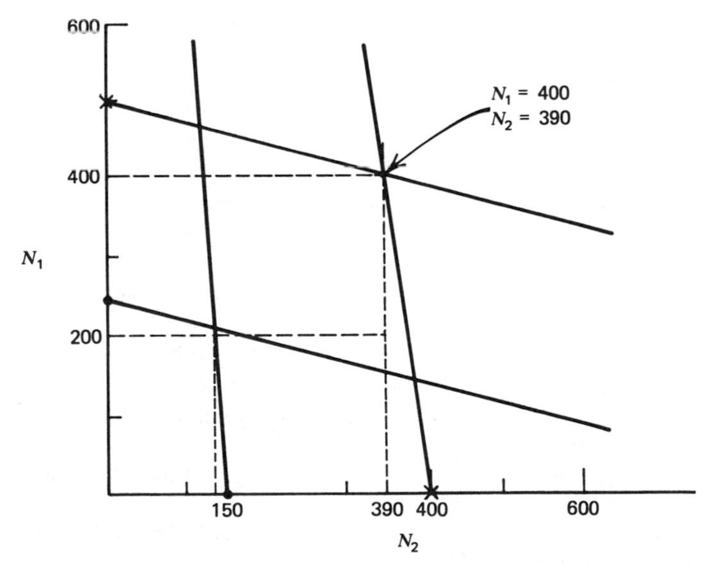

Figure 7.38

Because $N_1 = 400$, 200 of them exist in type A nests. Thus $400 - 200 = 200$ of the type B nests must be occupied by species 1. Because $n_2 = 390$, 100 of them exist in type C nests. Thus $390 - 100 = 290$ of the type B nests must be occupied by species 2. That makes $290 + 200$ type B nests. We seem to require more nests than are really there (i.e., we require 490 type B nests and there are only 300 in the first place).

REFERENCES

Ayala, F. J., M. E. Gilpin, and J. G. Ehrenfeld. 1973. Competition between species: theoretical models and experimental tests. *Theor. Popul. Biol.* **4**:331–356.

Bartlett, M. S. 1957. On theoretical medels for competitive and predatory biological systems. *Biometrika* **44**:27–42.

Beauchamp, R. S. A. and P. Ullyett, 1932. Competitive relationships between certain species of fresh-water triclads. *J. Ecol.* **20**:200–208.

Birch, L. C. 1957. The meanings of competition. *Am. Nat.* **91**:5–18.

Chiang, C. L. 1954. Competition and other interactions between species in O. Kempthorne, Ed. *Statistics and Mathematics in Biology*. Ames, Iowa: Iowa State College Press, pp. 197–215.

Cohen, J. E. 1968. *A model of simple competition*. Cambridge: Harvard University Press.

Cole, L. C. 1960. Competitive exclusion. *Science* **312**:348–349.

Connell, J. H. 1961. The influence of interspecific competition and other factors on the distribution of the barnacle *Chthamalus stellatus*. *Ecology* **42**:710–723.

Crombie, A. C. 1944. On intraspecific and interspecific competition in larvae of graminivorous insects. *J. Exp. Biol.* **20**:135–151.

Crombie, A. C. 1947. Interspecific competition. *J. Anim. Ecol.* **16**:44–73.

Crowell, K. L. 1962. Reduced interspecific competition among the birds of Bermuda. *Ecology* **43**:75–88.

DeBach, P. 1966. The competitive displacement and coexistence principles. *Annu. Rev. Entomol.* **11**:183–212.

Fujii, K. 1968. Studies on interspecies competition between the azuki bean weevil and the southern cowpea weevil: III. Some characteristics of strains of two species. *Res. Popul. Ecol.* **10**:87–98.

Gilpin, M. E. 1974. A Liapunov function for competition communities. *J. Theor. Biol.* **44**:35–48.

Gilpin, M. E. and F. J. Ayala. 1973. Global models of growth and competition. *Proc. Nat. Acad. Sci. USA* **70**:3590–3593.

Grant, P. R. 1972. Interspecific competition between rodents. *Annu. Rev. Ecol. Syst.* **3**:79–106.

Hardin, G. 1960. The competitive exclusion principle. *Science* **131**:1292–1297.

Haven, S. B. 1973. Competition for food between the identical gastropods *Acmaea scabra* and *Acmaea digitalis*. *Ecology* **54**:143–151.

Jaeger, R. G. 1970. Potential extinction through competition between two species of terrestrial salamanders. *Evolution* **24**:632–642.

Jaeger, R. G. 1971. Competitive exclusion as a factor influencing the distributions of two species of terrestrial salamanders. *Ecology* **53**:632–637.

Leon, J. A. 1974. Selection in contexts of interspecific competition. *Am. Nat.* **108**:739–757.

Leslie, P. H. 1962. A stochastic model for two competing species of *Tribolium* and its application to some experimental data. *Biometrika* **49**:1–25.

Miller, R. S. 1967. Pattern and process in competition. *Adv. Ecol. Res.* **4**:1–74.

Milne, A. 1961. Definitions of competition among animals. In F. L. Milthorpe, Ed. Mechanisms in Biological Competition. *Symp. Soc. Exp. Biol.* **15**:40–61.

Milthorpe, F. L., Ed. 1961. Mechanisms in biological competition. *Symp. Soc. Exp. Biol.* **15**. Cambridge: Cambridge University Press.

Park, T. 1948. Experimental studies of interspecific competition. I. Competition between populations of flour beetles *Tribolium confusum* Duvaal and *T. castaneum* Herbst. *Physiol. Zool.* **18**:265–308.

Park, T. 1954. Experimental studies of interspecific competition. II. Temperature, humidity, and competition in two species of *Tribolium. Physiol. Zool.* **27**:177–238.

Park, T. 1962. Beetles, competition, and populations. *Science* **138**:1369–1375.

Patten, B. C. 1961. Competitive exclusion. *Science* **134**:1599–1601.

Schoener, T. W. 1975. Competition and the form of habitat shift. *Theor. Popul. Biol.* **5**:265–307.

Vandermeer, J. H. 1975. Interspecific competition: A new approach to the classical theory. *Science* **188**:253–255.

Vandermeer, J. H. and D. Boucher. 1978. Varieties of mutualistic interactions in population models. *J. Theor. Biol.* **74**:549–558.

Wilbur, H. M. 1972. Competition, predation, and the structure of the *Ambystoma-Rana sylvatica* community. *Ecology* **53**:3–21.

8. Predator–Prey Theory

We now turn to the other major type of two-species interaction—that between predator and prey populations. As in Chapter 1, we begin with the simplest assumptions possible; that is, both populations grow according to the exponential equation

$$\frac{dN}{N\,dt} = b - d \tag{1}$$

where b is the birth rate and d, the death rate. Now we add the simple assumptions necessary to make the equations describe the process of a predator–prey interaction. The predator eats the prey. Let us suppose that the only source of death for the prey population is getting eaten by the predator. So, if the predator is not around, the prey population grows according to the simple exponential $dN_1/dt = b_1 N_1$, but if the predator is around each individual of the predator species decreases the per capita rate of increase of the prey population by a constant amount; that is, $d_1 = a_{12} N_2$. Similarly, assume that the only source of birth for the predator population is by eating the prey: $b_2 = a_{21} N_1$. Thus for the basic predator–prey equations we have

$$\begin{aligned} \frac{dN_1}{N_1\,dt} &= b - a_{12} N_2 \\ \frac{dN_2}{N_2\,dt} &= a_{21} N_1 - d \end{aligned} \tag{2}$$

which are called the Lotka–Volterra predator-prey equations.

☐ EXERCISES

1. Write the equilibrium solution for equations 1 and 2 in matrix form.
2. Express the equilibrium solution graphically (on a graph of N_1 against N_2).
3. For what part of the N_1, N_2 space will it be true that $dN_1/dt > 0$?
4. For what part of the N_1, N_2 space will it be true that $dN_2/dt > 0$?
5. Show all areas on the graph of N_1 against N_2 for the various conditions of the derivatives [i.e., areas for (1) $dN_1/dt > 0$ and $dN_2/dt > 0$, (2) $dN_1/dt > 0$ and $dN_2/dt < 0$, (3) $dN_1/dt < 0$ and $dN_2/dt > 0$, and (4) $dN_1/dt < 0$ and $dN_2/dt < 0$]. On another graph of N_1 against N_2 indicate the general dynamical flow of the system by using small arrows, as in Chapter 7. Qualitatively, what can you say about the system's dynamics? ☐

THE ANALYSIS OF GENERAL AUTONOMOUS SYSTEMS. At this time we must take a short side trip to develop some analytical skills before pursuing predator–prey theory. In general, the following are called simultaneous autonomous equations:

$$\frac{dX_1}{dt} = f_1(x_1, x_2)$$
$$\frac{dX_2}{dt} = f_2(x_1, x_2) \tag{3}$$

The crucial point to recognize is that the independent variable of the derivatives (t = time) does not appear on the right-hand side of the equations. It is this property that makes the equations "autonomous."

Given a set of autonomous differential equations, such as (3), it is possible to generate a diverse set of qualitative behavior patterns. In particular, we examine isolated singular points (i.e., those for which $dN_1/dt = dN_2/dt = 0$) and ask about the behavior of the system near one of them. This procedure is called a "neighborhood stability analysis" because we are concerned only with the behavior of the system in the "neighborhood" of an equilibrium (singular) point. Neighborhood stability analysis is to be contrasted with "global stability analysis." Whereas neighborhood analysis is concerned with an isolated singularity, global analysis is concerned with the behavior of the system over the entire N_1, N_2 space. In this section we discuss neighborhood analysis.

We can classify singularities in terms of the dynamical behavior of the system, near the singularity. In Figure 8.1 a classification of this kind is presented. In Figure 8.1 the three figures on the left are nonoscillatory; the three on the right are oscillatory.

☐ EXERCISES

6 For cases 2 and 4 in Chapter 7 (in 2 both species coexit; in 4 one or the other wins, depending on starting conditions) to which of the cases in Figure 8.1 do they correspond?

7 Draw rough graphs to indicate the behavior of X_1 and X_2 over time for b, d, and f in Figure 8.1 (plot X_1 and X_2 versus time). ☐

We now wish to develop analytical tools for use in a particular system that corresponds to the general system of equations 3 to determine to which of the cases in Figure 8.1 it belongs; that is, we want to be able to say which case (or cases) applies by examining f_1 and f_2 only.

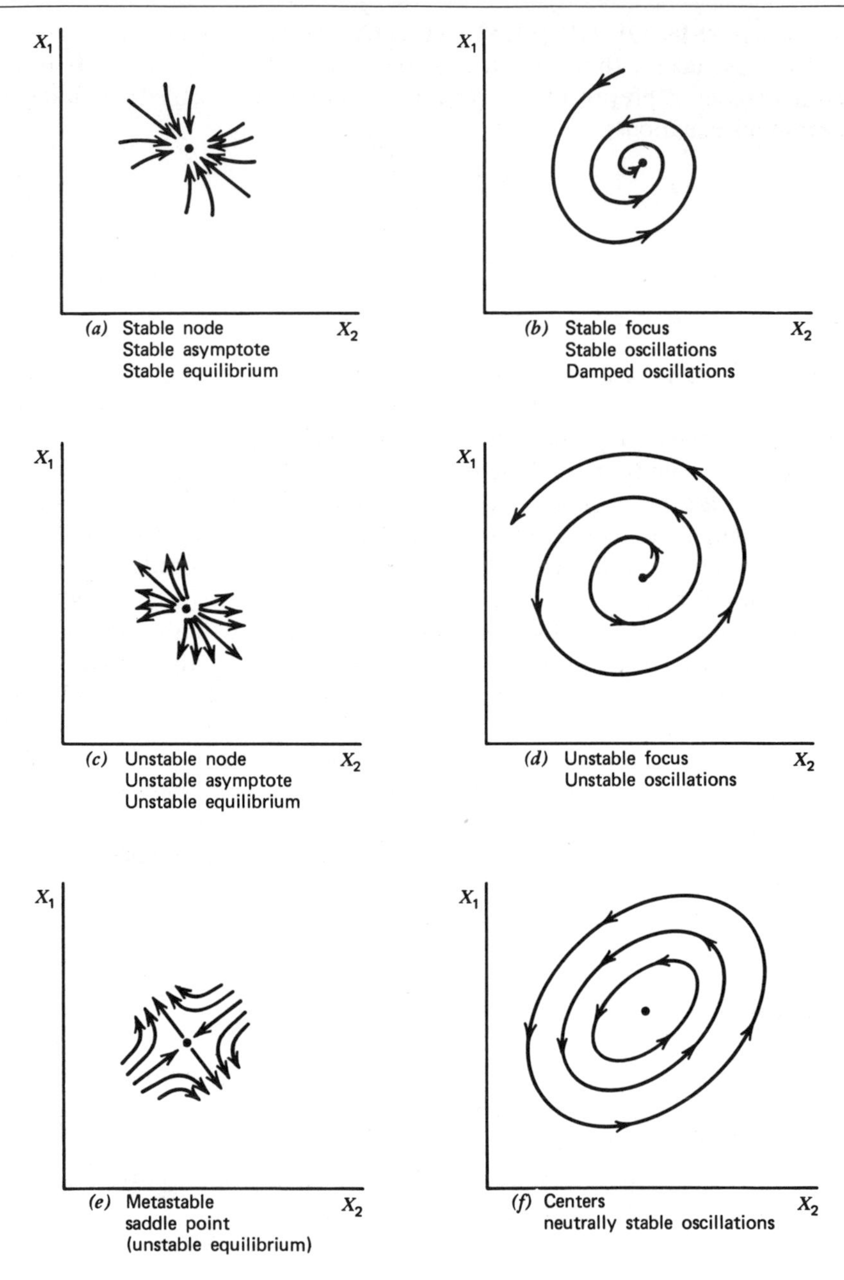
X_1
X_2
(a) Stable node
Stable asymptote
Stable equilibrium
(b) Stable focus
Stable oscillations
Damped oscillations
(c) Unstable node
Unstable asymptote
Unstable equilibrium
(d) Unstable focus
Unstable oscillations
(e) Metastable
saddle point
(unstable equilibrium)
(f) Centers
neutrally stable oscillations

Figure 8.1

The first three steps are

(a) find the singularity (equilibrium point);
(b) compute all partial derivatives at that equilibrium point;
(c) construct the Jacobian matrix and find its eigenvalues (see below).

The first step is accomplished, as in Chapter 7, by setting the derivatives equal to zero and solving for X_1 and X_2 in either of the equations. Sometimes only one point (X_1^*, X_2^*) will satisfy the equilibrium equations. At other times multiple points will satisfy the equilibrium equations. Remember that the neighborhood stability analysis is undertaken in the "neighborhood" of a single equilibrium point.

Next, all partial derivatives are computed *and* solved for the values of X_1 and X_2 at the equilibrium point.

Finally, compute the Jacobian matrix

$$J = \begin{vmatrix} \left(\dfrac{\partial f_1}{\partial x_1}\right)_0 & \left(\dfrac{\partial f_1}{\partial x_2}\right)_0 \\ \left(\dfrac{\partial f_2}{\partial x_1}\right)_0 & \left(\dfrac{\partial f_2}{\partial x_2}\right)_0 \end{vmatrix}$$

where $(\partial f_i/\partial x_j)_0$ means "partial of f_i with respect to X_j, evaluated at the point 0."

The eigenvalues are, by definition, the values of λ that satisfy the equation

$$\text{Det}\,(J - I\lambda) = 0$$

☐ EXERCISES

8 Given the following matrix

$$A = \begin{vmatrix} a & b \\ c & d \end{vmatrix}$$

what is the matrix $(A - \lambda I)$, where λ is a scalar.

9 For this example what is Det $(A - \lambda I)$?

10 Given the following matrix

$$B = \begin{vmatrix} 2 & 0 \\ 0 & 3 \end{vmatrix}$$

what is Det $(B - \lambda I)$?

11 What are the values of λ that satisfy Det $(B - \lambda I) = 0$? What are the roots of Det $(B - \lambda I) = 0$? What are the eigenvalues of the matrix B (as in exercise 10)?

12 Find the eigenvalues of the following matrix:

$$\begin{vmatrix} 5 & 4 \\ 1 & 2 \end{vmatrix}$$

13 For the Lotka–Volterra *competition* equations compute $\partial f_1/\partial x_1$, $\partial f_1/\partial x_2$, $\partial f_2/\partial x_1$, and $\partial f_2/\partial x_2$.

14 Find the equilibrium values of N_1 and N_2 and evaluate the derivatives in exercise 13 at that equilibrium point.

15 Write out the Jacobian matrix for the Lotka–Volterra competition equations.

16 Express the Jacobian matrix in exercise 15 in terms of $(\partial f_1/\partial N_1)_0$, $(\partial f_2/\partial N_2)_0$, α_{12}, and α_{21}.

17 For the system

$$\frac{dX_1}{dt} = X_1 - 0.01X_2$$

$$\frac{dX_2}{dt} = 2X_2 - 0.05X_1$$

compute the Jacobian matrix and its eigenvalues.

18 For the system

$$\frac{dX_1}{dt} = 0.5X_1 - 0.001X_1X_2$$

$$\frac{dX_2}{dt} = 0.005X_1X_2 - 0.1X_2$$

compute the Jacobian matrix and its eigenvalues. □

The eigenvalues of the Jacobian matrix tell us which of the six cases in Figure 8.1 we are dealing with. Remember that because we are dealing with two equations, we will always have two eigenvalues. Each of the eigenvalues has the following general form:

$$\lambda = a + bi \tag{4}$$

where a and b are constants, i is $\sqrt{-1}$, a is called the real part, and bi is the imaginary part. Based on the configuration of the two eigenvalues, we can tell which type of system behavior in Figure 8.1 applies to a particular system of differential equations. If the imaginary part of both eigenvalues is zero, we have case *a*, *c*, or *e*; that is, if the imaginary part of the eigenvalues is zero, we have a nonoscillatory system. If the system is nonoscillatory (b in equation 4 is zero), examine the signs of the real parts of the eigenvalues (a in equation 4). If both real parts are negative, we have a stable equilibrium (case *a*, stable node, stable asymptote). If the real parts of both eigenvalues are positive, we have an unstable equilibrium (case *c*, unstable node, unstable asymptote). If one eigenvalue is positive (real part) and the other, negative, we have a saddle point (case *e*, sometimes incorrectly called an unstable equilibrium).

All of the foregoing assumed zero imaginary parts for the eigenvalues. Now we consider the case in which both eigenvalues have nonzero imaginary parts (b in equation 4 is not equal to zero). With nonzero imaginary parts we are dealing with an oscillatory system ($b \neq 0$). We next look at the real parts of the eigenvalues. If both real parts are negative, we have case *b* in Figure 8.1. If both real parts are positive, we have case *d*. If both real parts are zero, we have case *f*.

If the two eigenvalues are symbolized as

$$\lambda_1 = a_1 + b_1 i$$

$$\lambda_2 = a_2 + b_2 i$$

we can construct the following table:

a_1, a_2	b_1, b_2	Case
Negative	Zero	Stable node (case *a*)
Positive	Zero	Unstable node (case *c*)
One positive One negative	Zero	Saddle point (case *e*)
Negative	Nonzero	Stable focus (case *b*)
Positive	Nonzero	Unstable focus (case *d*)
Zero	Nonzero	Centers (case *f*)

☐ EXERCISES

19 Compute the Jacobian matrix for equations 2 and determine to which case in Figure 8.1 they correspond.

20 Modify equations 2 so that the prey is density-dependent in a linear fashion (i.e., each individual of N_1 decreases the per capita growth rate of N_1 by a constant amount). Which case from Figure 8.1 do you get? (Assume that a_{11} is small, an important point.)

21 Modify equations 2 so that the predator is density-dependent in a linear fashion (i.e., each individual of N_2 decreases the per capita growth rate of N_2 by a constant amount). Which case from Figure 8.1 do you get? (Assume that a_{22} is small.)

22 Which case in Figure 8.1 do the following equations represent?

$$\frac{dN_1}{dt} = 5N_1 - 0.05N_1^2 - 0.08N_1N_2$$

$$\frac{dN_2}{dt} = 4N_2 - 0.05N_2^2 - 0.05N_1N_2$$

23 Add a positive density-dependent term to the prey equation (equations 2); that is, suppose that the basic per capita rate of increase of the prey itself is increased by the introduction of a prey individual. Which case does this represent? (Assume that a_{11} is small.) □

PREDATOR–PREY THEORY. With the foregoing theory (and exercises) we can discuss, in general terms, simple predator–prey theory. At its most elementary level we dealt with equations 2, the classical Lotka–Volterra predator–prey equations, which assumed that the prey was the only food source for the predator and the only source of death for the prey was the predator. There was no intraspecific competition—no density-dependent feedback on the predator or prey populations. The community matrix for equations 2 was

$$\begin{vmatrix} 0 & -a_{12} \\ a_{21} & 0 \end{vmatrix}$$

and the system behaved like Figure 8.1 f, neutrally stable oscillations. This result is *not* reasonable. It says that no matter where the system starts it will return repeatedly to that point. If the system is going through these permanent oscillations and we perturb it by adding predator or prey individuals to the system, it will begin to experience totally different, yet just as permanent, oscillations. Clearly the situation is biologically unrealistic.

One piece of biological realism we might add to the system is to presume that the prey population undergoes density-dependent regulation. This was done in exercise 20. The community matrix for that situation was

$$\begin{vmatrix} -a_{11} & -a_{12} \\ a_{21} & 0 \end{vmatrix}$$

The result was that the neutrally stable condition became a stable system (a stable oscillation if a_{11} were small and a stable node if a_{11} were large).

Alternatively, we might add density dependence to the predator population. This was done in exercise 21. The community matrix for that situation was

$$\begin{vmatrix} 0 & -a_{12} \\ a_{21} & -a_{22} \end{vmatrix}$$

Again the result was that the neutrally stable condition became stable.

In fact, we can easily generalize the above. Consider the following general equations:

$$\frac{dN_1}{N_1\,dt} = f_1(N_1, N_2)$$

$$\frac{dN_2}{N_2\,dt} = f_2(N_1, N_2)$$

The basic structure of a predator–prey situation requires $\partial f_1/\partial N_2 < 0$ and $\partial f_2/\partial N_1 > 0$ if N_1 is prey and N_2 is predator. If there is no intraspecific dependence on density at all, $\partial f_1/\partial N_1 = 0$ and $\partial f_2/\partial N_2 = 0$. Thus the Jacobian matrix is

$$J = \begin{vmatrix} 0 & \dfrac{\partial f_1}{\partial N_2} \\ \dfrac{\partial f_2}{\partial N_1} & 0 \end{vmatrix}$$

and Det $(J - \lambda I)$ is

$$\text{Det}\begin{vmatrix} -\lambda & \dfrac{\partial f_1}{\partial N_2} \\ \dfrac{\partial f_2}{\partial N_1} & -\lambda \end{vmatrix}$$

which is

$$(-\lambda)^2 - \left(\frac{\partial f_1}{\partial N_2}\right)\left(\frac{\partial f_1}{\partial N_2}\right)$$

and, when set equal to zero, gives $\lambda = \pm\sqrt{(\partial f_1/\partial N_2)(\partial f_2/\partial N_1)}$. Because $\partial f_1/\partial N_2 < 0$ and $\partial f_2/\partial N_1 > 0$, the term under the radical is negative. Thus both eigenvalues have zero real parts and nonzero imaginary parts and we have a neutrally stable system.

If we add intraspecific density dependence (i.e., $\partial f_1/\partial N_1 < 0$ and/or $\partial f_2/\partial N_2 < 0$), the Jacobian matrix is

$$J = \begin{vmatrix} \dfrac{\partial f_1}{\partial N_1} & \dfrac{\partial f_1}{\partial N_2} \\ \dfrac{\partial f_2}{\partial N_1} & \dfrac{\partial f_2}{\partial N_2} \end{vmatrix}$$

whence

$$\text{Det}\,(J - \lambda I) = \text{Det}\begin{vmatrix} \dfrac{\partial f_1}{\partial N_1} - \lambda & \dfrac{\partial f_1}{\partial N_2} \\ \dfrac{\partial f_2}{\partial N_1} & \dfrac{\partial f_2}{\partial N_2} - \lambda \end{vmatrix}$$

$$= \lambda^2 - \left(\frac{\partial f_1}{\partial N_1} + \frac{\partial f_2}{\partial N_2}\right)\lambda + \left(\frac{\partial f_1}{\partial N_1}\frac{\partial f_2}{\partial N_2} - \frac{\partial f_1}{\partial N_2}\frac{\partial f_2}{\partial N_1}\right)$$

Thus the eigenvalues are

$$\lambda = \frac{1}{2}\left(\frac{\partial f_1}{\partial N_1} + \frac{\partial f_2}{\partial N_2}\right) \pm \frac{1}{2}\sqrt{\left(\frac{\partial f_1}{\partial N_1} + \frac{\partial f_2}{\partial N_2}\right)^2 - 4\left(\frac{\partial f_1}{\partial N_1}\frac{\partial f_2}{\partial N_2} - \frac{\partial f_1}{\partial N_2}\frac{\partial f_2}{\partial N_1}\right)} \quad (5)$$

Because $\partial f_1/\partial N_1$ and $\partial f_2/\partial N_2$ are both negative, the system must be stable (either a stable oscillation or a stable node). In fact, if $\partial f_1/\partial N_1$ *or* $\partial f_2/\partial N_2$ were zero, the system would still be stable. Clearly, the biological result is that if predator or prey or both are subject to density-dependent population regulation, a stable system will be generated. Because it is thought that most populations in nature exhibit some sort of density control, does it not follow that all predator–prey systems should be stable?

What if one of the populations had a positive density-dependent response (i.e., for every individual added to the system the per capita rate of its species increased)? An examination of equation 5 reveals that it is at least possible to generate an unstable system (i.e., unstable oscillations or unstable nodes) if $\partial f_1/\partial N_1 > |\partial f_2/\partial N_2|$, where $\partial f_1/\partial N_1 > 0$ and $\partial f_2/\partial N_2 < 0$. This is a generalization of the specific case in exercise 23 and actually forms the basis for the next section.

THE METHOD OF ROSENZWEIG AND MACARTHUR. We begin with a quick review of the biological implications of the theory as developed so far. The basic assumption that predator eats prey and prey is limited only by the predator, with no intraspecific density dependence, led to neutrally stable oscillations, clearly an unreasonable biological result. If any sort of density-dependent negative feedback is added, the system becomes stable. This seems to be a reasonable result. On further inspection it is not an entirely reasonable result. In fact, many laboratory experiments with predator–prey systems are not in any sense stable. To generate the sort of instability that is frequently observed in the laboratory we had to add positive feedback; that is, one of the species had to show a positive change in the per capita growth rate as the result of adding an individual of its own species. This is an important result.

Rosenzweig and MacArthur (1963) used this basic result in formulating an elegant analysis of predator–prey interactions. The rest of this chapter is based on that method.

We begin by assuming that predator–prey dynamics may be described by the following familiar equations:

$$\frac{dN_1}{dt} = f_1(N_1, N_2)$$

$$\frac{dN_2}{dt} = f_2(N_1, N_2)$$

where N_1 is prey and N_2 is predator. Consider the equilibrium condition of these equations:

$$0 = f_1(N_1, N_2)$$
$$0 = f_2(N_1, N_2)$$

As in Chapter 7, these equations can be plotted on a graph of N_1 against N_2. For the present development it is convenient to ask what f_1 and f_2 might look like on a graph of N_1 against N_2 under various biological assumptions; for example, consider the case of interspecific competition. If we hold the first competitor (N_1) constant at zero, we may ask in what direction will the system move along the N_2 axis? Clearly, the second competitor should behave exactly like the logistic equation because $N_1 = 0$. Thus, if we indicate the dynamics of the system under this condition of $N_1 = 0$, we will have Figure 8.2:

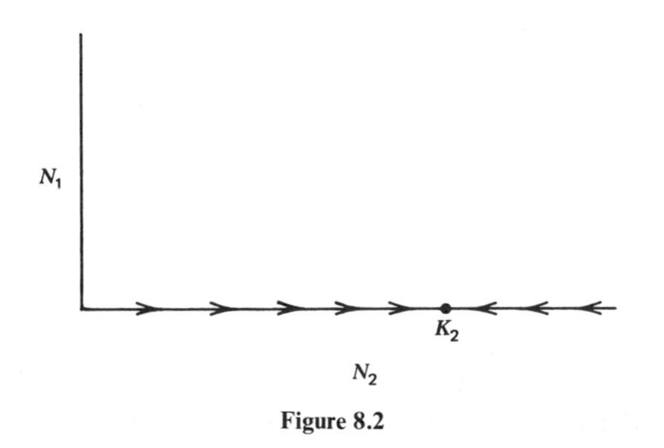

Figure 8.2

Now suppose that we allow N_1 to be slightly greater than zero. We presume, for ease of exposition, that N_1 will stay at a constant density—that we somehow will have the ability to keep N_1 from increasing or decreasing even though it is below K_1. What now will be the behavior of N_2? It will still increase if it is at a low density and decrease if it is at a high density. Also, there will be some population density for which it will neither increase nor decrease, but that density will be at least slightly less than the carrying capacity because of the competitive effect of species 1. Thus we obtain Figure 8.3:

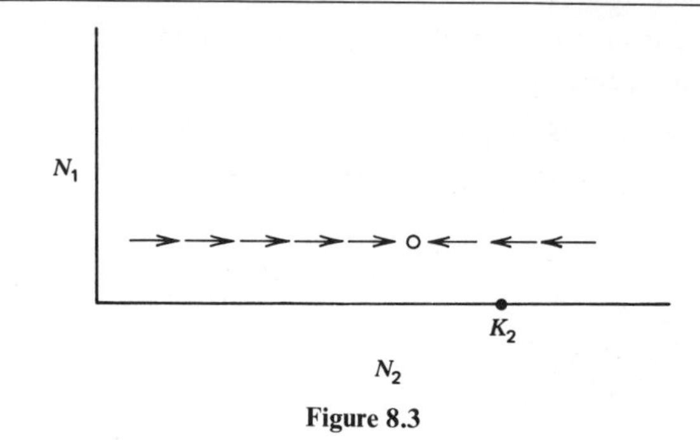

Figure 8.3

If we repeat the experiment but fix N_1 at a slightly higher density repeat it again at an even higher density, and continue on up, we would obtain something like Figure 8.4:

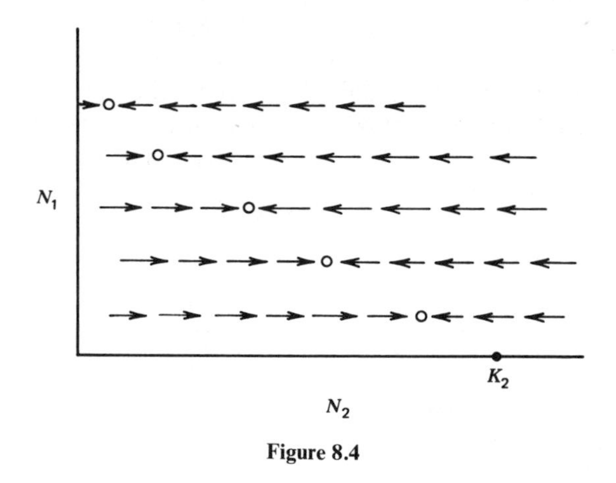

Figure 8.4

By connecting the points for which N_2 neither increases nor decreases we obtain the line that is identical to the graph of $dN_2/dt = 0$ in the Lotka–Volterra competition equations. This line is frequently referred to as an isocline or a zero isocline. It represents the locus of points for which one of the populations neither increases nor decreases.

□ EXERCISES

24 Draw the isocline for species 1 in the competition example. Use the reasoning given in the text.

25 Suppose we are dealing with a classic predator–prey interaction in which the death rate of the prey is due only to the predator and the birth rate of the predator is due only to the prey. (Remember that the birth rate of the prey and the death rate of the predator are constants.) Using this qualitative reasoning, construct the isoclines for both species.

26 Repeat exercise 25 but include negative density-dependent feedback for the prey.

27 Repeat exercise 25 but include negative density-dependent feedback for the predator.

28 Repeat exercise 25 but include positive density-dependent feedback for the prey. □

In these exercises you have seen how easy it is to construct isoclines, at least qualitatively. We now construct a somewhat more complicated isocline. Consider the prey isocline. Let us suppose that at low prey density the prey exhibits positive density-dependent feedback but at very high density it exhibits negative density dependence. Furthermore, presume that there is some minimal population density below which the prey population could not increase in numbers even if the predator were not there (this is sometimes called the Allee effect). Without predators the dynamics would like like Figure 8.5:

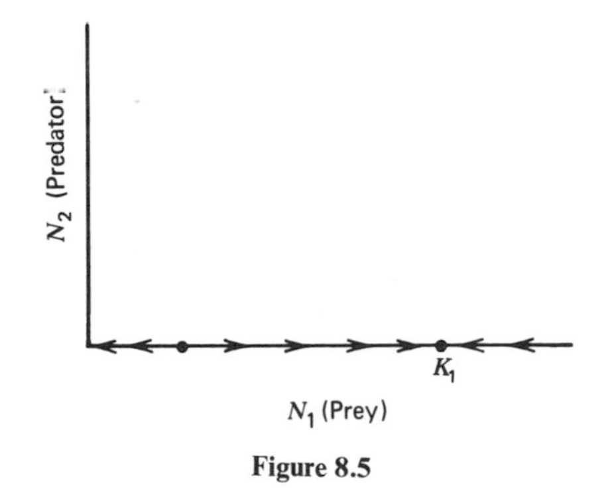

Figure 8.5

Now suppose we add a small number of predators. Using the reasoning in exercises 26 and 28, we find that the prey density below which the prey cannot persist is larger than it was without that extra predator pressure. Furthermore, the prey density below which the prey increases is smaller than it was without that extra predator pressure. Thus we obtain Figure 8.6:

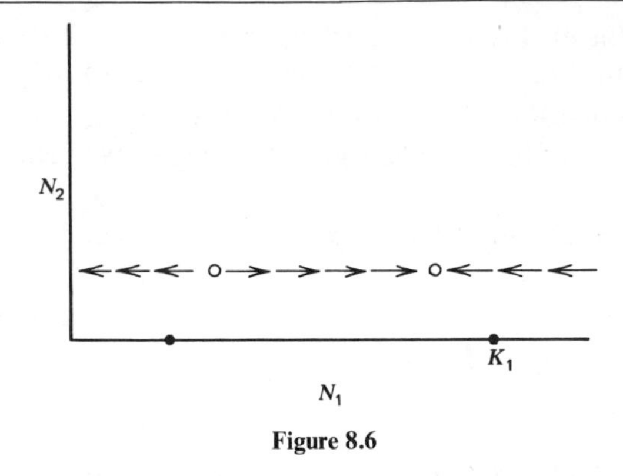

Figure 8.6

If we continue to increase the predator density step by step, we will obtain Figure 8.7:

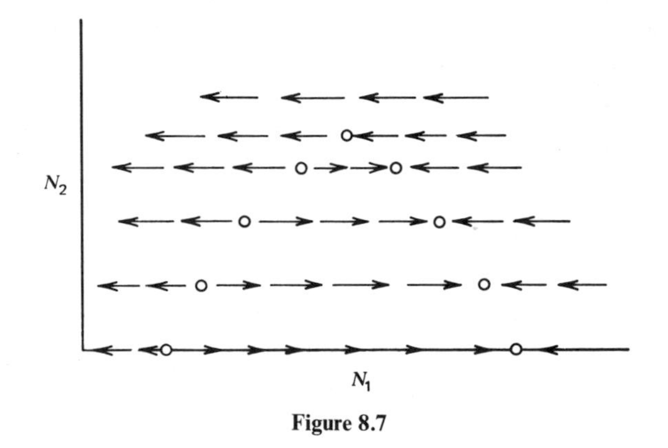

Figure 8.7

Clearly the isocline has a "hump." It need not be symmetrical, for it may look like any of the following (Figure 8.8) or virtually any other conceivable shpae as long as it has a hump.

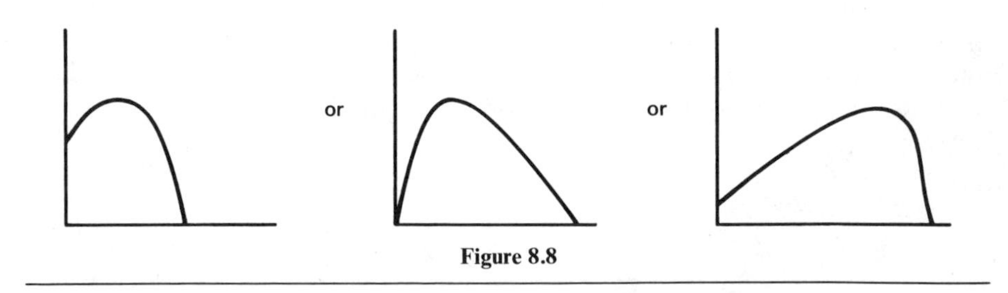

Figure 8.8

It should be clear that the "hump" results from a mixture of positive and negative density-dependent effects. What will happen if this humped curve is put together with the basic density-independent predator isocline?

Figure 8.9 shows three qualitatively interesting positions that the predator isocline might take:

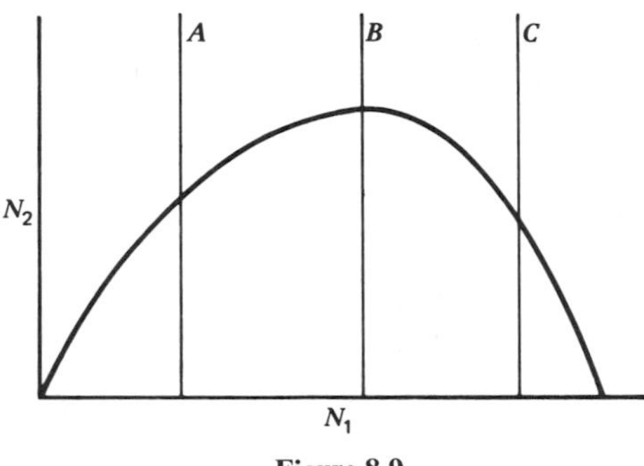

Figure 8.9

Position A lies to the left of the "hump" and crosses the prey isocline on a point at which the prey isocline is ascending (exhibit positive density dependence). Position C is to the right of the hump and crosses the prey isocline on a point at which the prey isocline is descending (exhibits negative density dependence). Position B occurs *exactly* on the point at which the prey isocline moves from positive to negative density dependence. We are concerned only with positions A and C.

Consider position C first. If we add the dynamic arrows that we used to construct the isoclines in the first place, we obtain Figure 8.10:

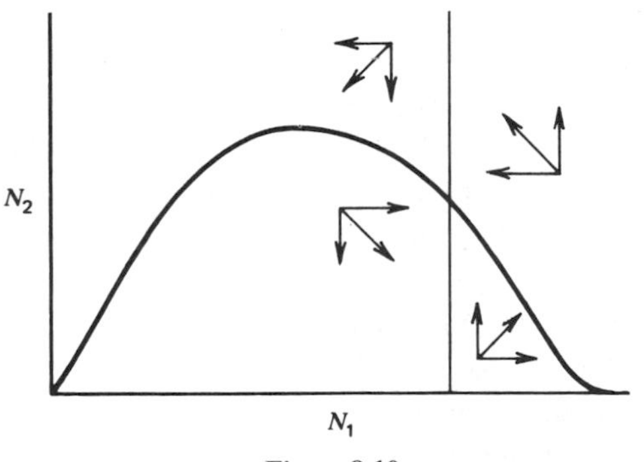

Figure 8.10

This is an oscillatory situation. Recall from the preceding analysis that when the prey exhibits negative density dependence and the predator is density independent the system is a stable focus or stable asymptote. In this case we must have a stable focus because on the point at which the two isoclines cross the prey is negatively density dependent; that is, if the predator isocline crosses the prey isocline to the right of the hump of the prey isocline, the system is stable.

Consider the predator isocline in position A. Again adding dynamic arrows, we obtain Figure 8.11:

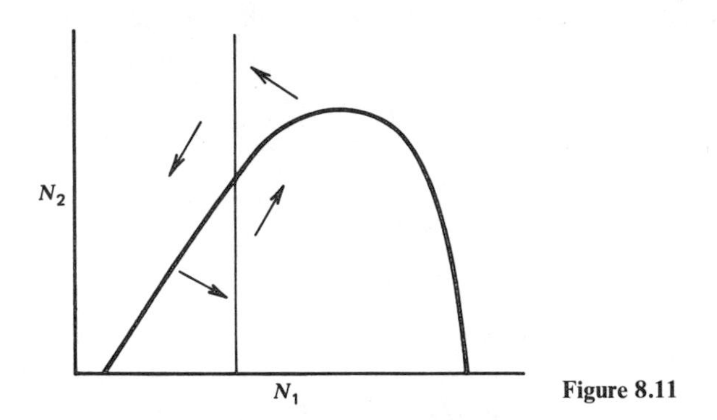

Figure 8.11

This is another oscillatory situation. Again recall from the preceding analysis that when the prey exhibits positive density dependence and the predator is density independent, the system is unstable. In this case, then, we must have an unstable focus; that is, if the predator isocline crosses the prey isocline to the left of the hump of the prey isocline, the system will be unstable.

What we have developed is a model that accounts for both stable and unstable oscillations. At least at a qualitative level, this model seems to explain most of the general patterns observed in nature, and apparently the theory generally acceptable for predator–prey interactions.

□ EXERCISES

29 Assume that the predator does not undergo density-dependent regulation until it reaches a very high density, at which point the per capita growth rate is decreased suddenly to zero. Add this feature to the basic Rosenzweig and MacArthur graph and show what its dynamic consequences are for (a) when the predator isocline crosses to the right of the hump and (b) when the predator isocline crosses to the left of the hump.

30 Assume that there is some refugium in which a small number of prey individuals can hide; that is, below some critical value of N_1 the predator can no longer catch any prey and the prey will increase in numbers. (Assume that the critical number is larger than the critical number for the Allee effect). What are the dynamic effects? □

LIMIT CYCLES. Exercises 29 and 30 provided examples of the generation of permanent oscillations in predator–prey communities. These permanent oscillations are different from the permanent oscillations generated in a neutrally stable system (case f in Figure 8.1). The difference is best illustrated by drawing the trajectories of a neutrally stable system and a stable limit cycle (Figure 8.12):

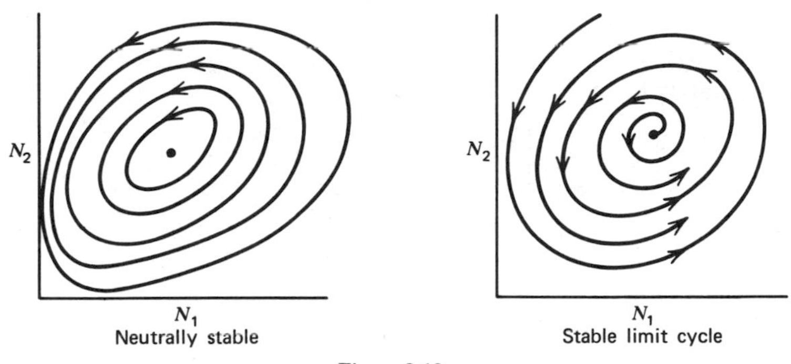

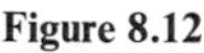

Figure 8.12

The difference between the two seems obvious. The stable limit cycle has a single "connected" trajectory. If it begins inside that trajectory, the system will spiral out until it reaches that trajectory and continue on it thereafter. If it begins outside the connected trajectory, it will spiral inward until it reaches that trajectory. The system is stable in the same sense that systems were stable or unstable before but this time the equilibrium "point" is a ring of points. In contrast, wherever the system starts in the neutrally stable system it will continue in the same trajectory forever. The stable limit cycle represents a system of permanent oscillations much more realistically than centers do. Most people agree that any permanently oscillating predator–prey system in nature probably represents a stable limit cycle.

□ EXERCISES

31 What might an unstable limit cycle look like?

32 What might a metastable limit cycle look like (i.e., the analogue to a saddle point)? □

ANSWERS TO EXERCISES

1

$$\frac{dN_1}{N_1 dt} = b - a_{12}N_2$$

$$\frac{dN_2}{N_2 dt} = a_{12}N_1 - d$$

at equilibrium (i.e., $dN_1/dt = dN_2/dt = 0$)

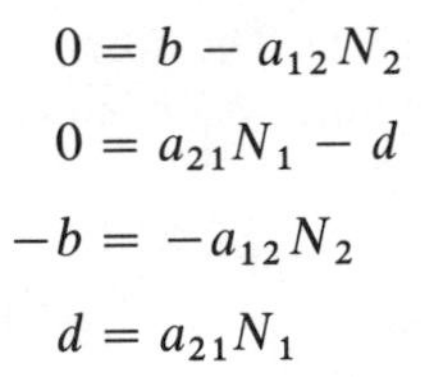

$$0 = b - a_{12}N_2$$

$$0 = a_{21}N_1 - d$$

$$-b = -a_{12}N_2$$

$$d = a_{21}N_1$$

and in matrix form

$$\begin{vmatrix} -b \\ d \end{vmatrix} = \begin{vmatrix} 0 & -a_{12} \\ a_{21} & 0 \end{vmatrix} \begin{vmatrix} N_1 \\ N_2 \end{vmatrix}$$

2

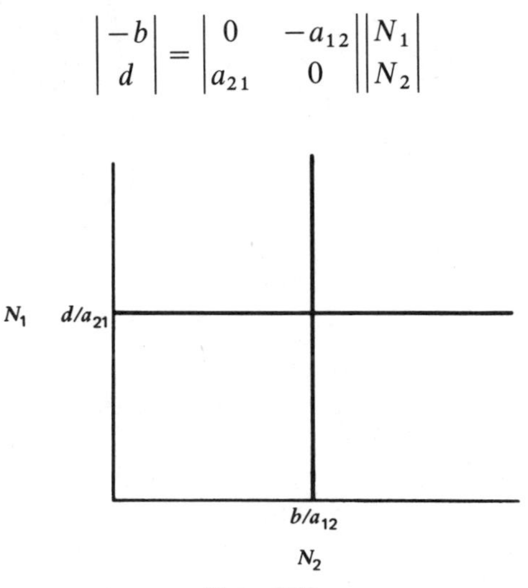

Figure 8.13

3

$$\frac{dN_1}{dt} > 0 \Rightarrow b - a_{12}N_2 > 0$$

$$N_2 < \frac{b}{a_{12}}$$

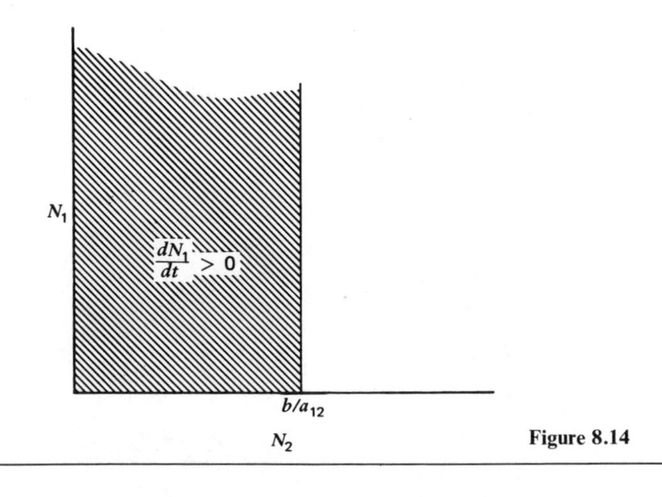

Figure 8.14

4
$$\frac{dN_2}{dt} > 0 \Rightarrow N_1 a_{21} - d > 0$$

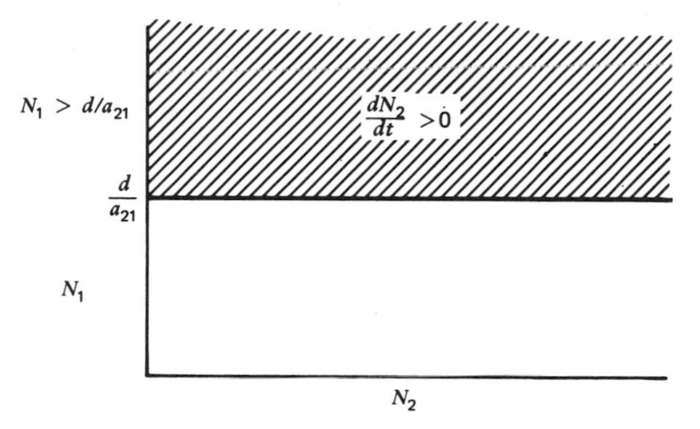

Figure 8.15

5

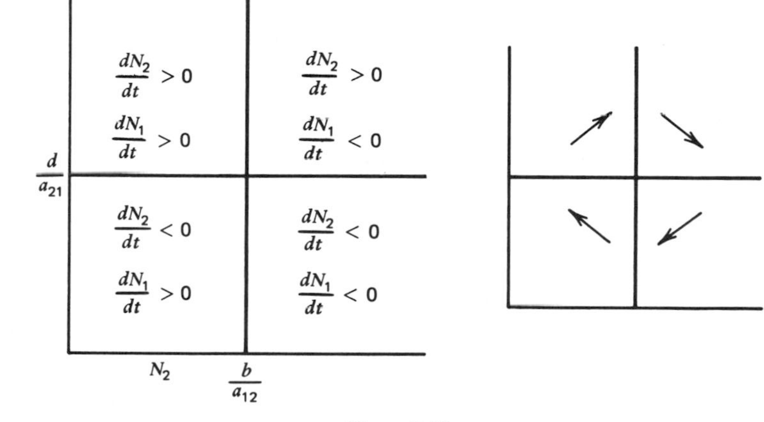

Figure 8.16

The system obviously is oscillatory.

6 Case 2 corresponds to stable node or stable asymptote. Case 4 corresponds to saddle point.

7 For Figure 8.1*b*

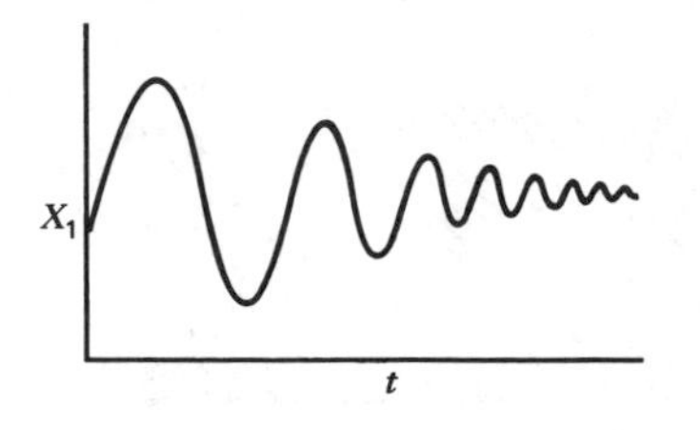

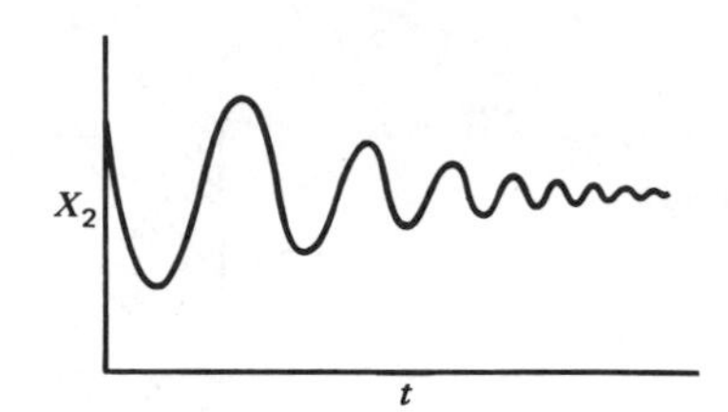

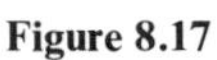

Figure 8.17

For Figure 8.1*d*

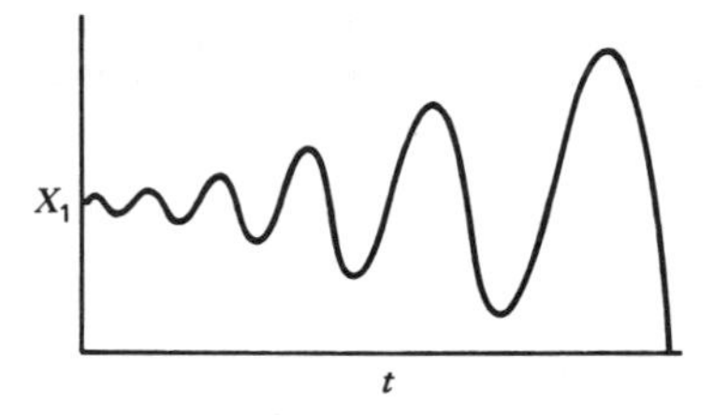

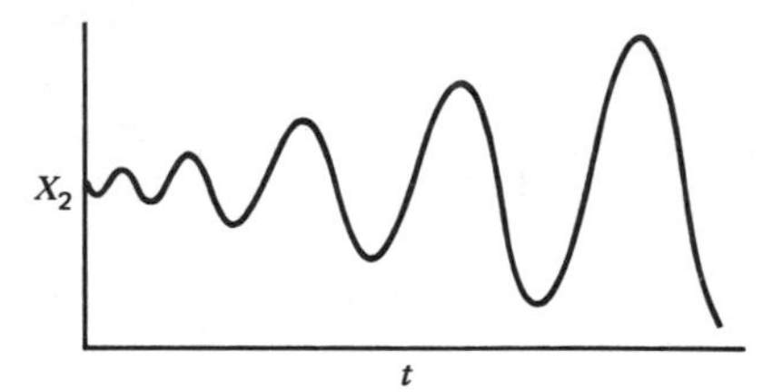

Figure 8.18

For Figure 8.1*f*

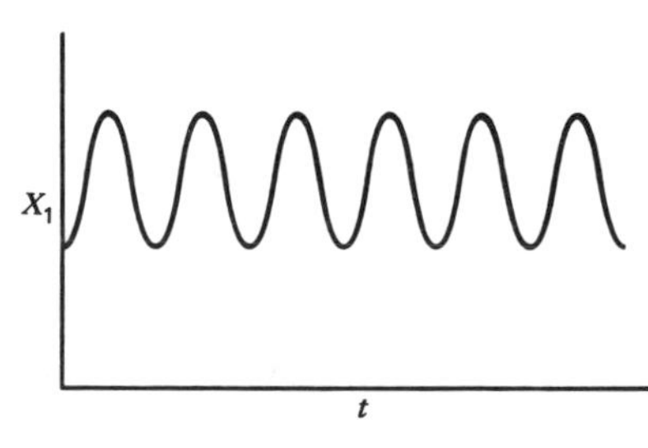

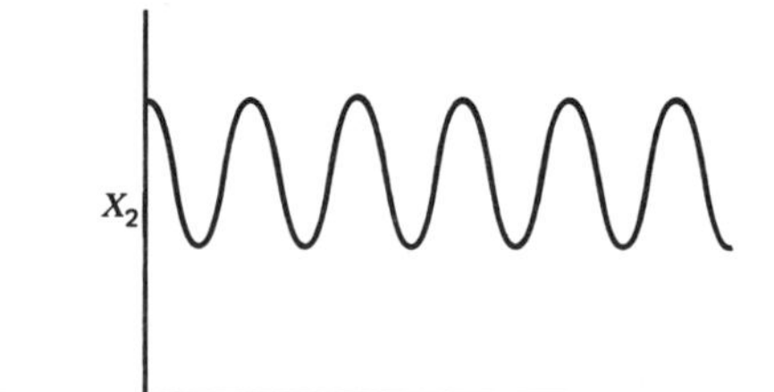

Figure 8.19

8 The matrix λI is

$$\lambda \begin{vmatrix} 1 & 0 \\ 0 & 1 \end{vmatrix} = \begin{vmatrix} \lambda & 0 \\ 0 & \lambda \end{vmatrix}$$

Thus

$$A - \lambda I = \begin{vmatrix} a & b \\ c & d \end{vmatrix} - \begin{vmatrix} \lambda & 0 \\ 0 & \lambda \end{vmatrix}$$

$$= \begin{vmatrix} a - \lambda & b \\ c & d - \lambda \end{vmatrix}$$

9 The matrix $A - \lambda I$ (from exercise 8) is

$$\begin{vmatrix} a - \lambda & b \\ c & d - \lambda \end{vmatrix}$$

Thus

$$\text{Det } (A - \lambda I) = \text{Det} \begin{vmatrix} a - \lambda & b \\ c & d - \lambda \end{vmatrix} = (a - \lambda)(d - \lambda) - bc$$

$$= ad - a\lambda - d\lambda + \lambda^2 - bc$$

$$= \lambda^2 - (a + d)\lambda + ad - bc$$

10

$$B - \lambda I = \begin{vmatrix} 2 & 0 \\ 0 & 3 \end{vmatrix} - \begin{vmatrix} \lambda & 0 \\ 0 & \lambda \end{vmatrix} = \begin{vmatrix} 2 - \lambda & 0 \\ 0 & 3 - \lambda \end{vmatrix}$$

$$\text{Det } (B - \lambda I) = \text{Det} \begin{vmatrix} 2 - \lambda & 0 \\ 0 & 3 - \lambda \end{vmatrix} = (2 - \lambda)(3 - \lambda) = 6 - 5\lambda + \lambda^2$$

11 From exercise 10 Det $(B - \lambda I) = \lambda^2 - 5\lambda + 6 = 0$. Applying the quadratic formula

$$\lambda = \frac{-b \pm \sqrt{b^2 - 4ac}}{2a} = \frac{5 \pm \sqrt{25 - 4(6)}}{2}$$

$$= \frac{5 \pm \sqrt{1}}{2} = \tfrac{5}{2} \pm \tfrac{1}{2} = 2.5 \pm 0.5$$

Thus $\lambda_1 = 3$, $\lambda_2 = 2$, are the two values of λ that satisfy the equation; they are the roots of the equation and the eigenvalues of the matrix B.

12

$$\begin{vmatrix} 5 & 4 \\ 1 & 2 \end{vmatrix} - \begin{vmatrix} \lambda & 0 \\ 0 & \lambda \end{vmatrix} = \begin{vmatrix} 5 - \lambda & 4 \\ 1 & 2 - \lambda \end{vmatrix}$$

$$\text{Det} \begin{vmatrix} 5 - \lambda & 4 \\ 1 & 2 - \lambda \end{vmatrix} = (5 - \lambda)(2 - \lambda) - 4 = 6 - 7\lambda + \lambda^2 = 0$$

$$\lambda = \tfrac{7}{2} \pm \tfrac{1}{2}\sqrt{49 - 4(6)} = \tfrac{7}{2} \pm \tfrac{5}{2}$$

$$\lambda_1 = \frac{7 + 5}{2} = 6$$

$$\lambda_2 = \frac{7 - 5}{2} = 1$$

13 For the Lotka–Volterra equations

$$f_1 = r_1 N_1 - \frac{r_1}{K_1} N_1^2 - \frac{r_1}{K_1} \alpha_{12} N_1 N_2$$

$$f_2 = r_2 N_2 - \frac{r_2}{K_2} N_2^2 - \frac{r_2}{K_2} \alpha_{21} N_1 N_2$$

$$\frac{\partial f_1}{\partial N_1} = r_1 - \frac{2r_1}{K_1} N_1 - \frac{r_1}{K_1} \alpha_{12} N_2$$

$$\frac{\partial f_1}{\partial N_2} = -\frac{r_1}{K_1} \alpha_{12} N_1$$

$$\frac{\partial f_2}{\partial N_1} = -\frac{r_2}{K_2} \alpha_{21} N_2$$

$$\frac{\partial f_2}{\partial N_2} = r_2 - 2\frac{r_2}{K_2} N_2 - \frac{r_2}{K_2} \alpha_{21} N_1$$

14 From Chapter 7

$$N_1 = \frac{\text{Det } A_1}{\text{Det } A} = \frac{K_1 - K_2 \alpha_{12}}{1 - \alpha_{12}\alpha_{21}}$$

$$N_2 = \frac{\text{Det } A_2}{\text{Det } A} = \frac{K_2 - K_1 \alpha_{21}}{1 - \alpha_{12}\alpha_{21}}$$

$$\left(\frac{\partial f_1}{\partial N_1}\right)_0 = r_1 - \frac{2r_1}{K_1}\left(\frac{K_1 - K_2\alpha_{12}}{1 - \alpha_{12}\alpha_{21}}\right) - \frac{r_1}{K_1}\alpha_{12}\left(\frac{K_2 - K_1\alpha_{21}}{1 - \alpha_{12}\alpha_{21}}\right)$$

$$= \frac{r_1 \alpha_{12} \dfrac{K_2}{K_1} - r_1}{1 - \alpha_{12}\alpha_{21}}$$

$$\left(\frac{\partial f_1}{\partial N_2}\right)_0 = \frac{\alpha_{12}[(\alpha_{12} r_1)/K_1)(K_2 - r_1)]}{1 - \alpha_{12}\alpha_{21}}$$

$$\left(\frac{\partial f_2}{\partial N_2}\right)_0 = \frac{(r_2 \alpha_{21} K_1/K_2) - r^2}{1 - \alpha_{12}\alpha_{21}}$$

$$\left(\frac{\partial f_2}{\partial N_1}\right)_0 \quad \frac{\alpha_{21}(r_2 \alpha_{22} K_1 - r_2)}{1 - \alpha_{12}\alpha_{21}}$$

15

$$J = \begin{vmatrix} \dfrac{(r_1 \alpha_{12} K_2/K_1) - r_1}{1 - \alpha_{12}\alpha_{21}} & \alpha_{12}\left[\dfrac{(r_1 \alpha_{12} K_2/K_1) - r_1}{1 - \alpha_{12}\alpha_{21}}\right] \\ \alpha_{21}\left[\dfrac{(r_2 \alpha_{21} K_1/K_2) - r_2}{1 - \alpha_{12}\alpha_{21}}\right] & \dfrac{(r_2 \alpha_{21} K_1/K_2) - r_2}{1 - \alpha_{12}\alpha_{21}} \end{vmatrix}$$

16

$$J = \frac{1}{1 - \alpha_{12}\alpha_{21}} \begin{vmatrix} \dfrac{r_1\alpha_{12}K_2}{K_1} - r_1 & 0 \\ 0 & \dfrac{r_2\alpha_{21}K_1}{K_2} - r_2 \end{vmatrix} \begin{vmatrix} 1 & \alpha_{12} \\ \alpha_{21} & 1 \end{vmatrix}$$

or

$$J = \begin{vmatrix} \left(\dfrac{\partial f_1}{\partial N_1}\right)_0 & 0 \\ 0 & \left(\dfrac{\partial f_2}{\partial N_2}\right)_0 \end{vmatrix} \begin{vmatrix} 1 & \alpha_{12} \\ \alpha_{21} & 1 \end{vmatrix}$$

17

$$f_1 = X_1 - 0.01X_2$$

$$f_2 = 2X_2 - 0.05X_1$$

$$\frac{\partial f_1}{\partial X_1} = 1 \qquad \frac{\partial f_1}{\partial X_2} = -0.01 \qquad \frac{\partial f_2}{\partial X_2} = 2 \qquad \frac{\partial f_2}{\partial X_1} = -0.05$$

Because neither X_1 nor X_2 appear in the general expression of the partials, we do not need to evaluate the partials at the equilibrium point. Thus

$$J = \begin{vmatrix} 1 & -0.01 \\ -0.05 & 2 \end{vmatrix}$$

The eigenvalues are the roots to the equation

$$\text{Det}\,(J - \lambda I) = 0$$

$$\text{Det}\left\{\begin{vmatrix} 1 & -0.01 \\ -0.05 & 2 \end{vmatrix} - \begin{vmatrix} \lambda & 0 \\ 0 & \lambda \end{vmatrix}\right\} = 0$$

$$\text{Det}\begin{vmatrix} 1 - \lambda & -0.01 \\ -0.05 & 2 - \lambda \end{vmatrix} = 0$$

$$(1 - \lambda)(2 - \lambda) - (0.01)(0.05) = 0$$

$$\lambda^2 - 3\lambda + 1.9995 = 0$$

$$\lambda = \frac{3 + \sqrt{9 - 4(1.9995)}}{2}$$

$$\lambda_1 = \frac{4.001}{2} = 2.0005$$

$$\lambda_2 = \frac{1.999}{2} = 0.9995$$

18

$$f_1 = 0.5X_1 - 0.001X_1X_2$$

$$f_2 = 0.005X_1X_2 - 0.1X_2$$

$$\frac{\partial f_1}{\partial X_1} = 0.5 - 0.001X_2$$

$$\frac{\partial f_1}{\partial X_2} = -0.001X_1$$

$$\frac{\partial f_2}{\partial X_1} = 0.005X_2$$

$$\frac{\partial f_2}{\partial X_2} = 0.005X_1 - 0.1$$

Now the partials must be evaluated at equilibrium:

$$0 = 0.5X_1 - 0.001X_1X_2$$

$$0 = 0.005X_1X_2 - 0.1X_2$$

$$0.001X_2 = 0.5$$

$$X_2 = \frac{0.5}{0.001} = 500$$

$$X_1 = \frac{0.1}{0.005} = 20$$

Thus

$$\left(\frac{\partial f_1}{\partial X_1}\right)_0 = 0.5 - (0.001)(500) = 0.5 - 0.5 = 0$$

$$\left(\frac{\partial f_1}{\partial X_2}\right)_0 = -0.001(20) = -0.02$$

$$\left(\frac{\partial f_2}{\partial X_1}\right)_0 = 0.005(500) = 2.5$$

$$\left(\frac{\partial f_2}{\partial X_2}\right)_0 = (0.005)(20) - 0.1 = 0.1 - 0.1 = 0$$

Thus

$$J = \begin{vmatrix} 0 & -0.02 \\ 2.5 & 0 \end{vmatrix}$$

$$J - \lambda I = \begin{vmatrix} -\lambda & -0.02 \\ 2.5 & -\lambda \end{vmatrix}$$

$$\text{Det}\,(J - \lambda I) = (-\lambda)^2 + (0.02)(2.5)$$

Eigenvalues are roots of

$$\lambda^2 + 0.05 = 0$$

$$\lambda = \frac{0 \pm \sqrt{-4(0.05)}}{2} = \pm\frac{0.447}{2}\sqrt{-1}$$

$$= \pm 0.224i$$

where $i = \sqrt{-1}$.

19

$$f_1 = N_1 b - a_{12}N_1N_2$$

$$f_2 = a_{21}N_1N_2 - dN_2$$

$$\frac{\partial f_1}{\partial N_1} = b - a_{12}N_2$$

$$\frac{\partial f_1}{\partial N_2} = -a_{12}N_1$$

$$\frac{\partial f_2}{\partial N_1} = a_{21}N_2$$

$$\frac{\partial f_2}{\partial N_2} = a_{21}N_1 - d$$

Equilibrium values:

$$N_1 = \frac{d}{a_{21}} \qquad N_2 = \frac{b}{a_{12}}$$

$$\left(\frac{\partial f_1}{\partial N_1}\right)_0 = b - a_{12}\frac{b}{a_{12}} = 0$$

$$\left(\frac{\partial f_1}{\partial N_2}\right)_0 = -a_{12}\frac{d}{a_{21}} = -\frac{a_{12}}{a_{21}}d$$

$$\left(\frac{\partial f_2}{\partial N_1}\right)_0 = a_{21}N_2 = \frac{ba_{21}}{a_{12}}$$

$$\left(\frac{\partial f_2}{\partial N_2}\right)_0 = a_{21}N_1 - d = 0$$

$$J = \begin{vmatrix} 0 & -\dfrac{a_{12}}{a_{21}}d \\ b\dfrac{a_{21}}{a_{12}} & 0 \end{vmatrix}$$

$$\text{Det}\,(J - \lambda I) = \text{Det}\begin{vmatrix} -\lambda & -\dfrac{a_{12}}{a_{21}}d \\ \dfrac{a_{21}}{a_{12}}b & -\lambda \end{vmatrix}$$

$$= \lambda^2 + bd$$

Eigenvalues:

$$\lambda = \frac{0 \pm \sqrt{-4bd}}{2}$$

Clearly, because b and d are positive,

$$\lambda_1 = +\sqrt{bd}\,i$$

$$\lambda_2 = -\sqrt{bd}\,i$$

20 The basic equations are

$$\frac{dN_1}{N_1\,dt} = b - a_{12}N_2 - a_{11}N_1$$

$$\frac{dN_2}{N_2\,dt} = a_{21}N_1 - d$$

$$\frac{\partial f_1}{\partial N_1} = b - a_{12}N_2 - 2a_{11}N_1$$

$$\frac{\partial f_1}{\partial N_2} = -a_{12}N_1$$

$$\frac{\partial f_2}{\partial N_1} = a_{21}N_2$$

$$\frac{\partial f_2}{\partial N_2} = a_{21}N_1 - d$$

Equilibrium values:

$$N_1 = \frac{\alpha}{a_{21}}$$

$$N_2 = \frac{a_{21}b - a_{11}d}{a_{12}a_{21}}$$

$$\left(\frac{\partial f_1}{\partial N_1}\right)_0 = -\frac{a_{11}}{a_{21}}d$$

$$\left(\frac{\partial f_1}{\partial N_2}\right)_0 = -a_{12}\left(\frac{d}{a_{21}}\right)$$

$$\left(\frac{\partial f_2}{\partial N_1}\right)_0 = a_{21}\left(\frac{a_{21}b - a_{11}d}{a_{12}a_{21}}\right) = \frac{a_{21}}{a_{12}}b - \frac{a_{11}}{a_{12}}d$$

$$\left(\frac{\partial f_2}{\partial N_2}\right)_0 = a_{21}N_1 - d = a_{21}\left(\frac{d}{a_{21}}\right) - d = 0$$

$$J = \begin{vmatrix} -\frac{a_{11}}{a_{21}}d & -a_{12}\left(\frac{d}{a_{21}}\right) \\ \frac{a_{21}}{a_{12}}b - \frac{a_{11}}{a_{12}}d & 0 \end{vmatrix}$$

whence the eigenvalues are

$$\lambda = \frac{-(a_{11}/a_{21})d \pm \sqrt{(a_{11}/a_{21})^2d^2 - 4(a_{12}/a_{21})d[(a_{21}/a_{12})b - (a_{11}/a_{12})d]}}{2}$$

By inspection, if a_{11} is small, the term under the radical is negative and the system remains oscillatory. The real parts of the eigenvalues, however, are no longer zero; they are negative. Therefore we have case b of Figure 8.1.

21 The modified equations are

$$\frac{dN_1}{N_1\,dt} = b - a_{12}N_2$$

$$\frac{dN_2}{N_2\,dt} = a_{21}N_1 - d - a_{22}N_2$$

Partial derivatives:

$$\frac{\partial f_1}{\partial N_1} = b - a_{12}N_2$$

$$\frac{\partial f_1}{\partial N_2} = -a_{12}N_1$$

$$\frac{\partial f_2}{\partial N_1} = a_{21}N_2$$

$$\frac{\partial f_2}{\partial N_2} = a_{21}N_1 - d - 2a_{22}N_2$$

Equilibrium:

$$N_1 = \frac{d}{a_{21}} + \frac{a_{22}}{a_{21}}\frac{b}{a_{21}} = \frac{da_{12} + a_{22}b}{a_{12}a_{21}}$$

$$N_2 = \frac{b}{a_{12}}$$

$$\left(\frac{\partial f_1}{\partial N_1}\right)_0 = b - a_{12}\left(\frac{b}{a_{12}}\right) = 0$$

$$\left(\frac{\partial f_1}{\partial N_2}\right)_0 = -a_{12}\left(\frac{da_{12} + a_{22}b}{a_{12}a_{21}}\right)$$

$$\left(\frac{\partial f_2}{\partial N_1}\right)_0 = a_{21}\left(\frac{b}{a_{12}}\right)$$

$$\left(\frac{\partial f_2}{\partial N_2}\right)_0 = a_{21}\left(\frac{da_{12} + a_{22}b}{a_{12}a_{21}}\right) - d - 2a_{22}\left(\frac{b}{a_{12}}\right) = \frac{-a_{22}}{a_{12}}b$$

$$J = \begin{vmatrix} 0 & -\left(\dfrac{da_{12} + a_{22}b}{a_{21}}\right) \\ a_{21}\left(\dfrac{b}{a_{12}}\right) & -\dfrac{a_{22}}{a_{12}}b \end{vmatrix}$$

$$\text{Det}\,(J - \lambda I) = \lambda^2 + \frac{a_{22}}{a_{12}}b\lambda + bd + \frac{a_{22}}{a_{12}}b^2$$

Eigenvalues:

$$\lambda = \frac{-(a_{22}/a_{12})b \pm \sqrt{(a_{22}/a_{12})^2b^2 - 4(bd + (a_{22}/a_{12})b^2)}}{2}$$

Clearly we no longer have centers. As long as a_{22} is relatively small, the system will be case b of Figure 8.1.

22

$$\frac{\partial f_1}{\partial N_1} = 5 - 2(0.05)N_1 - 0.08N_2$$

$$\frac{\partial f_1}{\partial N_2} = -0.08N_1$$

$$\frac{\partial f_2}{\partial N_1} = -0.05N_2$$

$$\frac{\partial f_2}{\partial N_2} = 4 - 2(0.05)N_2 - 0.05N_1$$

Equilibrium densities:

$$N_1 = 47 \qquad N_2 = 33$$

$$\left(\frac{\partial f_1}{\partial N_1}\right)_0 = 5 - 2(0.05)47 - 0.08(33) = -2.34$$

$$\left(\frac{\partial f_1}{\partial N_2}\right)_0 = -0.08(47) = -3.76$$

$$\left(\frac{\partial f_2}{\partial N_1}\right)_0 = -0.05(33) = \quad 1.65$$

$$\left(\frac{\partial f_2}{\partial N_2}\right)_0 = 4 - 2(0.05)(33) - 0.05(47) = -1.65$$

Jacobian matrix:

$$J = \begin{vmatrix} -2.34 & -3.76 \\ -1.65 & -1.65 \end{vmatrix}$$

$$\begin{aligned} \text{Det}\,(J - \lambda I) &= \text{Det}\begin{vmatrix} -2.34 - \lambda & -3.76 \\ -1.65 & -1.65 - \lambda \end{vmatrix} \\ &= \lambda^2 + 2.34\lambda + 1.65\lambda - 2.33 \\ &= \lambda^2 + 3.99\lambda - 2.33 = 0 \end{aligned}$$

Eigenvalues:

$$\lambda = \frac{-3.99 \pm \sqrt{3.99^2 + 4(2.33)}}{2}$$

$$\lambda_1 = 4.5166$$

$$\lambda_2 = -0.5166$$

Therefore the equations represent a saddle point (case *e* in Figure 8.1).

23 Basic equations become

$$\frac{dN_1}{N_1\,dt} = b - a_{12}N_2 + a_{11}N_1$$

$$\frac{dN_2}{N_2\,dt} = a_{21}N_1 - d$$

$$\frac{\partial f_1}{\partial N_1} = b - a_{12}N_2 + 2a_{11}N_1$$

$$\frac{\partial f_1}{\partial N_2} = -a_{12}N_1$$

$$\frac{\partial f_2}{\partial N_1} = a_{21}N_2$$

$$\frac{\partial f_2}{\partial N_2} = a_{21}N_1 - d$$

Equilibrium values:

$$N_1 = \frac{d}{a_{21}} \qquad N_2 = \frac{b}{a_{12}} + \frac{a_{11}d}{a_{12}a_{21}}$$

$$\left(\frac{\partial f_1}{\partial N_1}\right)_0 = b - a_{12}\left(\frac{b}{a_{12}} + \frac{a_{11}d}{a_{12}a_{21}}\right) + \frac{2a_{11}d}{a_{21}} = \frac{a_{11}d}{a_{21}}$$

$$\left(\frac{\partial f_1}{\partial N_2}\right)_0 = -\frac{a_{12}}{a_{21}}d$$

$$\left(\frac{\partial f_2}{\partial N_1}\right)_0 = \frac{a_{21}b + a_{11}d}{a_{21}}$$

$$\left(\frac{\partial f_2}{\partial N_2}\right)_0 = 0$$

Jacobian:

$$J = \begin{vmatrix} \dfrac{a_{11}d}{a_{21}} & -\dfrac{a_{12}d}{a_{21}} \\ \dfrac{a_{21}b + a_{11}d}{a_{12}} & 0 \end{vmatrix}$$

$$\text{Det}\,(J - \lambda I) = \lambda^2 - \frac{a_{11}}{a_{21}}d\lambda + bd + \frac{a_{11}}{a_{21}}d^2 = 0$$

$$\lambda = \frac{(a_{11}/a_{21})d \pm \sqrt{(a_{11}/a_{21})^2d^2 - 4(bd + (a_{11}/a_{21})d^2)}}{2}$$

By inspection, if a_{11} is small, the term under the radical will be negative, meaning that the system is oscillatory; but the real parts of the eigenvalues are positive. Thus we have case *d* in Figure 8.1, an unstable oscillation.

24

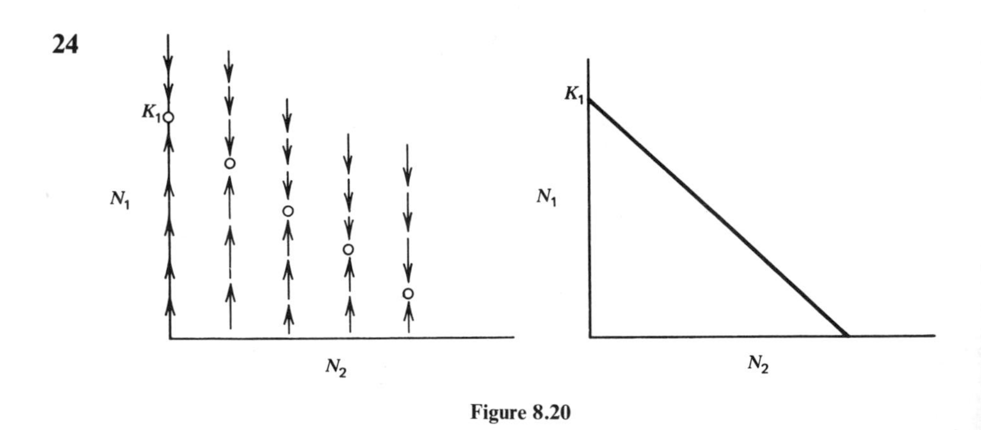

Figure 8.20

25 Let N_1 = prey and N_2 = predator. Because the death rate of the predator is a constant, the predator will always decrease in number if the prey population is below a critical value. That critical value is the population density of the prey which would just counterbalance the constant death rate. Thus for the predator we have Figure 8.21:

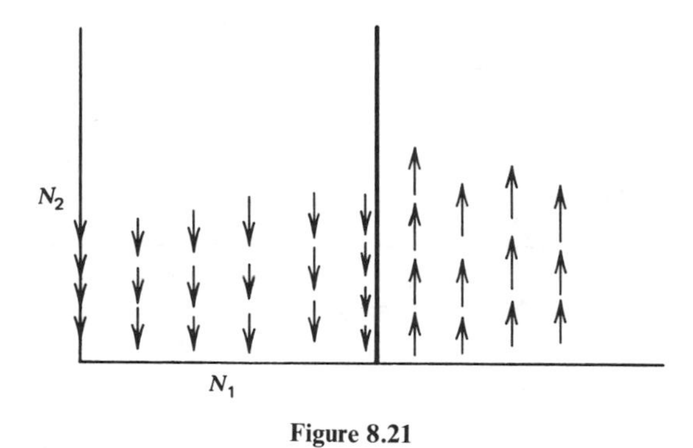

Figure 8.21

The predator isocline is a straight vertical line. Similarly, the birth rate of the prey is a constant. Below a critical density of predator the prey will increase. Above that critical density the prey will decrease (Figure 8.22):

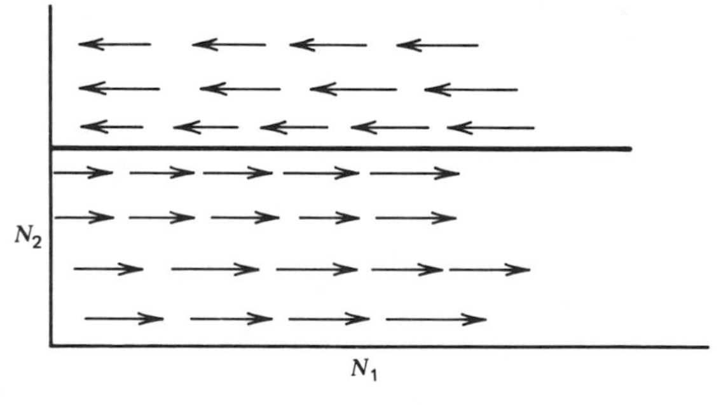

Figure 8.22

Thus the prey isocline is a straight horizontal line.

26

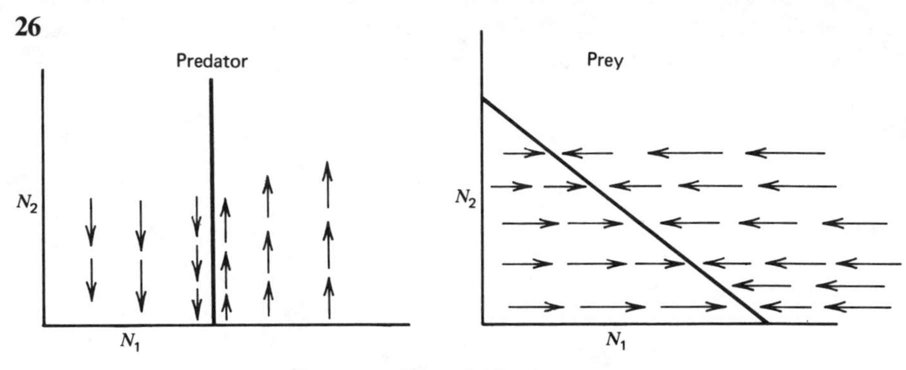

Figure 8.23

27

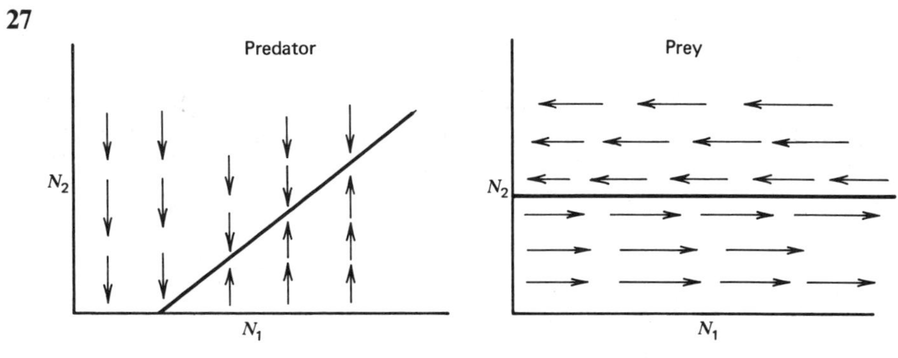

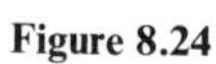
Figure 8.24

28

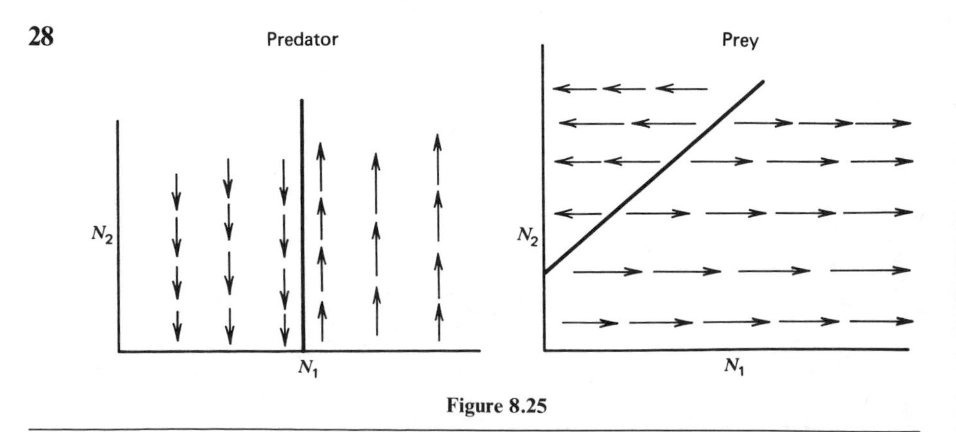

Figure 8.25

29

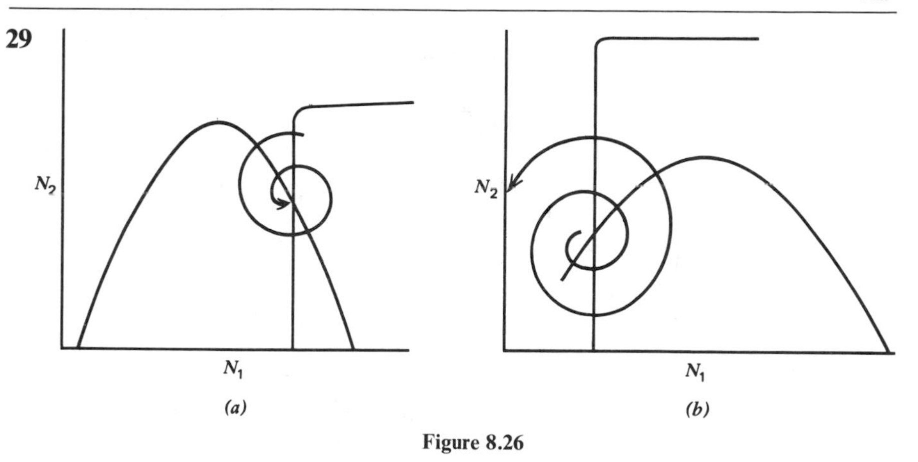

Figure 8.26

In (*a*) and (*b*) nothing has changed qualitatively. To the left of the hump unstable oscillations eventually lead to extinction and to the right of the hump we obtain stable oscillations. Suppose the critical density of the predator is somewhat less than in (*b*). We obtain Figure 8.27:

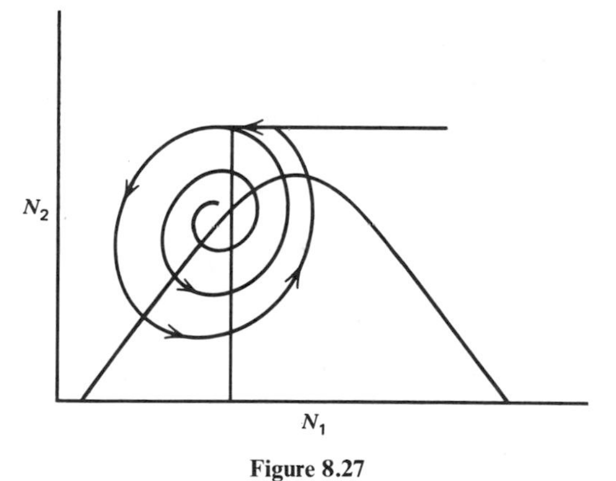

Figure 8.27

Clearly the system spirals out away from the equilibrium until it hits point P_0. It is then, in a sense, driven out of its outward spiral. The net result is that a permanent cycle is established in which, if the system begins inside the cycle, it spirals outward until it hits point P_0, at which time it continues on the constant cycle forever.

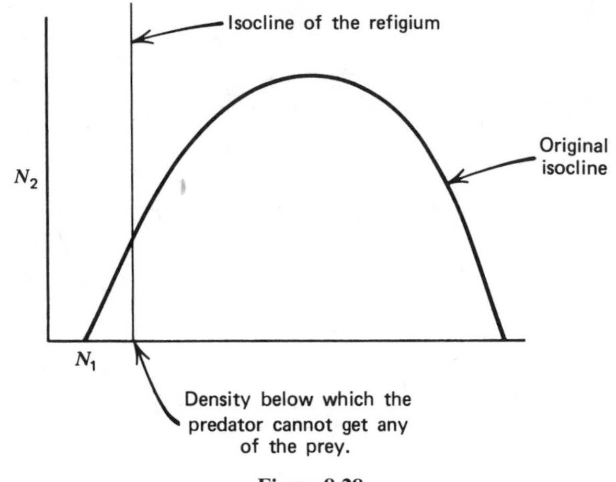

Figure 8.28

30

A composite isocline is shown in Figure 8.29:

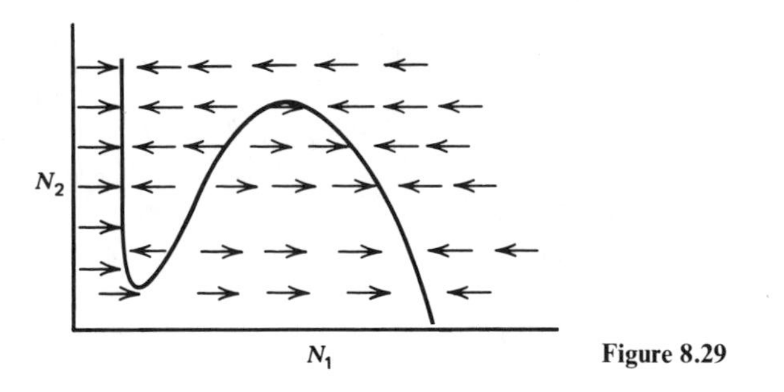

Figure 8.29

Putting the two isoclines together, we have Figure 8.30:

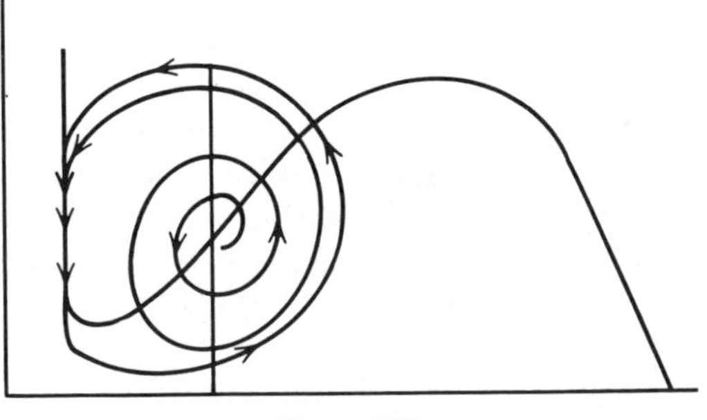

Figure 8.30

Thus the same thing happened as in exercise 29.

31

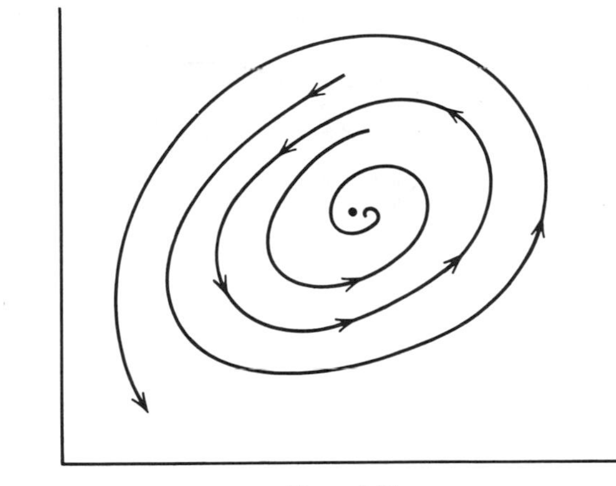

Figure 8.31

32 Like this:

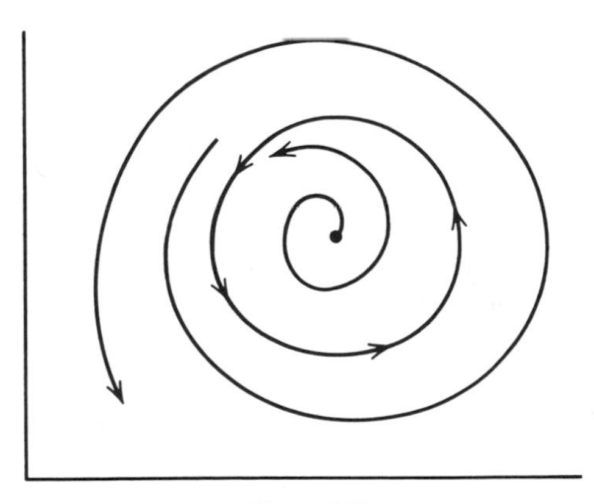

Figure 8.32

Like this:

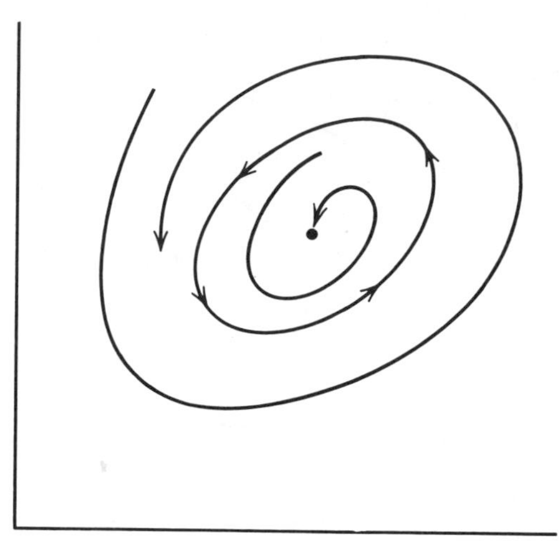

Figure 8.33

REFERENCES

Bartlett, M. S. 1960. Stochastic population models in ecology and epidemiology. New York: Wiley.

Beddington, J. R. 1975. Mutual interference between parasites or predators and its effect on searching efficiency. *J. Anim. Ecol.* **44**:331–340.

Beddington, J. R., M. P. Hassell, and J. H. Lawton. 1976. The components of arthropod predation, II. The predator rate of increase. *J. Anim. Ecol.* **45**:165–185.

Bradley, D. J. 1974. Stability in host-parasite systems. In M. B. Usher and M. H. Williamson, Eds. *Ecological Stability*. London: Chapman and Hall.

Crofton, H. D. 1971a. A quantitative approach to parasites. *Parasitology* **62**:179–193.

Crofton, H. D. 1971b. A model of host-parasite relationships. *Parasitology* **63**:343–364.

DeBach, P. and H. S. Smith. 1941a. Are population oscillations inherent in the host-parasite relations? *Ecology* **22**:363–369.

DeBach, P. and H. S. Smith. 1941b. The effect of host density on the rate of reproduction of entomophagous parasites. *J. Econ. Entomol.* **34**:741–745.

Dixon, A. F. G. 1959. An experimental study of the searching behaviour of the predatory coccinellid beetle Adalia decempunctata (L.). *J. Anim. Ecol.* **28**:259–281.

Gilpin, M. E. 1972. Enriched prey-predator systems: theoretical stability. *Science* **177**:902–904.

Griffiths, K. J. and C. S. Holling. 1969. A competition sub-model for parasites and predators. *Can. Entomol.* **101**:785–818.

Hassell, M. P. 1971. Mutual interferences between searching insect parasites. *J. Anim. Ecol.* **40**:473–486.

Luckinbill, L. S. 1973. Coexistence in Laboratory populations of *Paramecium aurelia* and its predator *Didinium Nasutum*. *Ecology* **54**:1320–1327.

Maly, E. J. 1969. A laboratory study of the interaction between the predatory rotifer *Asplanchna* and *Paramecium*. *Ecology* **50**:59–73.

May, R. M. 1972. Limit cycles in predator-prey communities. *Science*. **177**:900–902.

Mogi, M. 1969. Predation response of the larvae of *Harmonia axyridis* Pallas (Coccinellidae) to the different prey density. *Japn. J. Appl. Entomol. Zool.* **13**:9–16.

Murdie, G., and M. P. Hassill. 1973. Food distribution, searching success and predator-prey models. In M. S. Bartlett and R. W. Hiorns, Eds. *The Mathematical Theory of the Dynamics of Biological Populations*. London: Academic, pp. 87–101.

Murdoch, W. W. and A. Oaten, 1975. Predation and population stability. *Adv. Ecol. Res.* **9**:2–131.

Noy-Meir, I. 1975. Stability of grazing systems: an application of predator–prey graphs. *J. Ecol.* **63**:459–481.

Pan, C. T. 1965. Studies on the host-parasite relationship between *Schistosoma mansoni* and the snail *Australorbis glabratus*. *Am. J. Trop. Med. Hyg.* **14**:931–976.

Pimentel, D., S. A. Levin, and A. B. Soans. 1975. On the evolution of energy balance in some exploiter-victim systems. *Ecology* **56**:381–390.

Price, P. W. 1972. Parasitoids utilizing the same host: adaptive nature of differences in size and form. *Ecology* **53**:190–195.

Rogers, D. J. and M. P. Hassell, 1974. General models for insect parasite and predator searching behavior: interference. *J. Anim. Ecol.* **43**:239–253.

Rogers, D. J. and S. Hubbard. 1974. How the behaviour of parasites and predators promotes population stability. In M. B. Usher and M. H. Williamson, Eds. *Ecological Stability*. London: Chapman and Hall, pp. 99–119.

Rosenzweig, M. L. 1969. Why the prey has a hump. *Am. Nat.* **103**:81–87.

Rosenzweig. M. L. 1971. Paradox of enrichment: destabilization of exploitation ecosystems in ecological time. *Science*. **171**:385–387.

Rosenzweig, M. L. and R. H. MacArthur. 1963. Graphical representation and stability condition of prey-predator interactions. *Am. Nat.* **97**:209–223.

Royama, T. 1971. A comparative study of models for predation and parasitism. *Res. Popul. Ecol. Kyoto Suppl. I.* I–91.

Tanner, J. T. 1975. The stability and the intrinsic growth rates of prey and predator populations. *Ecology* **56**:855–867.

Thompson, D. J. 1975. Towards a predator-prey model incorporating age structure: the effects of predator and prey size on the predation of *Daphnia magna* by *Ichnura elegans*. *J. Amin. Ecol.* **44**:907–916.

Wangersky, P. J. and W. J. Cunningham. 1957. Time lag in prey-predator population models. *Ecology*. **38**:136–139.

Wratten, S. D. 1973. The effectiveness of the coccinellid beetle. *Adalia bipunctata* (L.), as a predator of the lime aphid, *Eucallipterus tiliae L. J. Anim. Ecol.* **42**:785–802.

9. Species Diversity

Naturalists have long been fascinated by the diversity of life—the vast array of color forms in a butterfly collection, the immense variety of birds at the edge of a tropical forest, or the seemingly endless variety of form in the mollusks on an intertidal rock face. Such fascination did not stop at aesthetic appreciation. Certain patterns seemed to occur repeatedly. Specifically, in almost any haphazard collection of individuals within a taxon (butterflies, birds, insects, herps) almost without exception one can observe one or two common species (morphotypes), a few of intermediate abundance, and a large number of rarities. This sort of casual observation led to an attempt at a more formal quantification of diversity.

The first notable attempt at such quantification was made in the 1940s by Fisher, Corbett, and Williams (1943). Noting the basic pattern of a small number of common species and a large number of rare species, these authors developed a model, the logarithmic series, which approximated many of the known data in the literature. Even with reasonably good correspondence of the data to this theory, it became obvious eventually that much of the actual deviation was consistently biased. In case after case there were fewer rare species than the theory predicted. This systematic bias led to the realization that many of the patterns could be explained as incomplete sampling from a lognormal distribution, an observation made originally by Preston (1948). Preston's analysis seems to be the most common method of community analysis applied today and about half of this chapter is based on it.

THE LOGNORMAL DISTRIBUTION. Suppose we have a collection of butterflies in hand and are sure that the collection was made at random with respect to the abundance of species. One of the first things we noticed about this collection is that few species make up the bulk of individuals in the collection. Furthermore, probably our first actual quantification was almost subconsciously done; that is, we probably noticed the orders of magnitude of abundances: species 1 is about twice as common as species 2, species 2 is about twice as common as species 3, and so on. This initial quantification is the basis for Preston's lognormal curve. How many rare species are there, how many species are twice as abundant as the rare ones, and how many species are twice as abundant as that? Thus we can construct a frequency diagram with the number of species on the ordinate and abundance (in orders of magnitude) on the abscissa.

We might graduate the orders of magnitude on the absissa in a variety of ways. Probably the most convenient mathematically would be in terms of natural logarithms. We shall follow Preston's lead and use a scale of "octaves"; that is, consider those species represented by zero to 1 individuals, 1 to 2 individuals, 2 to 4, 4 to 8, 8 to 16, and so on. Each of these intervals is called an octave; for example, if a species is represented in the collection by three individuals (say), that species belongs in the third octave (first octave is 0–1, second is 1–2, third is 2–4). When a species has an abundance that falls on a boundary of an octave (1, 2, 4, 8, 16), it is considered as 0.5 in the left octave and 0.5 in the right.

□ EXERCISES

1 Suppose the following numbers represent the number of individuals in each of 36 species: 1, 1, 2, 2, 3, 3, 3, 5, 5, 6, 6, 6, 7, 7, 9, 9, 10, 10, 11, 12, 13, 14, 19, 20, 23, 24, 25, 28, 29, 35, 50, 55, 60, 95, 110, 200. Plot number of species against octave. Draw a smooth curve through the points.

2 Suppose the following numbers represent the number of individuals in each of 42 species: 1, 1, 3, 3, 5, 6, 7, 9, 11, 12, 14, 17, 20, 25, 30, 35, 40, 42, 50, 53, 70, 75, 78, 83, 99, 130, 150, 178, 220, 260, 290, 320, 400, 600, 900, 950, 1100, 1300, 1900, 2500, 3000, 5000. Plot number of species against octave. Draw a smooth curve through the points.

3 Suppose the following numbers represent the number of individuals in each of 18 species: 1, 1, 1, 1, 1, 1, 1, 1, 1, 1, 1, 1, 1, 2, 2, 3, 4, 6. Plot number of species against octave. Draw a smooth curve through the points.

4 Suppose the following numbers represent the number of individuals in each of 55 species: 1, 1, 1, 1, 1, 1, 1, 1, 1, 1, 1, 1, 1, 1, 1, 1, 1, 1, 1, 1, 1, 1, 1, 1, 1, 1, 1, 1, 1, 1, 1, 1, 1, 1, 2, 2, 2, 2, 3, 3, 3, 3, 3, 3, 3, 4, 4, 5, 6, 7, 10, 15, 30. Plot number of species against octaves. Draw a smooth curve through the points. □

Exercises 1 and 2 were constructed to demonstrate how a community can have a distribution that at least looks normal on that kind of plot, but, in fact, exercises 3 and 4 are more like what would be obtained in an analysis of a real sample from nature. How can we maintain that communities are basically lognormal (as in exercises 1 and 2) when most communities are like exercises 3 and 4?

□ EXERCISES

5 Suppose we had taken a sample that was one-half the size of the sample that gave us the number in exercise 1; that is, each of the species would have

its abundance reduced by one-half (approximately) and one-half the singletons would not be represented. Plot such a situation (i.e., actually cut the abundance in half for each of the species) on a graph of number of species against octave. Draw a smooth curve through the points. Draw an arrow to indicate the mode and another arrow to indicate where the mode was in the example in exercise 1.

6 Cut the population densities of exercise 5 in half randomly (again one-half the singletons should go to zero). Plot the new data. Draw a smooth curve through the points. Draw arrows to indicate the mode of this exercise, exercise 5, and exercise 1.

7 Cut the population densities of exercise 6 in half. Plot the new data. Draw a smooth curve through the points. Draw arrows to indicate the mode of this exercise, exercise 6, exercise 5, and exercise 1. □

Exercises 5, 6, and 7 illustrate the effects of decreasing or increasing sample size. Indeed, what seems to happen is that an underlying lognormal curve is exposed as the sample size increases. If the sample size is small, only the right-hand side of the curve is visible. This is a mathematical property of sampling from a lognormal distribution; it has nothing to do with biological assumptions. Clearly this means that if a sample is graphed as in the above exercises, what would actually be seen is a part of the underlying curve. What we are actually dealing with is Figure 9.1:

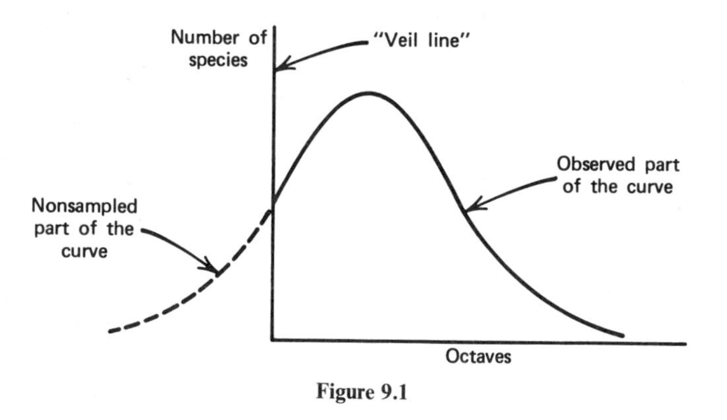

Figure 9.1

The line that separates the real part of the curve from the unsampled part is referred to as the veil line. Sampling, then, can be thought of as moving the veil line to the left. More exactly, if the sample size is doubled, the veil line will move exactly one octave to the left. This should be intuitively obvious from the examples in exercises 1, 5, 6, and 7.

Another minor but classically confusing point about the lognormal distribution has to do with the plotting of singletons (species represented by only one individual). It is perfectly valid to suppose that each species is counted as one-half a species in each octave if the species occurs on the boundary of an octave; however, a special problem is encountered in dealing with the first octave. As we go from high to low octaves, we proceed in a logarithmic fashion from high to low numbers, that is, 32, 16, 8, 4, 2, 1. What comes after 1? Until now I have referred to the first octave as containing 0 to 1 individual. In fact, this should be 0.5 to 1 individual. Thus we should include not only singletons in this octave but also all species represented by 0.5 individual. Because 0.5 individual obviously does not show up in any sample, the first octave (0.5–1) will *always* contain fewer species than it really should. Thus it is imperative that the veil line be drawn at exactly 1 at the end of the first octave.

This point has been vague in the literature ever since Preston first confused it. We frequently see lognormal curves with the veil line supposedly drawn at point 0 (actually it is at 0.5). This mistake frequently leads us to believe that the mode of the lognormal curve has already been exposed (the 0–1 or 0.5–1 octave *always* has fewer species than the 1–2 octave).

□ EXERCISES

8 Replot the data of exercises 1, 5, 6, and 7 with the veil line positioned at 1.

9 Remove at random one-half of all the populations in exercise 7. Plot. Draw a smooth curve.

10 Remove at random one-half of all the populations in exercise 9. Plot. Draw a smooth curve.

11 Graph the following equation (for $X = 0$ to 9):

$$Y = Y_0 e^{-(aX)^2}$$

for $Y_0 = 8$ and $a = 0.2$.

12 For this equation plot $\ln Y$ against X^2. □

The equation for the lognormal distribution is, most simply,

$$S(R) = S_0 e^{-(aR)^2}$$

where $S(R)$ is the number of species in the Rth octave from the modal octave, S_0 is the number of species in the modal octave, and a is a parameter related to the variance of the distribution. If the standard deviation of the distribution is σ, $a = 1/(\sigma\sqrt{2})$.

To fit the lognormal distribution to field data requires sophisticated technology. Preston fit his curves by eye, a practice that has become commonplace.

If the mode is clearly included in the sample, an approximate fit can be obtained by the method of linear regression if the following equation is used:

$$\ln S(R) = \ln S_0 - a^2 R^2$$

☐ EXERCISES

13 For the data in exercise 1 plot ln $S(R)$ against R^2 and estimate a. What is the value of σ? Show σ on the graph you produced in exercise 1.

14 Repeat exercise 13 for the data in exercise 2.

15 Repeat exercise 13 for the data in exercise 6. ☐

The total number of species in the community is given as

$$S_t = (\sqrt{2\pi}) S_0 \sigma$$

This result is obtained by integrating the equation for the lognormal, a somewhat tiresome task that is not presented here.

Quantities of field data have been analyzed by fitting the lognormal distribution. As long as the community is relatively large (usually more than 30 species, although Preston was really talking about hundreds) the lognormal distribution usually fits the data reasonably well. In fact, this is not really surprising. In any multiplicative sort of process (such as growing biological populations or compounding interest) in which a number of factors affect the growth of the elements in the process independently [(in this case the biological populations) a lognormal distribution is expected; for further discussion see May (1975)].

Until now we have talked about what appears to be a general biological law, expected theoretically and observed empirically. The lognormal distribution has been historically useful and will probably continue to be in pure and applied ecological research. It is a two-parameter distribution; that is, if S_0 (or S_r) and a (or σ) is specified, the entire distribution is specified. Problems with fitting it to real data are encountered mainly when the mode of the distribution is not included in a sample, a point to which we shall return.

Preston suggested a special case of the lognormal distribution in which a knowledge of only S_0 will specify the entire distribution. He termed this special case of the lognormal the "canonical" lognormal or simply canonical distribution. To understand the canonical distribution it is necessary to define a new distribution, the "individuals" curve.

☐ EXERCISES

16 For the raw data in exercise 1 plot the total number of individuals that belong to species found in particular octaves; for example, in the fourth octave (between four and eight individuals) we have two species that contain five individuals (a total of 10 individuals), three species that contain six individuals (a total of 18 individuals), and two species that contain seven individuals (a total of 14 individuals). Thus the fourth octave contains $10 + 18 + 14 = 42$ individuals.

17 Repeat exercise 16 for the data in exercise 2.

18 Construct the individuals curve for the data in exercise 1 by multiplying the number of species in an octave by the mean of the octave; for example the mean of the fourth octave is 6, [(4 + 8)/2] and there are seven species in that octave. Thus the number of individuals for that octave will be $7 \times 6 = 42$.

□

Thus we see that the individuals curve is estimated as the appropriate value on the species curve multiplied by the mean of the octave. So computed, it can be shown that the individuals curve itself is a lognormal curve, greatly displaced to the right. On reviewing a great quantity of data, Preston noted that many communities seem to have an individuals curve that terminates at its crest; that is, when the individuals curve is constructed, the modal octave occurs at exactly the same octave as the last octave of the species curve. When this happens—when the individuals curve terminates at its crest—the distribution is referred to as the canonical lognormal distribution.

This theory, coupled with the apparent empirical generalization that communities in nature tend to be canonical, is a powerful statement. It implies that we can say all sorts of things about community structure if we estimate only one parameter of the lognormal. Such power turns out to be somewhat illusory [see Preston, 1962, and May, 1975, for further discussion].

THE MEASUREMENT OF SPECIES DIVERSITY. This chapter began with a general statement about a search by naturalists for methods to describe certain patterns observed in nature. Among the numerous theoretical distributions proposed the lognormal seems to be the most useful. Some ecologists, however, have begun to question the fundamental reason for approaching community ecology this way in the first place. Rather than looking for a mathematical function that describes some universal law of nature, a more recent trend has been simply to describe patterns of species diversity in purposeful ignorance of any underlying theory.

The notion of species diversity has long been in the minds of field ecologists. Some areas contain many species, others few, but the number of species is not really the only thing of concern when a naturalist says one area is diverse and another depauperate; for example, if areas A and B both contain 100 species but almost all the individuals in area A are of a single species (the others being rare) and the individuals in area B are more or less evenly apportioned among the species, most ecologists would say intuitively that area B is more diverse. Clearly, the notion of species diversity confounds the area of "number of species" (species richness) and the way in which individuals are apportioned into those species (evenness). The goal is to develop a measurement that reflects these two notions: species richness and evenness.

The problem is similar to one we faced in deriving a measure of clumpedness. We must devise a measurement that is large when the number of species is large and/or the apportionment of individuals among the species, is relatively even and small when the number of species is small and/or the apportionment of individuals among the species is uneven. The most commonly used measure is the Shannon–Weiner function:

$$H = -\sum_{i=1}^{S} P_i \ln P_i$$

where p_i is the fraction of all individuals in the community contained in species i. In general, it takes on a large value when the number of species is large and/or when the individuals are apportioned evenly among the species.

☐ EXERCISES

19 From the data in exercise 1 compute the total number of species in the community, using the estimate of σ from exercise 13 in the equation given in the text. Compute H for the same data.

20 Repeat exercise 19 with data from exercise 6.

21 Compute H for the data from exercise 10.

22 (a) Suppose that 10 species are represented by the following number of individuals: 10, 10, 10, 10, 10, 10, 10, 10, 10, 10. Compute H.
(b) Suppose that 10 species are represented by the following numbers of individuals: 91, 1, 1, 1, 1, 1, 1, 1, 1, 1. Compute H.
(c) Suppose that 10 species are represented by the following numbers of individuals: 15, 15, 15, 15, 15, 5, 5, 5, 5, 5. Compute H.
(d) Suppose that 10 species are represented by the following numbers of individuals: 1, 3, 5, 7, 9, 11, 13, 15, 17, 19. Compute H.
(e) Suppose that 10 species are represented by the following numbers of individuals: 19, 19, 19, 19, 19, 1, 1, 1, 1, 1. Compute H.

23 (a) Suppose that five species are represented by the following numbers of individuals: 20, 20, 20, 20, 20. Compute H.
(b) Suppose that five species are represented by the following numbers of individuals: 96, 1, 1, 1, 1. Compute H.
(c) Suppose that five species are represented by the following numbers of individuals: 30, 30, 20, 10, 10. Compute H.

(d) Suppose that five species are represented by the following numbers of individuals: 4, 12, 20, 28, 36. Compute H.

(e) Suppose that five species are represented by the following numbers of individuals: 49, 48, 1, 1, 1. Compute H. □

It should be clear from these exercises that the Shannon–Weiner function behaves appropriately. As in our search for a measurement of pattern, a vast number of other measurements would also behave appropriately. The only reason for using the Shannon–Weiner function is its popularity. In dealing with species diversity it seems to have become standard practice to use this measure.

Many, however, contend that it is a useless measure. They usually argue that the concept of species diversity includes two interrelated components—species richness and evenness—and that it is absurd to try to combine them in any meaningful way. Rather, they argue, we should measure either or both but realize that they are separate concepts.

To this end we now derive a measure of evenness. Consider what H would be if all individuals were apportioned equally among the species in the community. If n_i is the number of individuals in the ith species, N is the total number of individuals in the community, and S is the total number of species in the community, we estimate p_i as n_i/N to obtain

$$H = -\sum p_i \ln p_i$$

But if all the species in the community have an equal number of individuals, $n_i = N/S$ for all n_i. Therefore

$$\begin{aligned} H_{\max} &= -\sum \frac{N/S}{N} \ln \frac{N/S}{N} \\ &= -\sum \frac{1}{S} \ln \frac{1}{S} \\ &= \ln S \end{aligned}$$

Thus we may define evenness as

$$J = \frac{H}{H_{\max}}$$

We now have two measurements that can be used to characterize species diversity: S, the total number of species (usually called species richness), and J, the evenness.

At this point we note that in going from the lognormal distribution to the measurement of species diversity we originally went from a two-parameter characterization (S_0 and σ) to a one-parameter characterization (H). With the added notion that species richness and evenness are separate concepts we are back to a two-parameter characterization. Indeed, the study of species diversity with the lognormal distribution is not very different from the study of species diversity in the Shannon–Weiner approach. As we demonstrate below, J from the Shannon–Weiner function and σ from the lognormal distribution seem to be measuring more or less the same thing.

□ EXERCISES

24 Compute evenness for all the examples in exercise 22.

25 Compute evenness for all the examples in exercise 23.

26 Compute J for the data from exercises 1 and 2.

27 For the following two communities compute σ (approximately) and J:

(a) 1, 1, 1, 1, 2, 2, 3, 3, 4, 4, 5, 6, 6, 7, 9, 11, 12, 13, 15, 17, 19, 21, 23, 25, 28, 35, 40, 45, 50, 55, 64, 64, 70, 80, 90, 110, 130, 150, 190, 240, 300, 400, 500, 600, 900, 1500.

(b) 1, 1, 2, 2, 3, 3, 3, 3, 3, 3, 5, 5, 5, 6, 6, 7, 7, 10, 14, 20.

28 Plot J against σ^2. Plot ($\ln S_0$) J against σ^2. (Use data from exercises 26 and 27.) □

In exercise 28 we can see that σ from the lognormal distribution is clearly related to J, the evenness from the Shannon–Weiner function. In fact, ($\ln S_0$) J is approximately linearly related to σ^2 for most practical purposes. Then $\sigma^2/\ln S_0$ is at least as good a measure of evenness as J.

THE RELATIONSHIP BETWEEN SPECIES NUMBER AND AREA. We began the discussion of the lognormal distribution with a consideration of sampling problems, noting that if a lognormally distributed community were randomly sampled, a truncated lognormal would be obtained. We now pursue further some aspects of this sampling problem. I note in passing that certain statistical problems are associated with the use of H in a community sample. Pielou (1966b) discusses these problems thoroughly; they are not repeated here.

We have already seen (by example) that doubling the sample size of a community shifts the observed part of the lognormal curve one octave (see exercises 5, 6, and 7), which means that the veil line moves to the left one octave for each doubling in sample size. We assume that doubling the sample size means doubling the area sampled. What then is the relationship between number of species and area?

□ EXERCISES

29 For the following lognormal distribution (Figure 9.2) suppose that an initial sample of one acre positions the veil line at position P. Consider what will happen as the sample size is doubled repeatedly. Plot the number of species against the area.

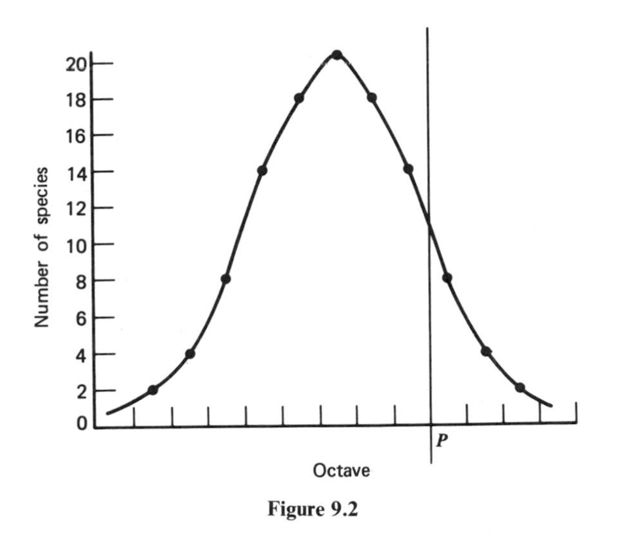

Figure 9.2

30 Plot the number of species against the log of the area for the data in exercise 29.

31 Plot ln S against ln A for the data of exercise 29.

32 Plot ln S against ln A for the following lognormal distribution (Figure 9.3) (for $A = 1$ suppose that the veil line is at P). Indicate on the graph what area represents the inclusion of the modal octave. If the sampling scheme included only those areas up to and including the modal octave, what might you conclude? □

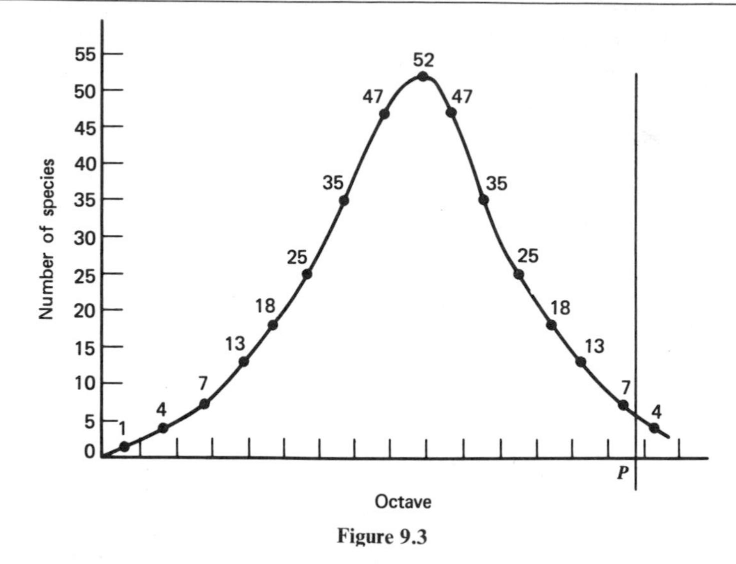

Figure 9.3

Thus the relationship between species richness and area can be interpreted as a sampling problem. In Chapter 10 we treat the problem in a more dynamic way (not necessarily more correct), but for now let us consider the species–area (or species–sampling unit) relationship as a sampling problem.

In exercises 29, 30, and 32 we see that a positive relationship generally occurs between number of species and area and that the increase in species number with increase in area decreases until the number of species asymptotes. This is a useful concept in sampling communities. With an incomplete sample one can estimate the true number of species in the community by a simple extrapolation. This procedure would be impossible if only a small fraction of the community had been sampled because the part of the curve that begins to approach the asymptote is not visible.

From exercise 32 we see that a sample that does not include the mode of the lognormal distribution appears to be linear in $\ln S$ versus $\ln A$. Indeed, this linear relationship is frequently observed when island biotas are studied. The log of the number of species on an island seems to be linearly related to the log of the area of the island. This common observation could be interpreted as a problem of incomplete sampling from a lognormal distribution. Another more popular interpretation seems to be in vogue right now. We begin Chapter 10 with an analysis of this other interpretation.

ANSWERS TO EXERCISES

1

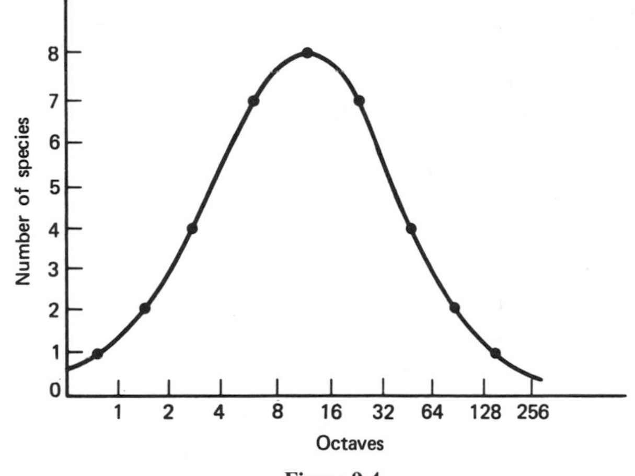

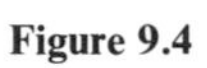

Figure 9.4

2

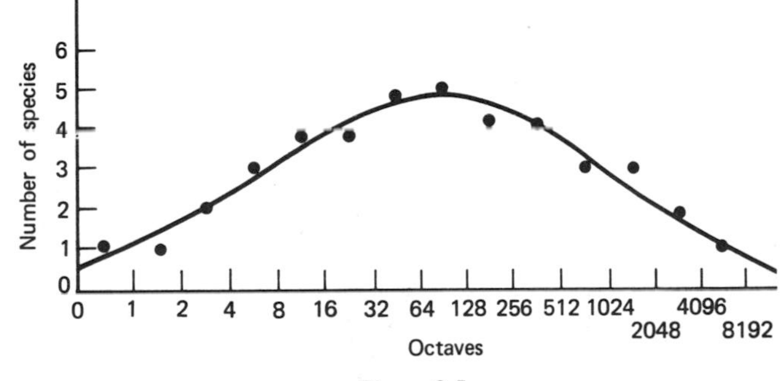

Figure 9.5

3

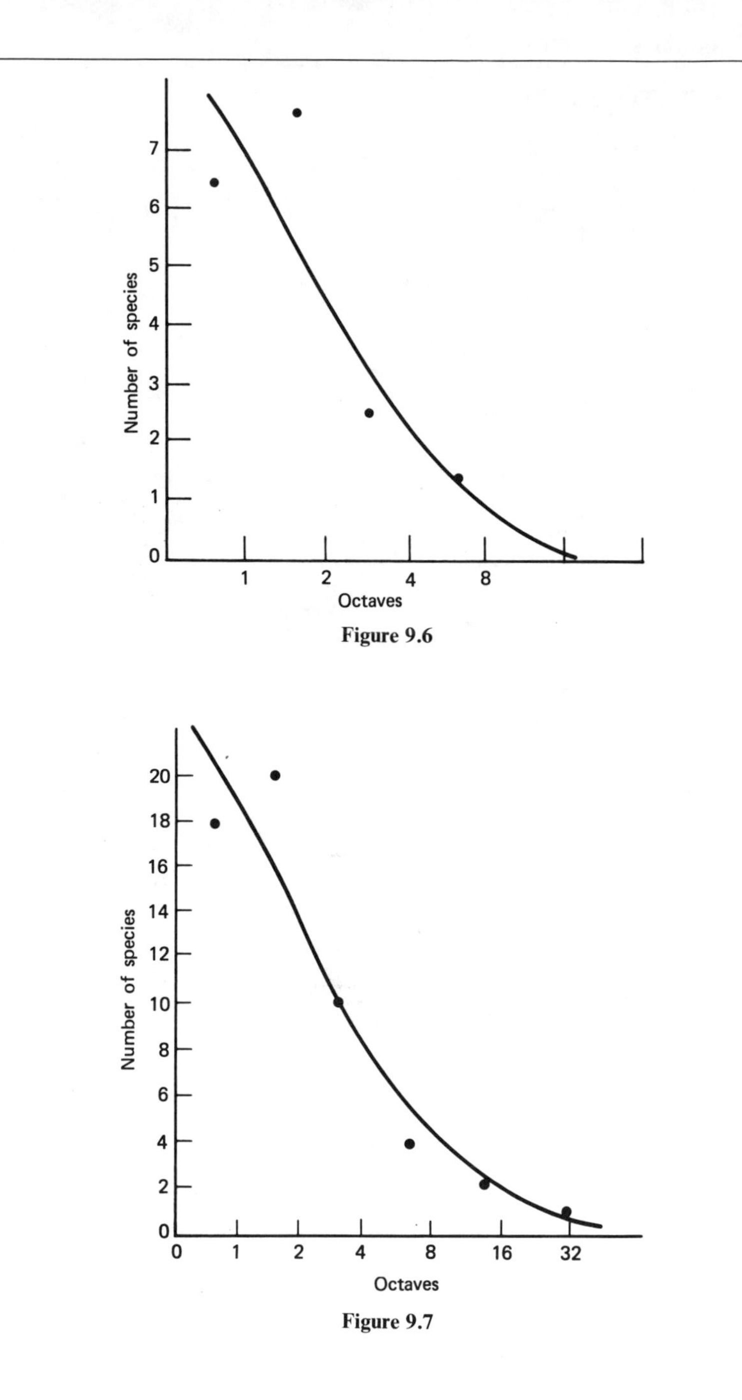

Figure 9.6

4

Figure 9.7

5 The number of individuals per species is 1, 1, 1, 1, 2, 2, 2, 3, 3, 3, 3, 3, 4, 4, 5, 5, 5, 6, 6, 6, 7, 8, 10, 11, 12, 13, 14, 15, 17, 25, 27, 30, 47, 55, 100 (Figure 9.8):

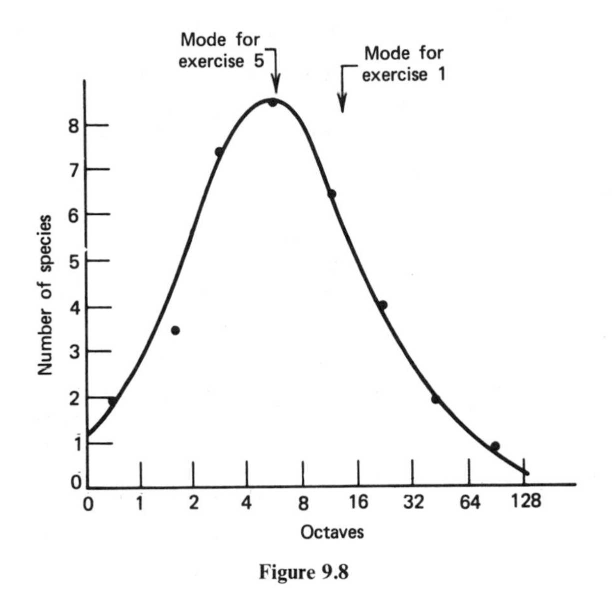

Figure 9.8

6 The number of individuals per species is 1, 1, 1, 1, 1, 1, 1, 2, 2, 2, 2, 2, 2, 3, 3, 3, 3, 3, 4, 4, 5, 5, 6, 7, 7, 7, 9, 12, 14, 15, 24, 27, 50 (Figure 9.9):

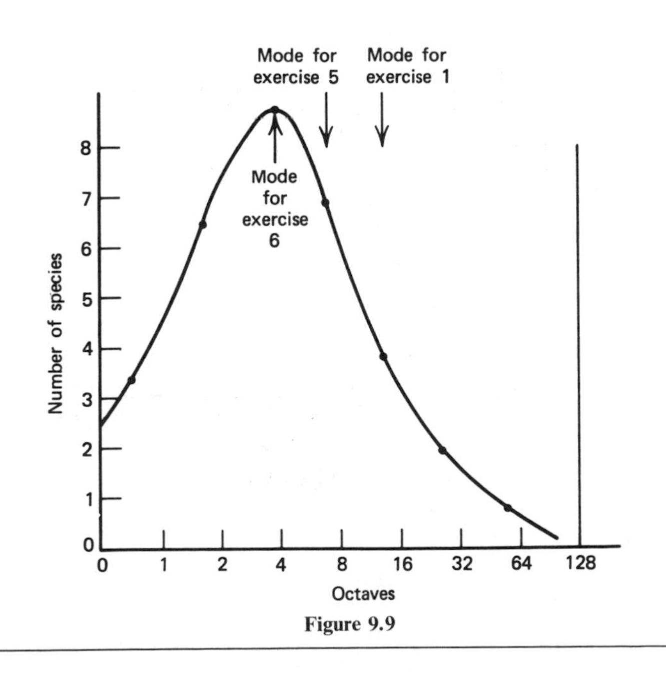

Figure 9.9

7 The number of individuals per species is 1, 1, 1, 1, 1, 1, 1, 1, 1, 1, 1, 1, 2, 2, 2, 2, 2, 2, 3, 3, 3, 4, 4, 4, 6, 7, 8, 12, 13, 25 (Figure 9.10):

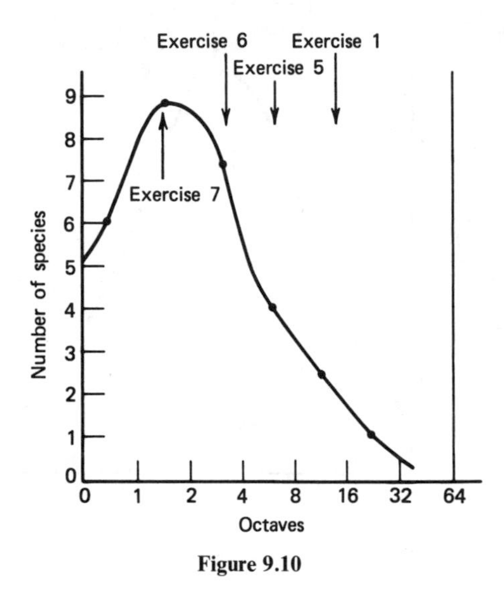

Figure 9.10

8

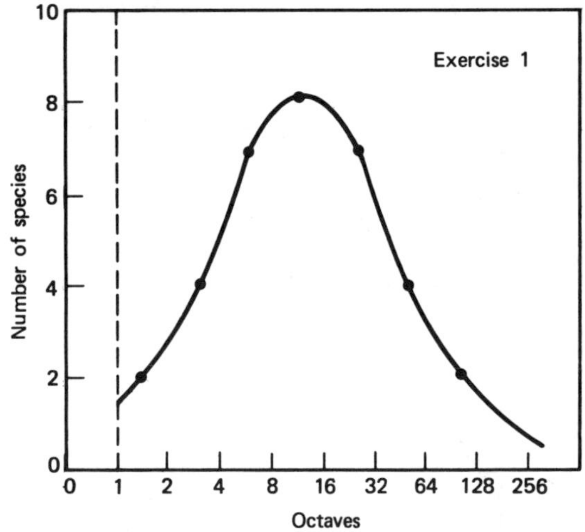

Figure 9.11

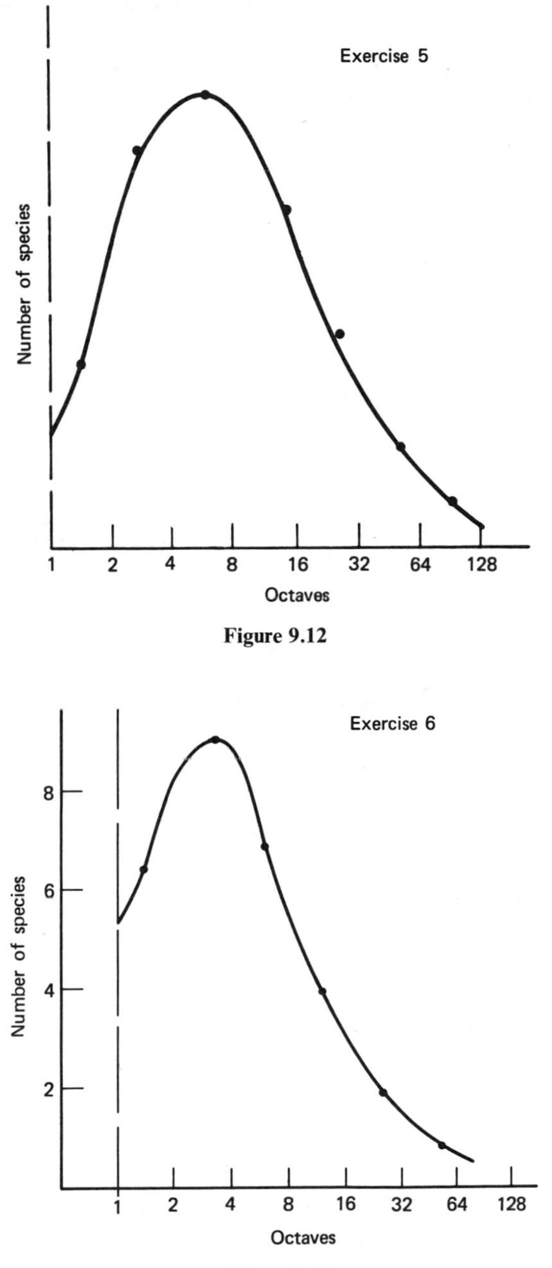

Figure 9.12

Figure 9.13

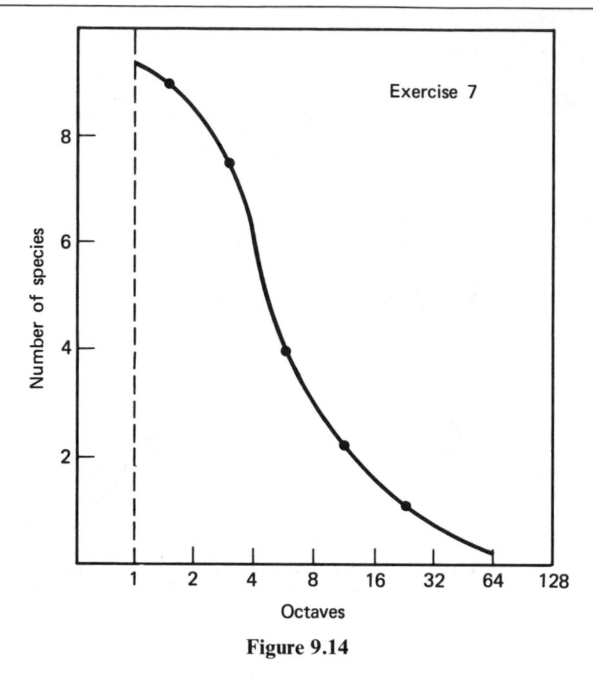

Figure 9.14

9 The number of individuals per species is, 1, 1, 1, 1, 1, 1, 1, 1, 1, 1, 1, 1, 1, 2, 2, 2, 2, 2, 3, 4, 4, 6, 7, 13 (Figure 9.15):

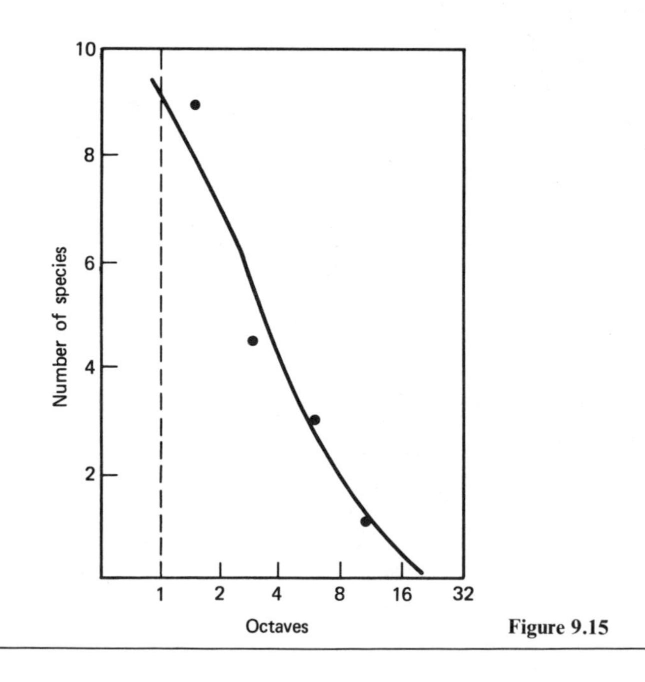

Figure 9.15

10 The number of individuals per species is, 1, 1, 1, 1, 1, 1, 1, 1, 1, 1, 1, 1, 1, 2, 2, 3, 4, 6 (Figure 9.16):

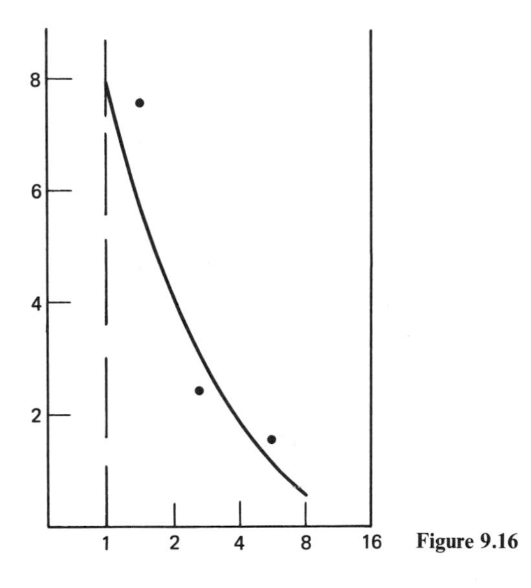

Figure 9.16

11 Construct the following table ($a^2 = 0.04$, $Y_0 = 8$):

X	X^2	a^2X^2	$e^{(aX)^2}$	$Y_0 e^{-(aX)^2}$
1	1	0.04	0.96	7.69
2	4	0.16	0.85	6.82
3	9	0.36	0.70	5.58
4	16	0.64	0.53	4.22
5	25	1.00	0.37	2.94
6	36	1.44	0.24	1.90
7	49	1.96	0.14	1.13
8	64	2.56	0.08	0.64
9	81	3.24	0.04	0.31

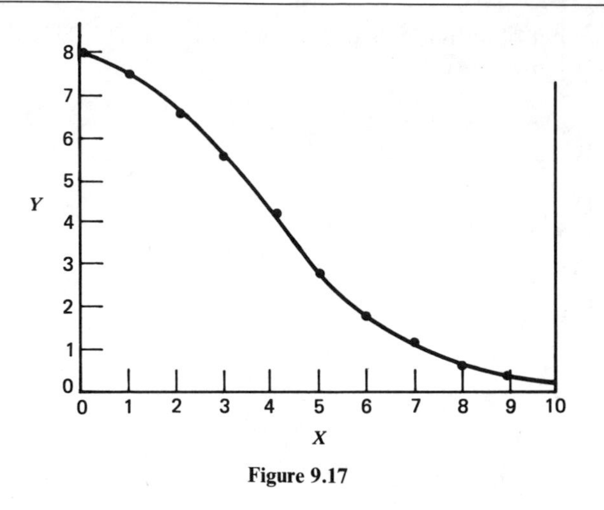

Figure 9.17

12 We already have the values for X^2 and Y both from exercise 11. We need only to find ln Y.

X^2	Y	ln Y
1	7.69	2.04
4	6.82	1.92
9	5.58	1.72
16	4.22	1.44
25	2.94	1.08
36	1.90	0.64
49	1.13	0.12
64	0.64	−0.45
81	0.31	−1.17

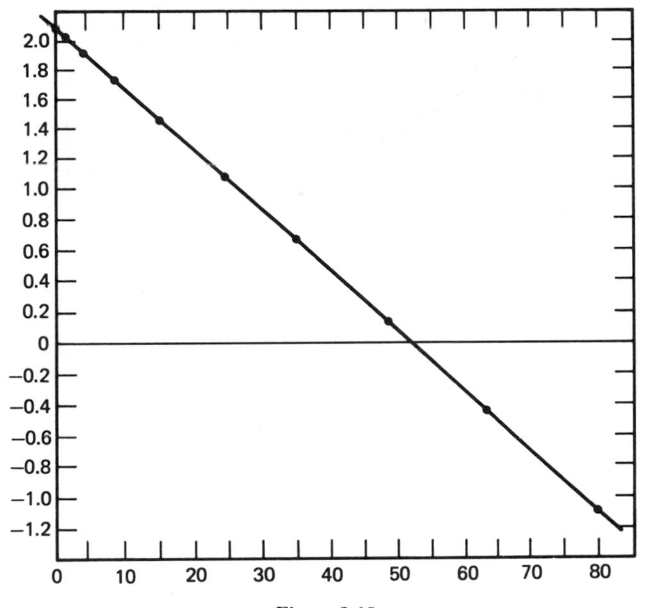

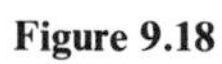
Figure 9.18

13

Octave	R	R^2	$S(R)$	$\ln S(r)$
1	4	16	1	0
2	3	9	2	0.69
3	2	4	4	1.39
3	1	1	7	1.95
5	0	0	8	2.08
6	1	1	7	1.95
7	2	4	4	1.39
3	3	9	2	0.69
9	4	16	1	0

$$a^2 = \frac{1.8}{12.5} = 0.14$$

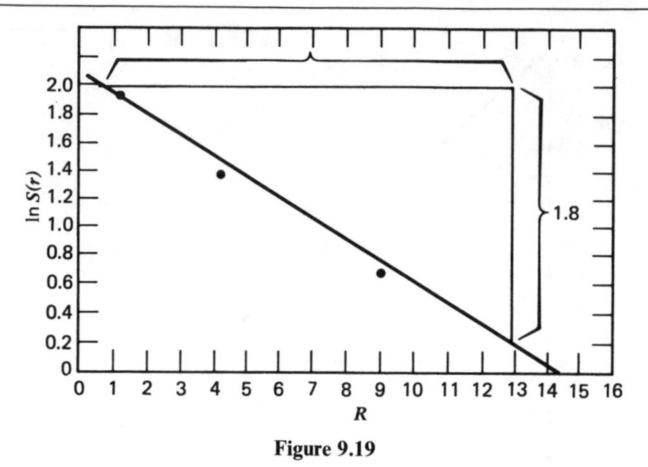

Figure 9.19

Because

$$a = \frac{1}{(\sigma\sqrt{2})},$$

$$\sigma\sqrt{2} = \frac{1}{a}$$

$$\sigma = \frac{1}{(a\sqrt{2})}$$

$$a^2 = 0.14 \qquad a = 0.37 \qquad \sigma = \frac{1}{(0.37)(1.41)} = 1.91$$

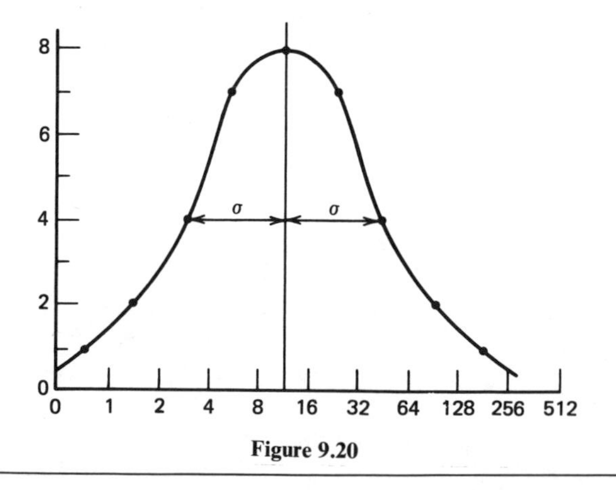

Figure 9.20

14

Octave	R	R^2	$S(R)$	$\ln S(R)$
1	6.5	42.25	1	0
2	5.5	30.25	1	0
3	4.5	20.25	2	0.69
4	3.5	12.25	3	1.10
5	2.5	6.25	4	1.39
6	1.5	2.25	4	1.39
7	0.5	0.25	5	1.61
8	0.5	0.25	5	1.61
9	1.5	2.25	4	1.39
10	2.5	6.25	4	1.39
11	3.5	12.25	3	1.10
12	4.5	20.25	3	1.10
13	5.5	30.25	2	0.69
14	6.5	42.25	1	0

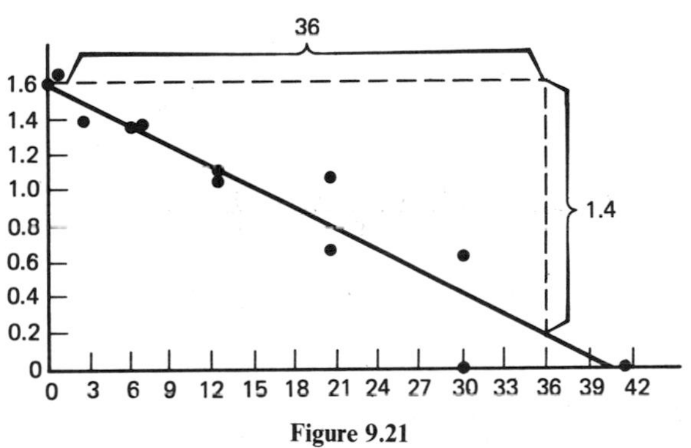

Figure 9.21

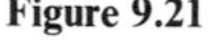

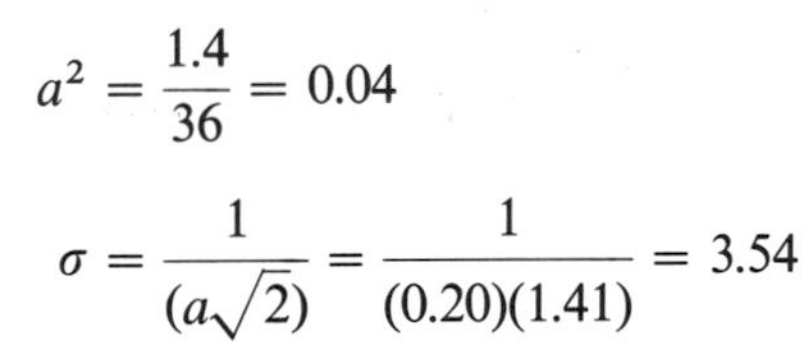

$$a^2 = \frac{1.4}{36} = 0.04$$

$$\sigma = \frac{1}{(a\sqrt{2})} = \frac{1}{(0.20)(1.41)} = 3.54$$

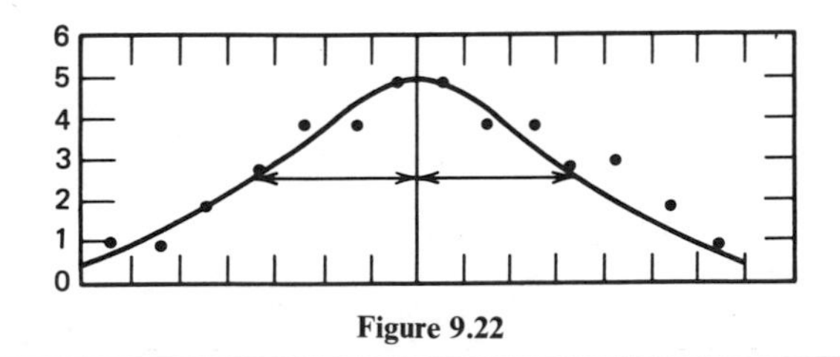

Figure 9.22

15

Octave	R	R^2	$S(R)$	$\ln S(R)$
1	2	4	3.5	1.25
2	1	1	6.5	1.87
3	0	0	9	2.20
4	1	1	7	1.95
5	2	4	4	1.39
6	3	9	2	0.69
7	4	16	1	0

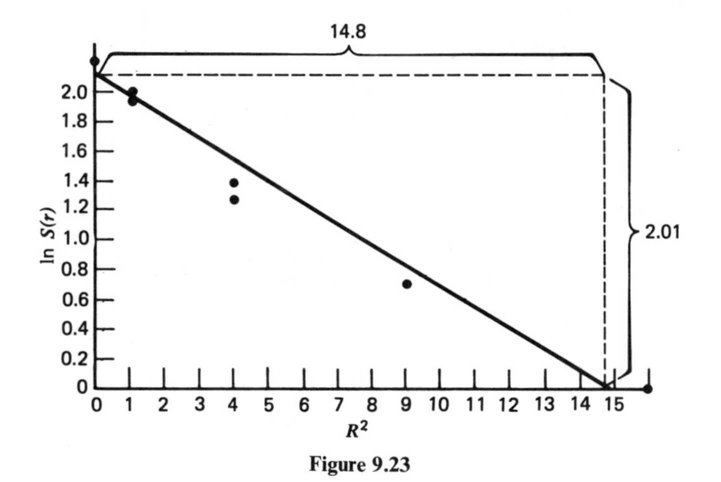

Figure 9.23

$$a^2 = \frac{2.01}{14.8} = 0.14 \qquad a = 0.37$$

$$\sigma = \frac{1}{a\sqrt{2}} = \frac{1}{(0.37)(1.41)} = 1.91$$

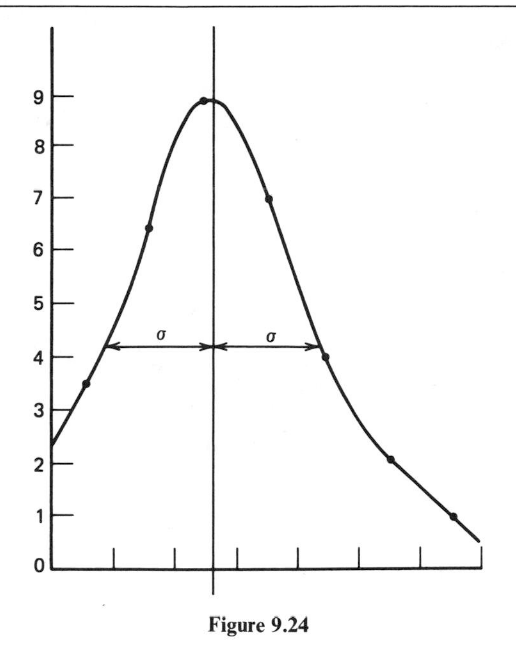

Figure 9.24

As expected, because these data were really a random sample from the data in exercise 1, the value of σ and/or a is the same as in exercise 13.

16

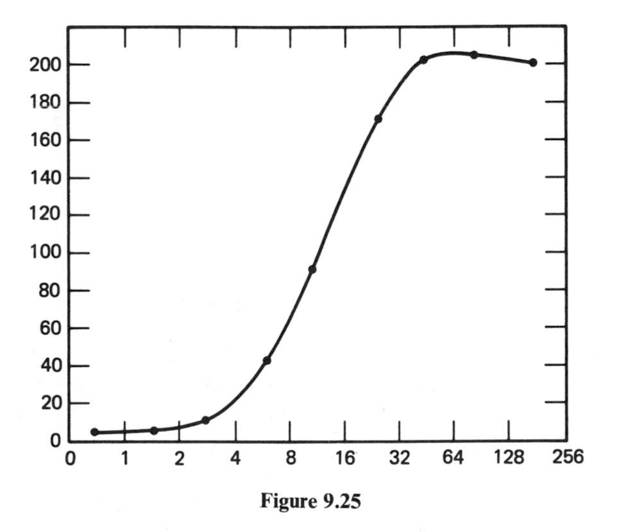

Figure 9.25

17

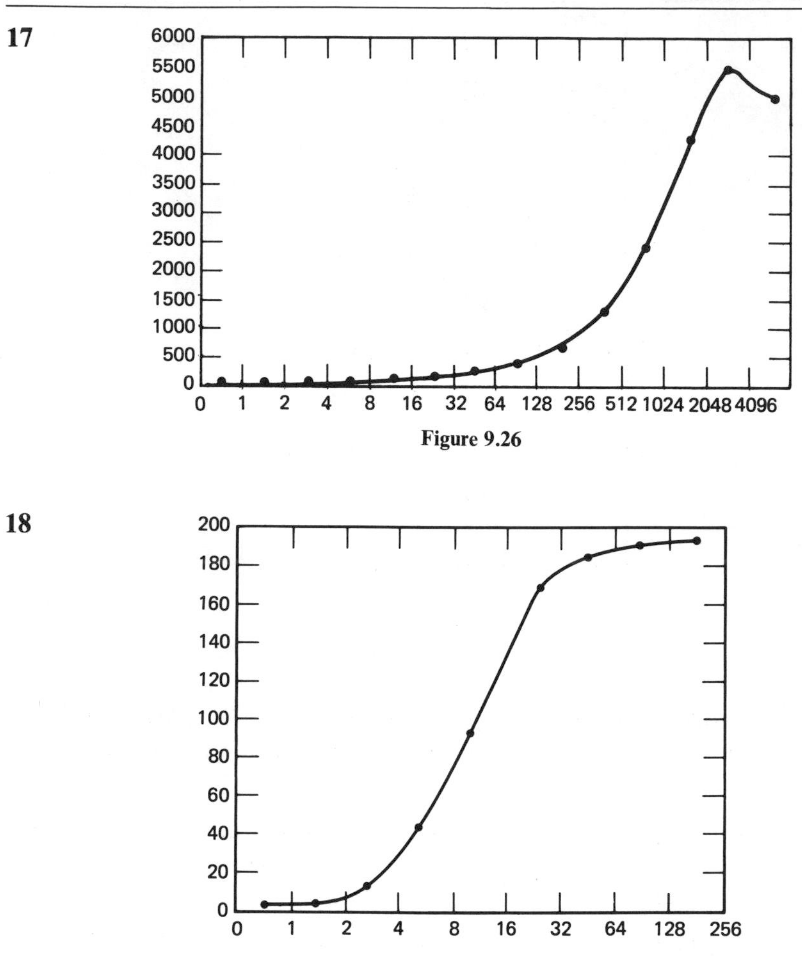

Figure 9.26

18

Figure 9.27

19 The computing equation is

$$S_T = (\sqrt{2\pi})S_0\sigma = [(3.1416)(2)]^{1/2}(8)(1.91) = 38$$

From exercise 1 the total number of individuals is 918. The values of p_i are 0.001, 0.001, 0.002, 0.002, 0.003, 0.003, 0.003, 0.005, 0.005, 0.006, 0.006, 0.006, 0.008, 0.008, 0.010, 0.010, 0.011, 0.011, 0.012, 0.013, 0.014, 0.015, 0.021, 0.022, 0.025, 0.026, 0.027, 0.030, 0.032, 0.038, 0.054, 0.060, 0.065, 0.100, 0.120, and 0.218.

$$-\sum p_i \ln p_i = 2.8471$$

20

$$S_T = 43$$

$$-\sum p_i \ln p_i = 2.866$$

21 $$-\sum p_i \ln p_i = 2.656$$

22 (a) 2.303
(b) 0.500
(c) 2.171
(d) 2.113
(e) 1.808

23 (a) 1.609
(b) 0.223
(c) 1.505
(d) 1.429
(e) 0.840

24 (a) $S = 10 \quad H_{\max} = -\ln\left(\frac{1}{S}\right) = \ln(S) = \ln(10) = 2.303$

From exercise 19 $H = 2.303$; therefore $J = 1.00$.

(b) $J = 0.217$
(c) 0.943
(d) 0.918
(e) 0.785

25 (a) $S = 5 \qquad H_{\max} = \ln(5) = 1.609 \qquad J = \frac{1.609}{1.609} = 1.00$

(b) 0.139
(c) 0.935
(d) 0.888
(e) 0.522

26 For exercise 1 from exercise 16, $H = 2.8471$. $H_{\max} = -\ln\left(\frac{1}{36}\right) = \ln(36) =$ 3.584.

$$J = \frac{2.8471}{3.584} = 0.794$$

For exercise 2 $H = 2.503$. $H_{\max} = \ln(42) = 3.738$, $J = 0.670$.

27

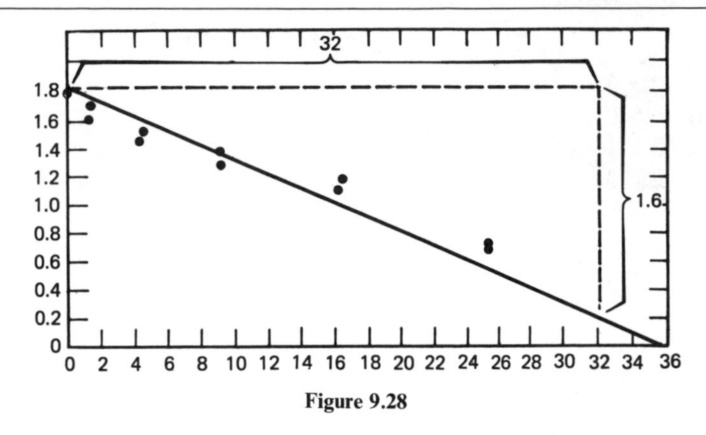

Figure 9.28

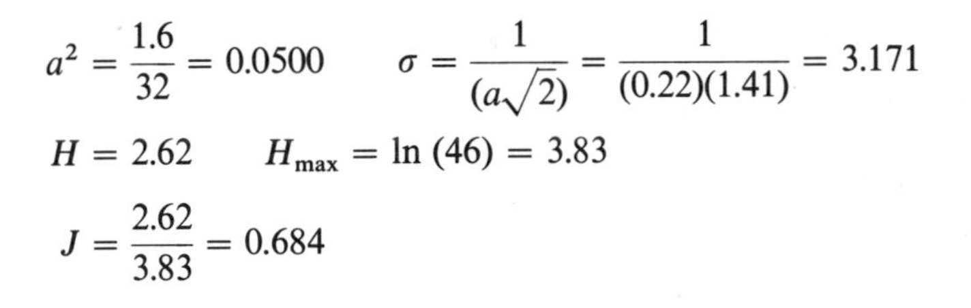

$$a^2 = \frac{1.6}{32} = 0.0500 \qquad \sigma = \frac{1}{(a\sqrt{2})} = \frac{1}{(0.22)(1.41)} = 3.171$$

$$H = 2.62 \qquad H_{\max} = \ln(46) = 3.83$$

$$J = \frac{2.62}{3.83} = 0.684$$

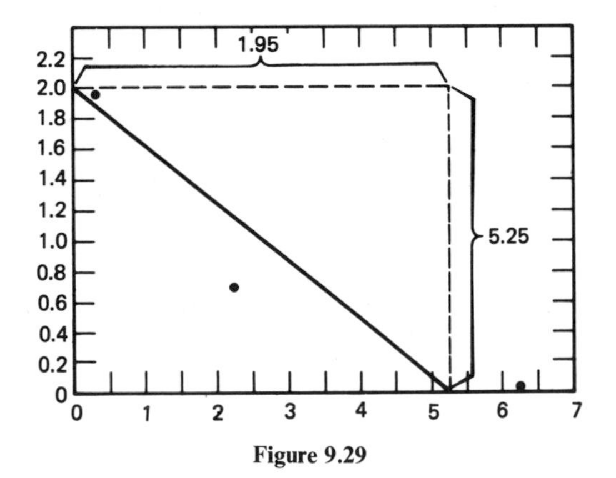

Figure 9.29

$$a^2 = \frac{1.95}{5.25} = 0.371 \qquad \sigma = \frac{1}{(a\sqrt{2})} = 1.164 \qquad H = 2.72 \qquad H_{\max} = 2.996$$

$$J = 0.91$$

28

Exercise	σ^2	J	S_0	$\ln S_0$	$(\ln S_0)J$
1	3.684	0.794	8	2.08	1.65
2	12.532	0.670	5	1.61	1.08
27a	10.055	0.684	6	1.79	1.22
27b	1.355	0.910	8	2.08	1.89

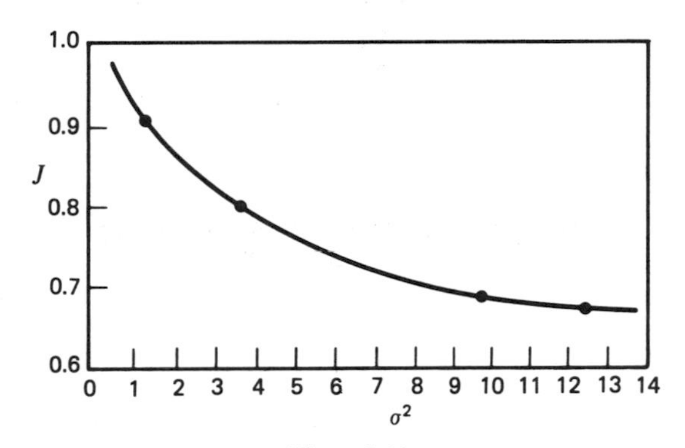

Figure 9.30

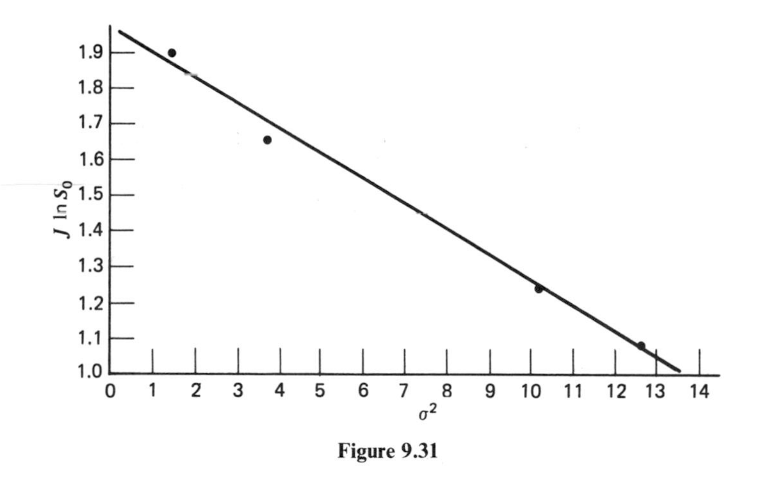

Figure 9.31

29 Clearly, when the veil line is at position P, the area is one acre and $S = 14$. When the sample size is doubled, the area becomes two acres and the species number becomes 28 because doubling the sample size moves the veil line one octave to the left. In this way we generate the following table:

Area	S
1	14
2	28
4	46
8	66
16	84
32	98
64	106
128	110
256	112

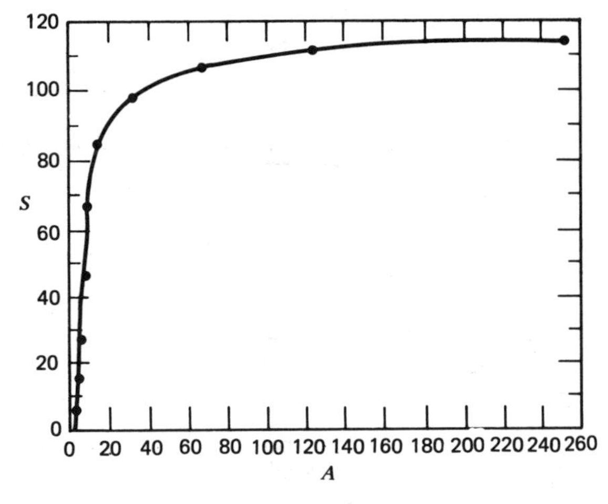

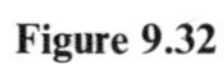
Figure 9.32

30

ln A	S
0	14
0.69	28
1.39	46
2.08	66
2.77	84
3.47	98
4.16	106
4.85	110
5.55	112

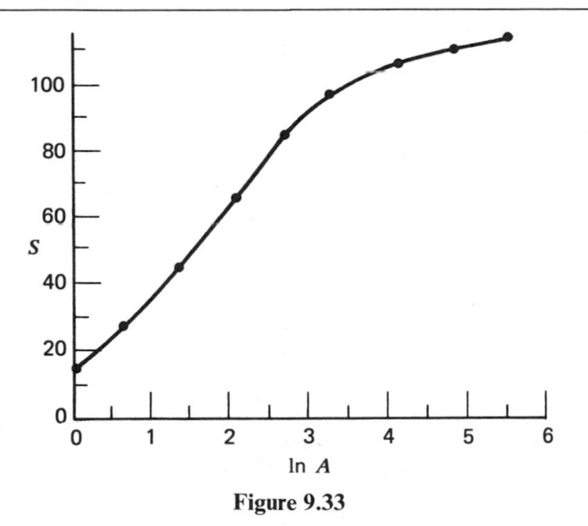

Figure 9.33

31

$\ln A$	$\ln S$
0	2.64
0.69	3.33
1.39	3.83
2.08	4.19
2.77	4.43
3.47	4.58
4.16	4.66
4.85	4.70
5.55	4.72

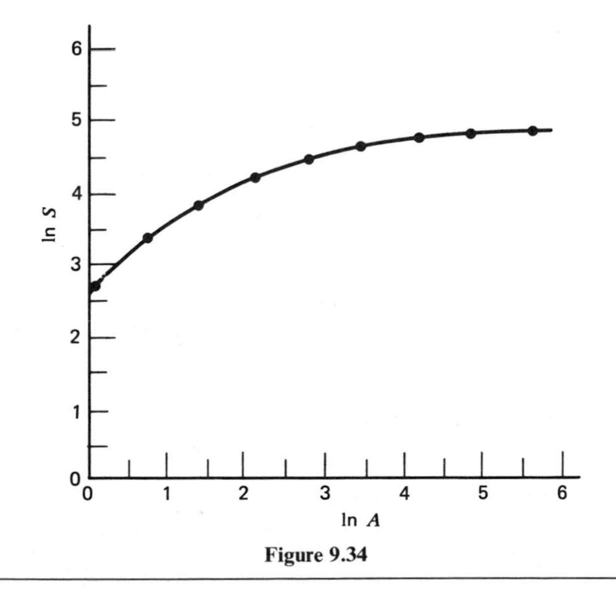

Figure 9.34

32

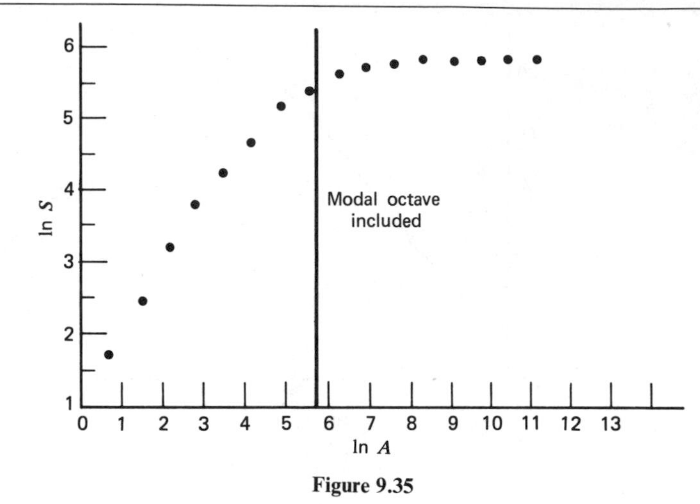

Figure 9.35

If the sampling included only up to the modal octave, we might be tempted to conclude that ln S and ln A were approximately related.

REFERENCES

Bazzaz, F. A. 1975. Plant species diversity in old-field successional ecosystems in southern Illinois. *Ecology* **56**:485–488.

Connell, J. H. 1972. Community interactions on marine rocky intertidal shores. *Annu. Rev. Ecol. Syst.* **3**:169–192.

Diamond, J. M. and E. Mayr. 1976. The species-area relation for birds of the Solomon Archipelago. *Proc. Nat. Acad. Sci. USA* **73**:262–266.

Fager, E. W. 1968. The community of invertebrates in decaying oak wood. *J. Anim. Ecol.* **37**:121–142.

Fisher, R. A., A. S. Corbet, and C. B. Williams. 1943. The relation between the number of species and the number of individuals in a random sample of an animal population. *J. Anim. Ecol.* **12**:42–58.

Hairston, N. G. 1969. Species abundance and community organization. *Ecology* **40**:404–416.

Hairston, N. G. and G. W. Byers. 1954. The soil arthropods of a field in southern Michigan: a study in community ecology. *Contrib. Lab. Vertebr. Biol. Univ. Mich.* **64**:1–37.

Hairston, H. G., J. D. Allan, R. K. Colwell, D. J. Futuyma, J. Howell, M. D. Lubin, J. Mathias, and J. H. Vandermeer. 1968. The relationship between species diversity and stability: an experimental approach with protozoa and bacteria. *Ecology* **49**:1091–1101.

Hutcheson, K. 1970. A test for comparing diversities based on the Shannon formula. *J. Theor. Biol.* **29**:151–154.

Hutchinson, G. E. 1959. Homage to Santa Rosalia, or why are there so many kinds of animals? *Am. Nat.* **93**:145–159.

Hutchinson, G. E. 1961. The paradox of the plankton. *Am. Nat.* **95**:137–145.

Janzen, D. J. 1973. Sweep samples of tropical foliage insects: effects of seasons vegetation types, elevation, time of day, and insularity. *Ecology* **54**:687–708.

Kilburn, P. D. 1966. Analysis of the species-area relation, *Ecology* **47**:831–843.

Lloyd, M., J. H. Zar, and J. R. Karr. 1968. On the calculation of information theoretical measures of diversity. *Am. Midl. Nat.* **79**:257–272.

May, R. M. 1975. Patterns of species abundance and diversity. In M. L. Cody and J. M. Diamond, Eds. *Ecology and Evolution of Communities*. Cambridge: Harvard University Press, pp. 81–120.

Patrick, R. 1973. Use of algae, especially diatoms, in the assessment of water quality. *ASTM, Spec. Tech. Publ.* **528**:76–95.

Patrick, R. 1975. Structure of stream communities. In M. L. Cody and J. M. Diamond, Eds. *Ecology and Evolution of Communities*. Cambridge: Harvard University Press, pp. 445–459.

Pielou, E. C. 1966a. Species-diversity and pattern-diversity in the study of ecological succession. *J. Theor. Biol.* **10**:370–383.

Pielou, E. C. 1966b. The measurement of diversity in different types of biological collections. *J. Theor. Biol.* **13**:131–144.

Preston, F. W. 1948. The commonness, and rarity, of species. *Ecology* **29**:254–283.

Preston, F. W. 1962. The canonical distribution of commonness and rarity. *Ecology* **43**:185–215 and 410–432.

Recher, H. F. 1969. Bird species diversity and habitat diversity in Australia and North America. *Am. Nat.* **103**:75–80.

Sanders, H. L. 1968. Marine benthic diversity: a comparative study. *Am. Nat.*, **102**:243–282.

Schoener, T. W. 1976. The species-area relation within archipelagos: models and evidence from island land birds. *Proc. 16th Int. Ornithol. Cong.* Canberra, Australia, August 1974.

Stout J. and J. Vandermeer. 1975. Comparison of species richness for stream-inhabiting insects in tropical and mid-latitude streams. *Am. Nat.* **109**:263–280.

10. Dynamics of Multiple Species Assemblages

THE EQUILIBRIUM THEORY OF ISLAND BIOGEOGRAPHY. As noted in Chapter 9, one of the commonest observations of community ecologists concerns the positive relationship between number of species and area, whether we are dealing with sampling or island areas. It has frequently been noted that this relationship is approximately linear when plotted on a log–log plot. Another common, albeit less ubiquitous, observation is that the number of species on an island decreases as the island is located farther from the source area. We now develop a theoretical framework that accounts for these two common observations elegantly. The theory is attributable to MacArthur and Wilson (1963).

Consider an island totally denuded of species. If we watch that island for a day (say), how many species of insects will fly or be blown to it? Suppose 10 species arrive. If we watch the same island for a second day, how many species will arrive? Again, suppose 10, but on that second day, some of the arrivals will already be present from the first day's flux of immigrants. If seven of the species arriving on the second day were already on the island as a result of the first day's immigrants, only three new species will have arrived on the second day.

☐ EXERCISES

1 Suppose that we begin with a totally denuded island and that each day 20 species arrive. On the second day 10 of the 20 arrivals were already on the island. On the third day 15 of the 20 were already on the island. On the fourth day 17 of the 20 were already on the island. On the fifth day 18 of the 20 were already on the island. On the sixth day 19 of the 20 were already on the island. Graph the number of new arrivals (Y axis) against the number of species already on the island (X axis) and assume that all arrivals remain.

2 The island in exercise 1 was 5 miles from the mainland source. Suppose we consider a second island, 10 miles from the mainland. Because this island is farther from the source of arrivals, we must expect fewer arrivals per day. Suppose that only one-half the number arriving on the first island arrive on the second island. If the fraction of arrivals already present is the same as in exercise 1, graph the number of new arrivals (Y axis) against the number of species already on the island (X axis). Again, assume that all arrivals stay on the island. Plot this graph on the plot used in exercise 1.

3 Consider a group of 10 islands all at exactly the same distance from the mainland source. All the islands are exactly the same except for age. Some are young and have not been colonized by many species; others are old and have been colonized by many species. Suppose that 30 species from the mainland arrive at each of the islands every year. Suppose, in addition, that all the species on the island are included in the group of species arriving; that is, if there are 20 species already on the island, only 10 of the 30 arriving will be new arrivals. If, in some particular year, the 10 islands contain 3, 8, 10, 13, 16, 19, 21, 24, 27, and 29 species, graph the number of new arrivals against the number already there for that year. □

We see from these exercises that, in general, a graph of the number of new species arriving on an island at some particular time, when graphed against the number already on the island, is monotonically decreasing; that is, the rate of arrival of new species, or simply the *immigration rate*, is a monotonically decreasing function of the number of species present on the island. We call that function the immigration function.

We also see, as in exercise 2, that the immigration function changes in a regular way, depending on the distance of the island from the mainland. The immigration function has a smaller slope and smaller intercept for islands that are far away from the mainland than for nearby islands. The general behavior of the immigration function is shown in Figure 10.1:

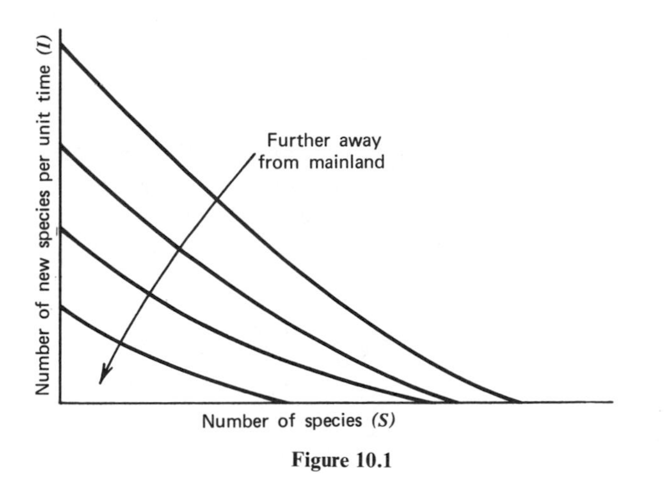

Figure 10.1

We must now examine the extinction rate. How many species become extinct in a unit time interval?

□ EXERCISES

4 Suppose that an island that originally contained 50 species is artificially isolated and no further immigration from the mainland is possible. If we keep track of the island for eight days, we will find that on the first day 20 species will become extinct; on the second day, 12 species; on the third day, 7 species; on the fourth day, 4 species; on the fifth day, 2 species; and on the sixth, seventh, and eighth days, 1 species will become extinct on each day. Graph the number of species that will become extinct per day against the number of species on the island.

5 In exercise 4 suppose that exactly 40% of the existing species died each day. Graph the result on the graph used in exercise 4.

6 Suppose that the island in exercise 5 had been 1/10th the size, or 1/100th the size. It is reasonable to suppose that if the island were significantly smaller, the species would become extinct more rapidly because in general the population sizes would be smaller and habitats less diverse. For the smaller island let us suppose that 60% of all existent species became extinct each day. Graph on the graph used in exercise 5.

7 Suppose we have a series of eight islands, all the same size. Suppose we isolate these islands and observe them for a year. Suppose that the islands contained 50, 48, 45, 40, 30, 20, 15, and 10 species at the beginning of the year and 36, 38, 39, 35, 27, 18, 13, and 9, respectively, at the end of the year. Graph the extinction rate against the number of species.

8 Suppose that the eight islands in exercise 8 were half the size. Suppose that this reduction in size had a concomitant reduction in survival rate. In particular, suppose that the extinction rate were doubled. Plot the result on the graph used in exercise 7. □

Clearly, the *extinction rate*, the number of species that become extinct per unit time, is a monotonically increasing function of the number of species present. That function is called the extinction function. As shown in exercises 6 and 8 the extinction function changes in a regular and predictable way as island areas change. This behavior is described in Figure 10.2:

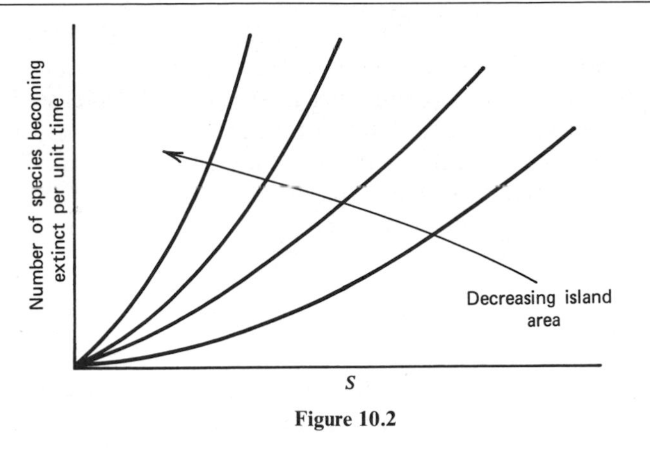

Figure 10.2

Extinction rates increase as island areas decrease.

We have now discussed two processes particularly associated with islands: immigration and extinction. What happens when the two concepts are put together? A graph of rate against number of species is presented in Figure 10.3:

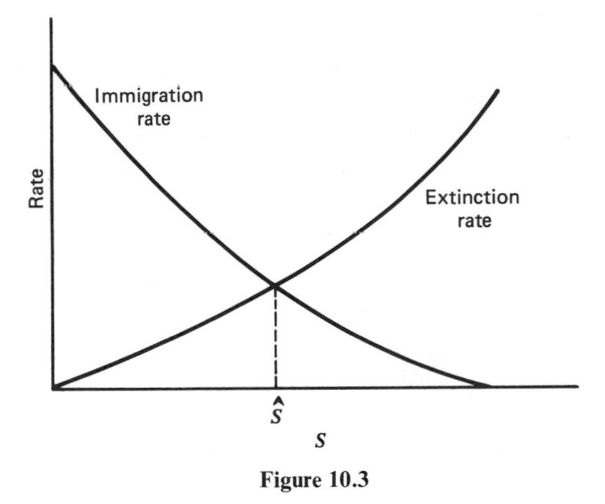

Figure 10.3

The curves must cross because the immigration function is monotonically decreasing and the extinction function is monotonically increasing. If we now concern ourselves with an island in which immigration and extinction occur simultaneously (as they most undoubtedly do in nature), we must ask whether the total number of species on the island is increasing or decreasing for various values of S. Suppose that the value of S is left of the point at which the two functions intersect. If so, the immigration rate will be greater than the extinction rate and the number of species will increase. Suppose that the value of S is right of the intersection of the two functions. If so, then the immigration rate is less than the extinction rate and the number of species will decrease. Suppose the value of S is exactly at $\hat{S}$ (where the two functions intersect). At that point the immigration rate is equal to the extinction rate and the number of species must remain constant. The composition of species can and will change but the number of species will remain constant. Thus the value $\hat{S}$ (see Figure 10.3) is called the equilibrium number of species.

□ EXERCISES

9 Compute the equilibrium number of species graphically for (a) the immigration function of exercise 1 and the extinction function of exercise 5; (b) the immigration function of exercise 1 and the extinction function of exercise 6; (c) the immigration function of exercise 1 and the extinction function of exercise 7. (Do all three on the same graph.)

10 Compute the equilibrium number of species graphically for (a) the immigration function of exercise 1 and the extinction function of exercise 8; (b) the immigration function of exercise 2 and the extinction function of exercise 8; (c) the immigration function of exercise 3 and the extinction function of exercise 8. (Do all three on the same graph.) □

We can most easily examine the biological meanings of some of the features of this model if we make some simplifying assumptions; namely, that the immigration and extinction functions are linear. This is almost certainly not true in nature, but it is convenient for discussion here. Assuming linearity, we obtain Figure 10.4:

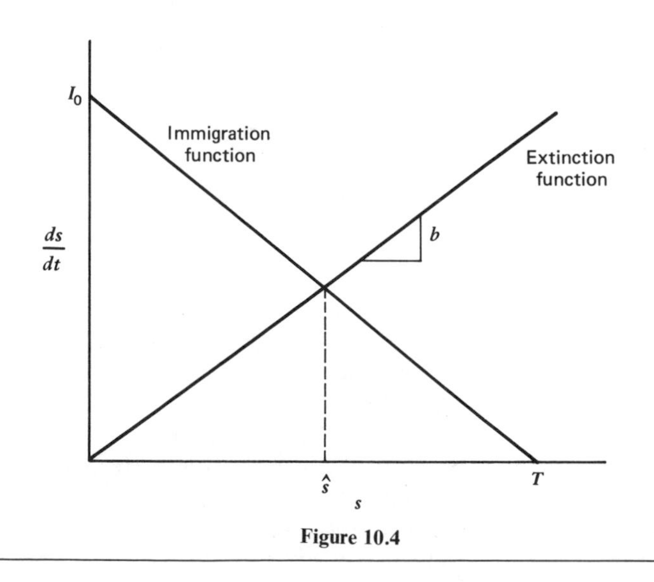

Figure 10.4

Here we have symbolized the rate as dS/dt, I_0 is the rate of immigration to an empty island, T is the total number of species in the species pool (those on the mainland that could migrate to the island), and b is the slope of the extinction function (the amount the extinction rate is increased by the introduction of one species onto the island).

□ EXERCISES

11 Write the equations that describe the immigration and extinction functions in Figure 10.4.

12 Solve for $\hat{S}$. □

THE RELATIONSHIP BETWEEN SPECIES AND AREA AGAIN. We have already said that the extinction function rises as the area of the island becomes smaller. This is equivalent (if the extinction function is linear) to the statement that $b = f(A)$, where $\partial f(A)/\partial A < 0$ and f is some function of area A. What will the function f look like? Qualitatively, it should be such that b is very large when the island is very small and very small when the island is very large; that is, $f(A)$ should look something like Figure 10.5:

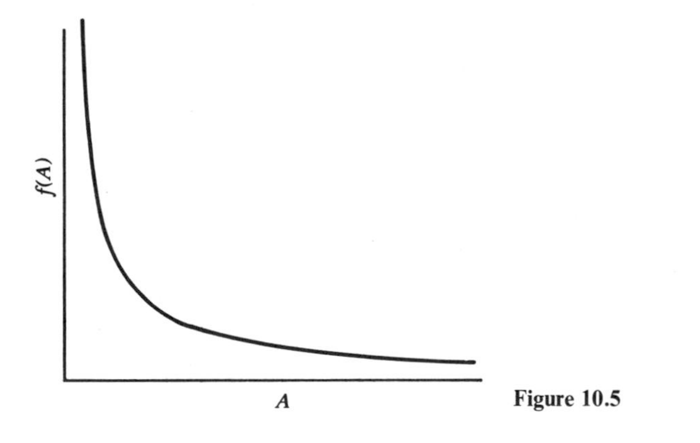

Figure 10.5

Perhaps the simplest function that does resemble Figure 10.5 is $f(A) = A^{-z}$, where z is some constant.

Recalling from exercise 12 that

$$\hat{S} = \frac{I_0}{b + I_0/T}$$

and substituting $b = f(A) = A^{-z}$

$$\hat{S} = \frac{I_0}{A^{-z} + I_0/T}$$

Biologically this function behaves quite appropriately. If A is very large, equation 1 will reduce to $S = T$, that is, when the area of the island approaches that of the mainland, the number of species on the island will equal the entire species pool. If the species pool is large in relation to the immigrant influx, we will obtain

$$\hat{S} = I_0 A^z$$

which is the form most commonly applied to actual island species lists (i.e., the linear relationship on a ln S versus ln A plot).

Therefore, at least qualitatively, we have a theoretical relationship between number of species and area that looks something like Figure 10.6

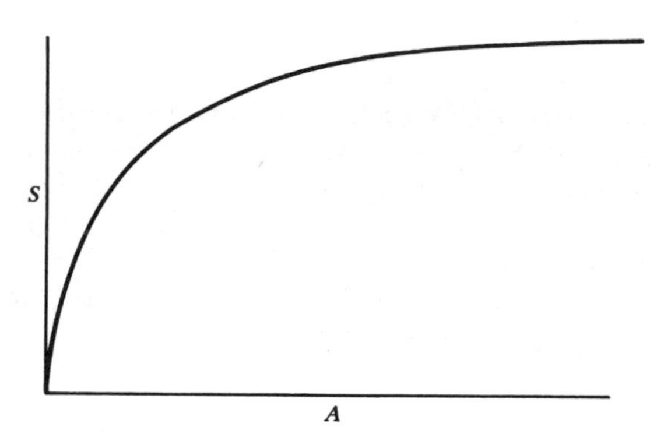

Figure 10.6

or Figure 10.7

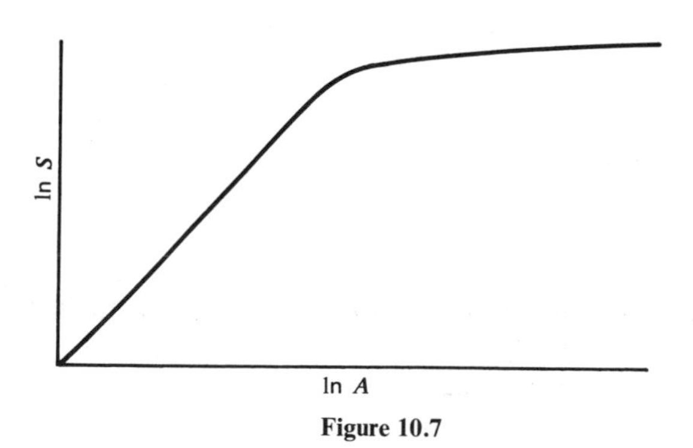

Figure 10.7

You may recall in Chapter 9 that it is possible to derive a direct species area relationship as a problem of sampling from a lognormal distribution. The sampling and island biogeographical approaches gives the same qualitative result, which is the general relationship between S and A.

☐ EXERCISE

13 Suppose that I_0 is linearly related to the distance from the mainland. What is $\hat{S}$ and how is it related to the distance from the mainland? ☐

INTRODUCTION TO POPULATION INTERACTIONS AND COMMUNITY STRUCTURE. Until now we discussed communities (multiple species assemblages, mixed populations) in a very general way. We have tried to examine patterns of diversity, richness, evenness, and so on, and have asked questions about general mechanisms that might explain these patterns. First was a presentation of the lognormal distribution, then the measurement of species diversity, which was followed by a discussion of the relationship between species numbers and areas. Then the equilibrium theory of island biogeography was presented to explain species numbers in terms of a balance between immigration and extinction rates. What we have neglected to show is an analysis of the population level phenomenon that might lead to various patterns.

In fact, some of the experimental results of the now classic studies of Wilson and Simberloff suggest an interplay of approaches such as the equilibrium island biogeography theory and a theory of population interactions. In particular, Wilson has suggested that in an early colonizing phase extinction rates are relatively low because the populations of the newly immigrated species are at low densities. As population densities increase, density-dependent effects, both intraspecific and interspecific, come into play. Competition and predation then act to increase the rate of extinction over what it was when all or most of the populations were at low densities. Graphically, we have Figure 10.8:

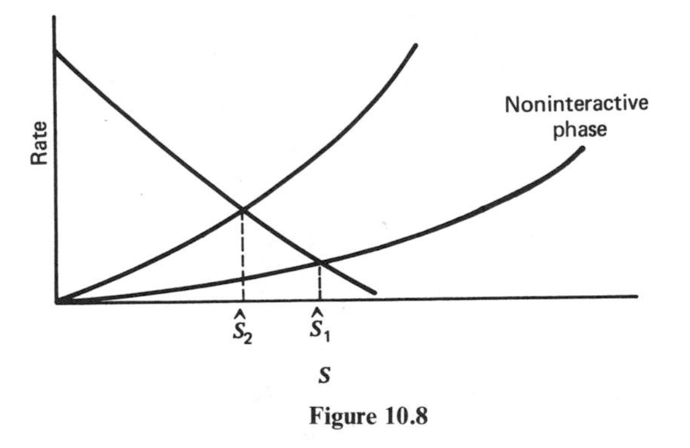

Figure 10.8

If we begin with a totally denuded island, the dynamic process of approaching an equilibrium number of species will act to push the system toward $\hat{S}_1$, but as all the species populations increase in density the system will shift to the interactive phase. Therefore, as a denuded island becomes colonized, we expect first an approach to $\hat{S}_1$, followed by an approach to $\hat{S}_2$. Because $\hat{S}_1 > \hat{S}_2$, we expect an overshoot, an accumulation of species greater than the final equilibrium value. This is exactly what was observed in some of the Wilson and Simberloff experiments.

It makes sense to try to understand community structure not only from the point of view of mathematical distributions or gross properties, such as rates of species immigrations and extinctions, but also from the more mechanistic view of population interactions.

THE COMMUNITY MATRIX. Developed by Levins and later greatly expanded by May, the community matrix shows promise of being one of the more useful theoretical tools for gaining an intuitive understanding of community structure. Its use in practical application to field situations remains to be convincingly demonstrated.

The community matrix was introduced in Chapter 7 (exercises 31 and 33). By way of review, if we have a two-species system,

$$\frac{dN_1}{dt} = r_1 N_1 \left(\frac{K_1 - N_1 - \alpha_{12} N_2}{K_1} \right)$$

$$\frac{dN_2}{dt} = r_2 N_2 \left(\frac{K_2 - N_2 - \alpha_{21} N_1}{K_2} \right)$$

At equilibrium ($dN_i/dt = 0$, $i = 1, 2$) we have

$$K_1 = N_1 + \alpha_{12} N_2$$

$$K_2 = \alpha_{21} N_1 + N_2$$

which we can express in matrix form as

$$K = AN$$

where K is the K-vector, N is the N-vector, and A is the community matrix. Similarly this notion can be extended to m species as

$$\begin{aligned} \frac{dN_1}{dt} &= r_1 N_1 \left(\frac{K_1 - \sum_{i=1}^{m} \alpha_{1i} N_i}{K_1} \right) \\ \frac{dN_2}{dt} &= r_2 N_2 \left(\frac{K_2 - \sum_{i=1}^{m} \alpha_{2i} N_i}{K_2} \right) \\ &\vdots \\ \frac{dN_m}{dt} &= r_m N_m \left(\frac{K_m - \sum_{i=1}^{m} \alpha_{mi} N_i}{K_m} \right) \end{aligned} \tag{1}$$

which at equilibrium is (remember $\alpha_{ii} = 1$)

$$K_1 = \sum_{i=1}^{m} \alpha_{1i} N_i$$

$$K_2 = \sum_{i=1}^{m} \alpha_{2i} N_i$$

$$\vdots$$

$$K_m = \sum_{i=1}^{m} \alpha_{mi} N_i$$

which can be expressed in matrix form as

$$K = AN$$

where K is the K-vector (this time a column of m numbers), N is the N-vector, and A is the community matrix (this time an $m \times m$ matrix). Clearly the community matrix has the following form:

$$\begin{vmatrix} 1 & \alpha_{12} & \alpha_{13} \cdots \alpha_{1m} \\ \alpha_{21} & 1 & \alpha_{23} \cdots \alpha_{2m} \\ \alpha_{31} & \alpha_{32} & 1 \ \cdots \alpha_{3m} \\ \alpha_{m_1} & \cdots & 1 \end{vmatrix}$$

that is, ones on the principal diagonal and competition coefficients everywhere else.

☐ EXERCISES

14 Consider the following set of equations:

$$\frac{dN_i}{N_i dt} = r_i - \sum_{j=1}^{m} a_{ij} N_j$$

for $i = 1, 2, \ldots, m$.
Transform these equations into equations 1 [i.e., express the parameters of equations 1(α_{ij}, K_i) in terms of the parameters of the equations in this exercise (r_i, a_{ij})].

15 If a three-species competitive system obeys the Lotka–Volterra competition equations and $r_1 = 0.1$, $r_2 = 0.2$, $r_3 = 0.3$, $a_{11} = 0.01$, $a_{22} = 0.002$,$a_{33} = 0.004$, $a_{12} = 0.002$, $a_{13} = 0.002$, $a_{21} = 0.005$, $a_{23} = 0.005$, $a_{31} = 0.008$, $a_{32} = 0.008$, what is the K-vector and the community matrix? ☐

Recall from Chapter 8 (exercise 16) that the form of the community matrix above is related to the Jacobian matrix as

$$\begin{vmatrix} \left(\frac{\partial f_1}{\partial N_1}\right)_0 & \left(\frac{\partial f_1}{\partial N_2}\right)_0 \\ \left(\frac{\partial f_2}{\partial N_1}\right)_0 & \left(\frac{\partial f_2}{\partial N_2}\right)_0 \end{vmatrix} = \begin{vmatrix} \left(\frac{\partial f_1}{\partial N_1}\right)_0 & 0 \\ 0 & \left(\frac{\partial f_2}{\partial N_2}\right)_0 \end{vmatrix} \begin{vmatrix} 1 & \alpha_{12} \\ \alpha_{21} & 1 \end{vmatrix}$$

which may be generalized as

$$\begin{vmatrix} \left(\frac{\partial f_1}{\partial N_1}\right)_0 & \left(\frac{\partial f_1}{\partial N_2}\right)_0 & \cdots & \left(\frac{\partial f_1}{\partial N_m}\right)_0 \\ \left(\frac{\partial f_2}{\partial N_1}\right)_0 & \left(\frac{\partial f_2}{\partial N_2}\right)_0 & \cdots & \frac{\partial f_2}{\partial N_m}\Big)_0 \\ \vdots & & & \vdots \\ \left(\frac{\partial f_m}{\partial N_1}\right)_0 & & \cdots & \left(\frac{\partial f_m}{\partial N_m}\right)_0 \end{vmatrix}$$

$$= \begin{vmatrix} \left(\frac{\partial f_1}{\partial N_1}\right)_0 & 0 & \cdots & 0 \\ 0 & \left(\frac{\partial f_2}{\partial N_2}\right)_0 & \cdots & 0 \\ \vdots & & & \vdots \\ 0 & \cdots & & \left(\frac{\partial f_m}{\partial N_m}\right)_0 \end{vmatrix} \begin{vmatrix} 1 & \alpha_{12} & \alpha_{13} & \cdots & \alpha_{1m} \\ \alpha_{21} & 1 & \alpha_{23} & \cdots & \alpha_{2m} \\ \vdots & & & & \vdots \\ \alpha_{m1} & & & \cdots & 1 \end{vmatrix}$$

or in matrix form

$$J = BA$$

where J is the Jacobian matrix, B is the diagonal matrix on the right-hand side, and A is the community matrix.

This terminology has recently been replaced by May who refers to the Jacobian matrix as the community matrix. May's terminology is generally more convenient and is used henceforth. When referring to the A matrix, I shall be specifying the "Levins community matrix." Otherwise, any reference to the community matrix will be to the Jacobian matrix. Be careful in reading the literature. It is somewhat confused on this point.

Recall the analytical development in Chapter 7. Based on the eigenvalues of the community matrix (Jacobian matrix), we were able to classify equilibrium points into several different categories. We are able to apply a similar method to the multiple species situation; however, the behavior of equilibrium points can be particularly complicated. It is useful to distinguish between two types of equilibrium, stable and unstable. A stable point is one in which any slight perturbation of the system results in the ultimate return of the system to that point. An unstable point is one in which any slight perturbation does not result in the ultimate return to the point. The detailed behavior of the system may be quite complicated, but qualitatively we need to deal only with stable versus unstable.

The eigenvalues of the Jacobian matrix of a two-series system are given as the roots to the following equation:

$$\text{Det}\,(J - \lambda I) = 0$$

(as developed in Chapter 8). The same procedure is used in an m species situation, the only difference being that we have m eigenvalues instead of two.

The basic rule that distinguishes the stable from the unstable is that, if the real parts of all the eigenvalues are negative, the system is stable. If one or more of the eigenvalues has a positive real part, the system is unstable.

May has developed several interesting models based on this sort of analysis. A particularly interesting approach, developed by Levins and myself, uses the determinant of the Levins community matrix to predict what might be the maximum number of species in a community (Levins 1968, Vandermeer, 1970, 1972a). Levins recently developed a technique he calls "loop analysis," whereby the relative values and positions of the α's in the Levins community matrix give some idea of the stability of the community (Levins, 1975).

These approaches are fascinating and at first glance seem to hold much promise for the study of community structure. They are also fairly difficult mathematically. More importantly, their use in the real world hinges on one practical problem that we seem to be far from solving. We now move to a development of that problem.

THE CONSUMER-RESOURCE EQUATIONS. The following development is due to one of the many insights of Robert MacArthur. We wish, at first, to consider two predator species and several prey species.

☐ EXERCISES

16 Write the Lotka–Volterra type equations for a system of two predators and two prey in which the prey do not compete with each other but both predator species eat both prey species. The predators should not have a self-damping term. The prey should have a self-damping term. (Let the predators be N_1 and N_2; let the prey be N_3 and N_4.)

17 What are the equilibrium values for the prey species? (Be sure you have the same answer for exercise 16 as in the text.

18 Put these equations (exercise 16) in the form of the classic Lotka–Volterra competition equations when the two competitors are N_1 and N_2. ☐

The equations in exercises 16, 17, and 18 are the *consumer-resource equations.* With their format we can compute the values of α_{12} and α_{21} (and by extension all alphas in a community) but only if we know the intrinsic rates of all four species, the interaction coefficients between all combinations of predator and prey, and the carrying capacities of the prey. We do not seem to be much better off than before.

MacArthur, however, provided powerful biological interpretations for the various parameters in the basic equations. Consider the term r_{13}. Clearly this is the amount the per capita rate of species 1 is increased by the addition of a single individual of species 3. Biologically, r_{13} must have something to do with the biomass of species 3, the probability that a particular individual of species 3 will be captured by species 1, and a conversion factor to tell how much biomass of species 1 will be made from a unit of biomass of species 3. Assume that the conversion rate of species 1 is the same for both prey species. Thus we may write $r_{13} = w_3 a_{13} C_1$, where w_3 is the biomass of an individual of species 3, a_{13} is the probability that an individual of species 3 will be eaten by an individual of species 1, and C_1 is the conversion ratio. We now make an important assumption; the reduction in per capita rate of a prey species due to the presence of one individual of the predator species is exactly equal to the probability of being eaten; for example, $\alpha_{31} = a_{13}$.

□ EXERCISES

19 Rewrite the consumer resources equations in terms of C, w, and a (the r and K of the prey species remain).

20 What are α_{12} and α_{21} in terms of a, w, K, r, and C? □

If we define

$$\mu_{ji} = a_{ji}\sqrt{K_i w_i / r_i}$$

as the degree of utilization of prey species i by predator species j, we will have

$$\alpha_{12} = \frac{\mu_{13}\mu_{23} + \mu_{14}\mu_{24}}{\mu_{13}^2 + \mu_{14}^2}$$

and

$$\alpha_{21} = \frac{\mu_{13}\mu_{23} + \mu_{14}\mu_{24}}{\mu_{23}^2 + \mu_{24}^2}$$

If we had had two predator species and S prey species, the equations would have been

$$\alpha_{12} = \frac{\sum_{j=1}^{s} \mu_{1j}\mu_{2j}}{\sum_{j=1}^{s} \mu_{1j}^2}$$

and

$$\alpha_{21} = \frac{\sum_{j=1}^{s} \mu_{1j}\mu_{2j}}{\sum_{j=1}^{s} \mu_{2j}^{2}}$$

These equations have been used, usually rather loosely, to estimate α's for a variety of situations. Indeed, if a particular system were at equilibrium (one of the assumptions made to derive the equations) and K, r, w, and a were actually known, the estimates would be totally accurate. It is usual, however, to estimate or approximate the utilization functions in somewhat dubious ways; for example, in the commonest method the foregoing formula is equated with some measure of niche overlap, as introduced in Chapter 6. Thus we let $u_{ij} \approx P_{ij}$, where P_{ij} is the proportion of the individuals of species i that occurs on resource j or in habitat j. These procedures have come under frequent criticism and currently are used only infrequently.

In general, the framework presented above is about all the theoretical framework we have available for estimating α's in a natural situation. In spite of the tremendous amount of literature that has accumulated in regard to this framework, I remain doubtful of its usefulness.

On the other hand, the more recent work of May and Levins holds more promise. It is not necessary to know the exact values of the α's but whether they are greater than, less than, or equal to zero. Several intriguing results emerge and although the method is still in its infancy it seems to hold some promise for the future. I have not included these new methods because of their complexity [Levins (1975); May (1973b)].

ANSWERS TO EXERCISES

1 On the first day there were zero species on the island and 20 new ones arrived. Thus the first point should be at $y = 20$ and $x = 0$. On the second day 10 of the 20 arrivals were already there, making the number of *new* arrivals 10. Because the number already on the island is 20, the second point is at $y = 10$, $X = 20$. According to the same reasoning, the third point is at $y = 5$, $x = 30$. Thus we have Figure 10.9;

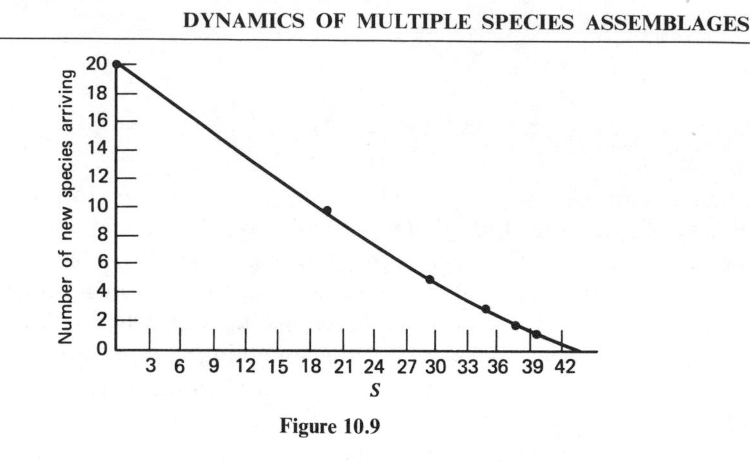

Figure 10.9

2 On the first day there were zero species on the island and 10 new ones arrived. On the second day 5 of the 10 arrivals were already there. On the third day 8 of the 10 were already there (actually 7.5, but we will round). On the fourth day 8 of the 10 were already there (actually 8.5). On the fifth day 9 of the 10 were already there. On the sixth day 10 of the 10 were already there. Thus we obtain Figure 10.10:

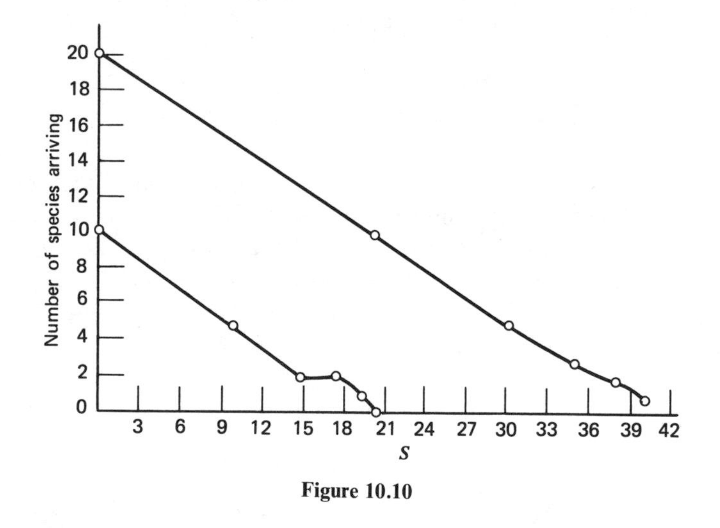

Figure 10.10

3 The first island contained three species. Because three of the 30 species were already on the island, 27 of the 30 must be new. Thus the first point on the graph is $y = 27$, $x = 3$. By following the same reasoning for the rest of the graph we obtain Figure 10.11:

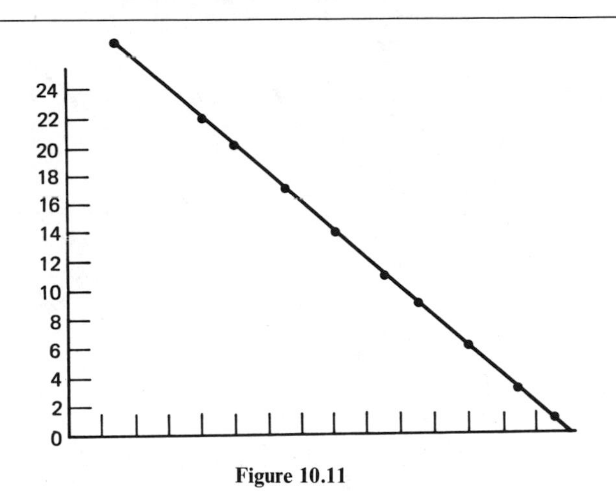

Figure 10.11

4

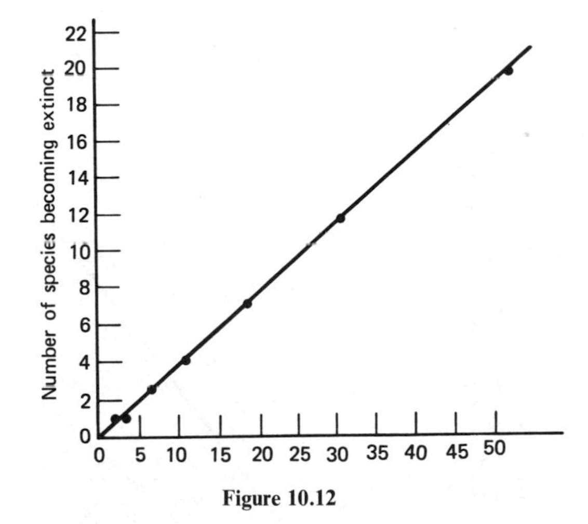

Figure 10.12

5

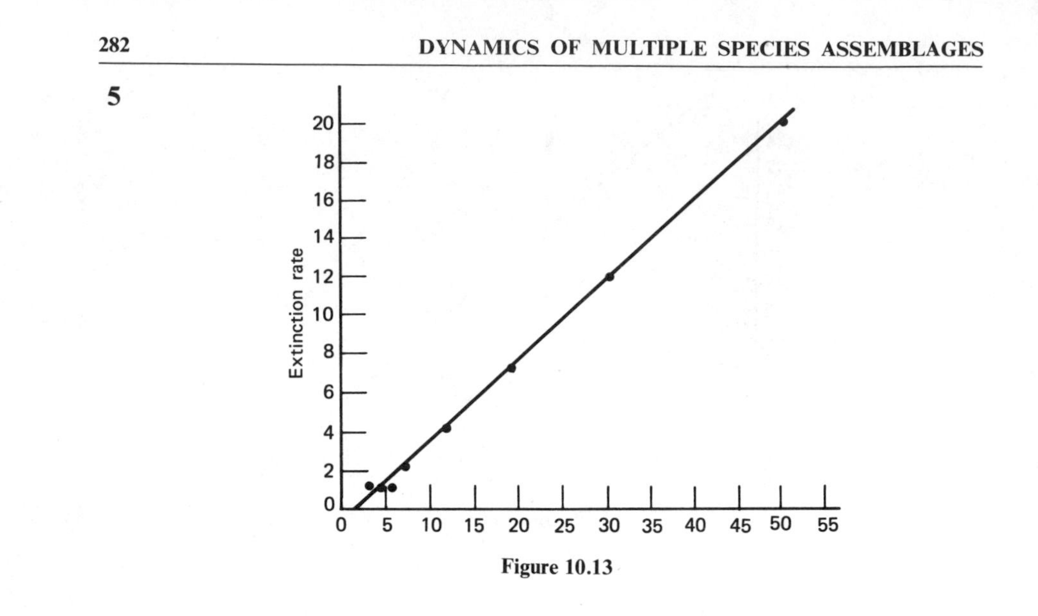

Figure 10.13

Note that the graph is exactly the same.

6

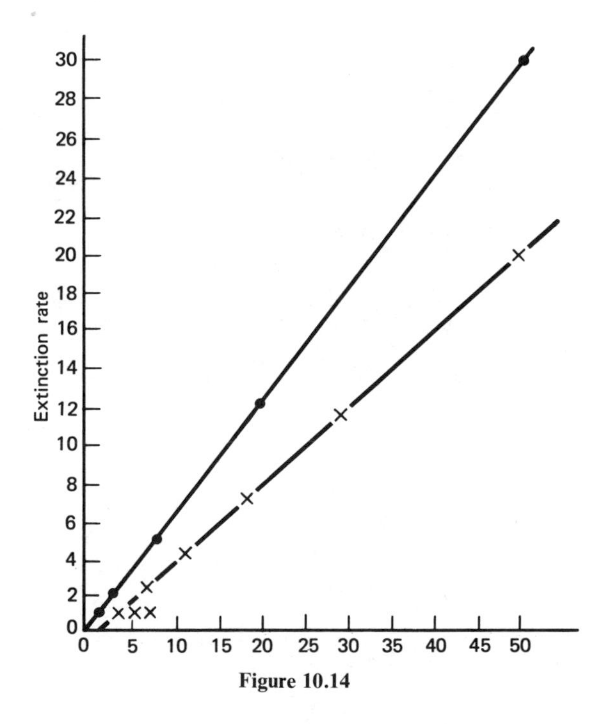

Figure 10.14

7

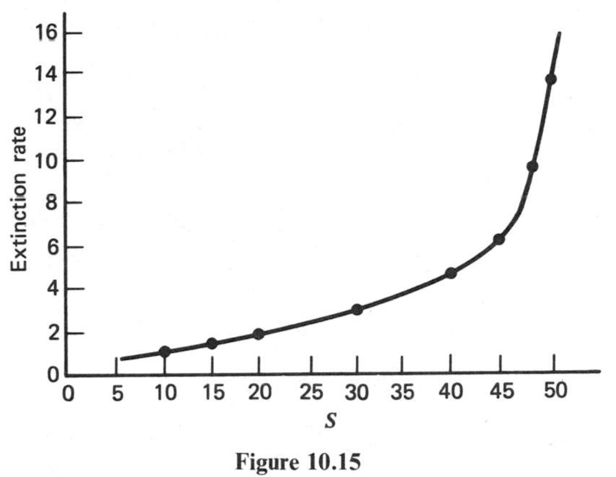

Figure 10.15

8

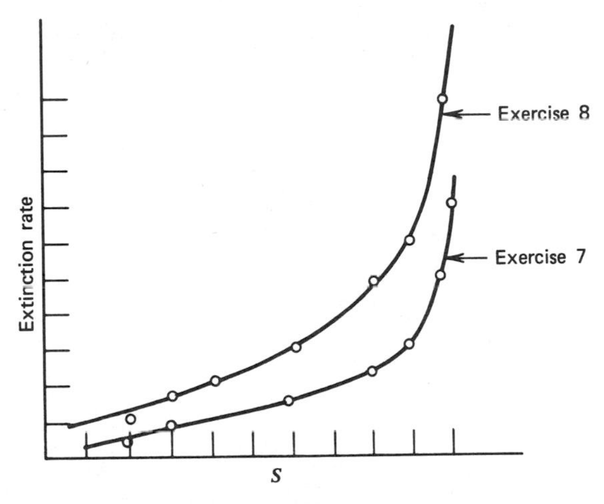

Figure 10.16

9

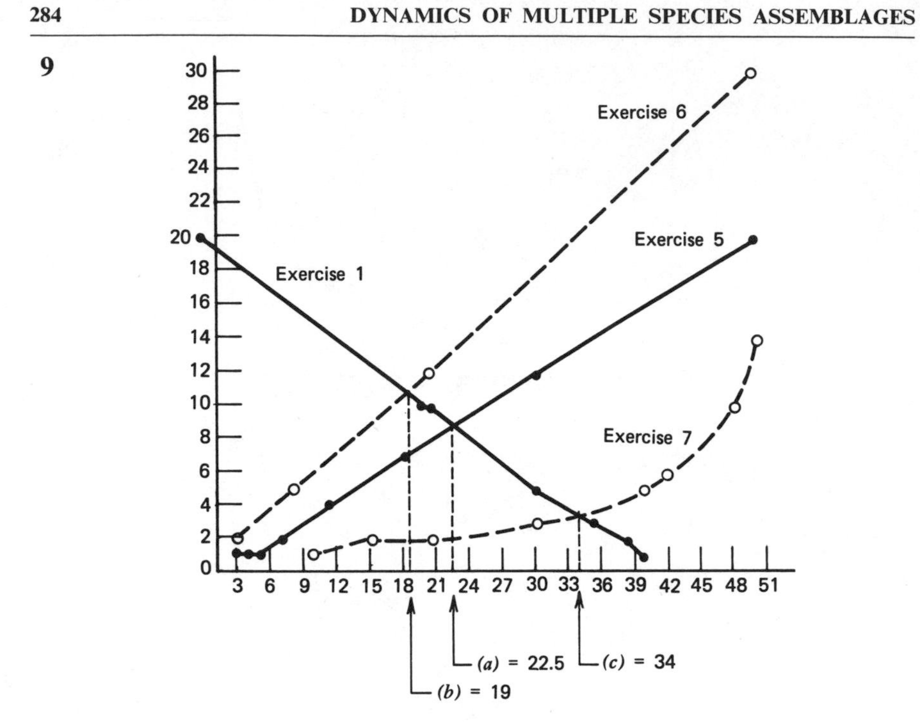

Figure 10.17

10

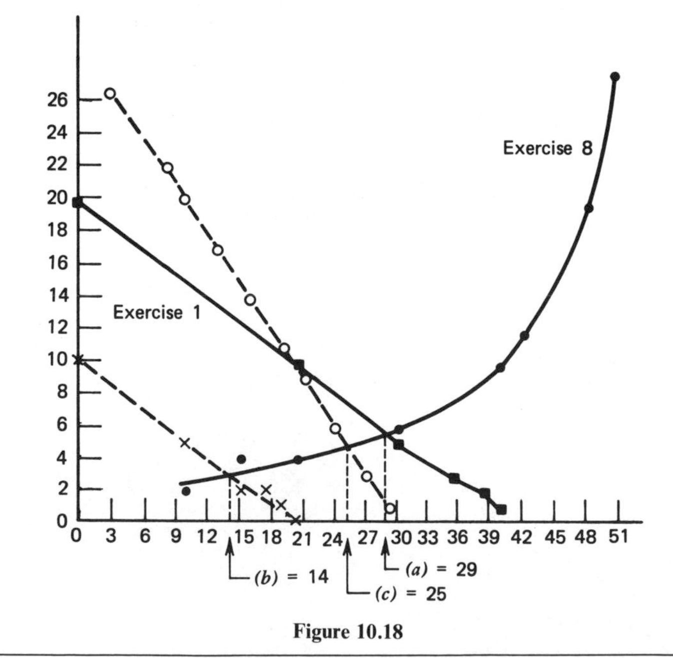

Figure 10.18

11 Immigration function $= \dfrac{ds}{dt} = I_0 - \dfrac{I_0}{T} s$

Extinction function $= \dfrac{ds}{dt} = bs$

12 Equilibrium when $I_0 - \dfrac{I_0}{T}\hat{s} = b\hat{s}$

Thus

$$\hat{s}\left(b + \frac{I_0}{T}\right) = I_0$$

$$\hat{s} = \frac{I_0}{b + I_0/T} = \frac{TI_0}{bT + I_0}$$

13 When the distance from the mainland is zero, the immigrant influx must be T. Thus

$$I_0 = T - \frac{T}{D_c} D$$

where D_c is the critical distance beyond which no species can immigrate. (An island greater than or equal to D_c from the mainland is too far away to receive any immigrants.) From exercise 12

$$\hat{s} = \frac{TI_0}{bT + I_0}$$

substitute for I_0

$$\hat{s} = \frac{T(T - TD/D_c)}{bT + T - (T/D_c)D} = \frac{T(1 - D/D_c)}{(b + 1) - D/D_c}$$

$$\hat{s} = \frac{T(D_c - D)}{D_c(b + 1) - D}.$$

14 Divide through by a_{ii}

$$\frac{1}{a_{ii}} \frac{dN_i}{N_i\, dt} = \frac{r_i}{a_{ii}} - \sum_{j=1}^{m} \alpha_{ij} N_j$$

because $\alpha_{ij} = a_{ij}/a_{ii}$. Recall that K_i (the carrying capacity of the ith species) is r_i/a_{ii}; therefore

$$\frac{dN_i}{N_i\, dt} = a_{ii}(K_i - \sum \alpha_{ij} N_j)$$

and, of course $a_{ii} = r_i/K_i$, so,

$$\frac{dN_i}{N_i\,dt} = r_i\left(\frac{K_i - \sum \alpha_{ij} N_j}{K_i}\right).$$

15

$$K_1 = \frac{r_1}{a_{11}} = \frac{0.1}{0.01} = 10$$

$$K_2 = \frac{r_2}{a_{22}} = \frac{0.2}{0.002} = 100$$

$$K_3 = \frac{r_3}{a_{33}} = \frac{0.3}{0.004} = 75$$

Thus the K-vector is

$$K = \begin{vmatrix} 10 \\ 100 \\ 75 \end{vmatrix}$$

$$\alpha_{12} = \frac{a_{12}}{a_{11}} = \frac{0.002}{0.01} = 0.2$$

$$\alpha_{13} = \frac{a_{13}}{a_{11}} = \frac{0.002}{0.01} = 0.2$$

$$\alpha_{21} = \frac{a_{21}}{a_{22}} = \frac{0.005}{0.002} = 2.5$$

$$\alpha_{23} = \frac{a_{23}}{a_{22}} = \frac{0.005}{0.002} = 2.5$$

$$\alpha_{31} = \frac{a_{31}}{a_{33}} = \frac{0.008}{0.004} = 2.0$$

$$\alpha_{32} = \frac{a_{32}}{a_{33}} = \frac{0.008}{0.004} = 2.0$$

Thus the community matrix is

$$A = \begin{vmatrix} 1.0 & 0.2 & 0.2 \\ 2.5 & 1.0 & 2.5 \\ 2.0 & 2.0 & 1.0 \end{vmatrix}$$

16

$$\frac{dN_1}{N_1\,dt} = r_{13}N_3 + r_{14}N_4 - \Delta_1$$

$$\frac{dN_2}{N_2\,dt} = r_{23}N_3 + r_{24}N_4 - \Delta_2$$

$$\frac{dN_3}{N_3\,dt} = r_3 - \alpha_{31}N_1 - \alpha_{32}N_2 - a_{33}N_3$$

$$\frac{dN_4}{N_4\,dt} = r_4 - \alpha_{41}N_1 - \alpha_{42}N_2 - u_{44}N_4$$

where N_1 and N_2 are predators and N_3 and N_4 are prey. The equations for N_3 and N_4 may be written equivalently

$$\frac{dN_3}{N_3\,dt} = r_3\left(\frac{K_3 - N_3}{K_3}\right) - \alpha_{31}N_1 - \alpha_{32}N_2$$

$$\frac{dN_4}{N_4\,dt} = r_4\left(\frac{K_4 - N_4}{K_4}\right) - \alpha_{41}N_1 - \alpha_{42}N_2$$

where $K_i = r_i/a_{ii}$.

17

$$0 = r_3 - \frac{r_3}{K_3}N_3 - \alpha_{31}N_1 - \alpha_{32}N_2$$

$$0 = r_4 - \frac{r_4}{K_4}N_4 - \alpha_{41}N_1 - \alpha_{42}N_2$$

$$N_3 = \frac{K_3}{r_3}(r_3 - \alpha_{31}N_1 - \alpha_{32}N_2)$$

$$N_4 = \frac{K_4}{r_4}(r_4 - \alpha_{41}N_1 - \alpha_{42}N_2)$$

18 Substitute into the predator equations (consider only N_1)

$$\begin{aligned}\frac{dN_1}{N_1\,dt} &= r_{13}\left[\frac{K_3}{r_3}(r_3 - \alpha_{31}N_1 - \alpha_{32}N_2)\right] \\ &\quad + r_{14}\left[\frac{K_4}{r_4}(r_4 - \alpha_{41}N_1 - \alpha_{42}N_2)\right] - \Delta_1 \\ &= r_{13}K_3 - \frac{r_{13}K_3}{r_3}\alpha_{31}N_1 - \frac{r_{13}K_3}{r_3}\alpha_{32}N_2 + r_{14}K_4 \\ &\quad - \frac{r_{14}K_4}{r_4}\alpha_{41}N_1 - \frac{r_{14}K_4}{r_4}\alpha_{42}N_2 - \Delta_1\end{aligned}$$

Recall the basic form

$$\frac{dN_1}{N_1\,dt} = r_1 - a_{11}N_1 - a_{12}N_2$$

Thus

$$\begin{aligned}\frac{dN_1}{N_1\,dt} &= (r_{13}K_3 + r_{14}K_4 - \Delta_1) - \left(\frac{r_{13}K_3}{r_3}\alpha_{31} + \frac{r_{14}K_4}{r_4}\alpha_{41}\right)N_1 \\ &\quad - \left(\frac{r_{13}K_3}{r_3}\alpha_{32} + \frac{r_{14}K_4}{r_4}\alpha_{42}\right)N_2\end{aligned}$$

Thus

$$K_1 = \frac{r_1}{a_{11}} = \frac{r_{13}K_3 + r_{14}K_4 - \Delta_1}{(r_{13}K_3/r_3)\alpha_{31} + (r_{14}K_4/r_4)\alpha_{41}}$$

$$\alpha_{12} = \frac{a_{12}}{a_{11}} = \frac{(r_{13}K_3/r_3)\alpha_{32} + (r_{14}K_4/r_4)\alpha_{42}}{(r_{13}K_3/r_3)\alpha_{31} + (r_{14}K_4/r_4)\alpha_{41}}$$

and similar equations for N_2.

19 From the text we have

$$r_{13} = \omega_3 a_{13} C_1.$$

Similar reasoning yields

$$r_{14} = \omega_4 a_{14} C_1$$

$$r_{23} = \omega_3 a_{23} C_2$$

$$r_{24} = \omega_4 a_{24} C_2$$

For the prey species $\alpha_{ij} = a_{ji}$. Thus the consumer resource equations are

$$\frac{dN_1}{N_1\,dt} = C_1(\omega_3 a_{13} N_3 + \omega_4 a_{14} N_4 - \Delta_1')$$

$$\frac{dN_2}{N_2\,dt} = C_2(\omega_3 a_{23} N_3 + \omega_4 a_{24} N_4 - \Delta_2')$$

$$\frac{dN_3}{N_3\,dt} = r_3\left(\frac{K_3 - N_3}{K_3}\right) - a_{13}N_1 - a_{23}N_2$$

$$\frac{dN_4}{N_4\,dt} = r_4\left(\frac{K_4 - N_4}{K_4}\right) - a_{14}N_1 - a_{24}N_2$$

where Δ_i from the original equation is now replaced by $C_i\Delta_i'$.

20 From exercise 18

$$\alpha_{12} = \frac{(r_{13}K_3/r_3)\alpha_{32} + (r_{14}K_4/r_4)\alpha_{42}}{(r_{13}K_3/r_3)\alpha_{31} + (r_{14}K_4/r_4)\alpha_{41}}$$

Substituting for r_{13} and r_{14} and α_{ij}, we have

$$\alpha_{12} = \frac{(K_3\omega_3 a_{13}C_1/r_3)a_{23} + (K_4\omega_4 a_{14}C_1/r_4)a_{24}}{(K_3\omega_3 a_{13}C_1/r_3)/a_{13} + (K_4\omega_4 a_{14}C_1/r_4)a_{14}}$$

$$= \frac{(K_3\omega_3/r_3)a_{13}a_{23} + (K_4\omega_4/r_4)a_{24}a_{14}}{(K_3\omega_3/r_3)a_{13}^2 + (K_4\omega_4/r_4)a_{14}^2}$$

Similar reasoning gives

$$\alpha_{21} = \frac{(K_3\omega_3/r_3)a_{13}a_{23} + (K_4\omega_4/r_4)a_{24}a_{14}}{(K_3\omega_3/r_3)a_{23}^2 + (K_4\omega_4/r_4)a_{24}^2}$$

REFERENCES

Abrams, P. 1976. Limiting similarity and the form of the competition coefficient. *Theor. Popul. Biol.* **8**:356–375.

Brown, J. H. 1971. Mammals on mountaintops: Non-equilibrium insular biogeography. *Am. Nat.* **105**:467–478.

Cody, M. 1968. On the methods of resource division in grassland bird communities. *Am. Nat.* **102**:107–148.

Cody, M. L. 1974. *Competition and the structure of bird communities.* Princeton: Princeton University Press.

Colwell, R. K. 1973. Competition and coexistence in a simple tropical community. *Am. Nat.* **107**:737–760.

Colwell, R. K. and E. R. Fuentes, 1975. Experimental studies of the niche. *Annu. Rev. Ecol. Syst.* **6**:281–310.

Colwell, R. K. and D. J. Futuyma, 1971. On the measurement of niche breadth and overlap. *Ecology* **52**:567–576.

Culver, D. C. 1970. Analysis of simple cave communities: niche separation and specie packing. *Ecology* **51**:949–958.

Diamond, J. M. 1969. Avifaunal equilibria and species turnover rates on the Channel Islands of California. *Proc. Nat. Acad. Sci. USA* **64**:57–63.

Diamond, J. M. 1971. Comparison of faunal equilibrium turnover rates on a tropical island and a temperate island. *Proc. Nat. Acad. Sci. USA* **68**:2742–2745.

Diamond, J. M. 1972. Biogeographic kinetics: Estimation of relaxation times for avifaunas of Southwest Pacific Island. *Proc. Nat. Acad. Sci. USA* **69**:3199–3203.

Diamond, J. M. 1975. Assembly of species communities. In M. L. Cody and J. M. Diamond, Eds. *Ecology and Evolution of Communities.* Cambridge: Harvard University Press, pp. 342–444.

Feldman, M. and J. Roughgarden. 1975. A population's stationary distribution and chance of extinction in a stochastic environment with remarks on the theory of species packing. *Theor. Popul. Biol.* **7**:197–207.

Fenchel, T. and F. B. Christiansen. 1976. *Theories of biological communities.* New York: Springer-Verlag.

Gilpin, M. E. 1975. Limit cycles in competition communities. *Am. Nat.* **109**:51–60.

Hairston, N. G. 1951. Interspecies competition and its probable influence upon the vertical distribution of Appalachian salamanders of the genus Plethodon. *Ecology* **32**:266–274.

Heatwole, H. and R. Levins. 1972. Trophic structure stability and faunal change during recolonization. *Ecology* **53**:531–534.

Hunt, G. J. Jr. and M. W. Hunt. 1974. Trophic levels and turnover rates: the avifauna of Santa Barbara Island, California. *Condor* **76**:363–369.

Koplin, J. R. and R. S. Hoffman. 1968. Habitat overlap and competitive exclusion in voles (*Microtus*). *Am. Midl. Nat.* **80**:494–507.

Levin, S. A. 1970. Community equilibria and stability, and an extension of the competitive exclusion principle. *Am. Nat.* **104**:413–423.

Levins, R. 1968. Evolution in changing environments. Princeton: Princeton University Press.

Levins, R. 1975. Evolution of Communities near equilibrium. In Ecology and Evolution of Communities, M. L. Cody, and J. M. Diamond (Eds), Cambridge: Belknap Press of Harvard University Press: pp. 16–50.

MacArthur, R. H. 1968. Theory of the niche. In R. C. Lewontin Ed. *Population biology and evolution*, Syracuse: Syracuse University Press, pp. 1159–1176.

MacArthur, R. H. 1969. Species packing and what competition minimizes. *Proc. Nat. Acad. Sci. USA* **64**: 1369–1371.

MacArthur, R. H. and R. Levins. 1967. The limiting similarity, convergence, and divergence of coexisting species. *Am. Nat.* **101**: 377–385.

MacArthur, R. H. and E. O. Wilson. 1963. An equilibrium theory of insular zoogeography. *Evolution* **17**: 373–387.

MacArthur, R. H. and E. O. Wilson. 1967. *The theory of Island biogeography*. Princeton: Princeton University Press.

May, R. M. 1973a. Time-delay versus stability in population models with two and three trophic levels. *Ecology* **54**: 315–325.

May, R. M. 1973b. Qualitative stability in model ecosystems. *Ecology* **54**: 638–641.

May, R. M. 1974. On the theory of niche overlap. *Theor. Popul. Biol.* **5**: 297–332.

May, R. M. 1975a. *Stability and complexity in model ecosystems*. Princeton: Princeton University Press.

May, R. M. 1975b. Some notes on estimating the competition matrix. *Ecology* 737–741.

May, R. M. and W. J. Leonard. 1975. Nonlinear aspects of competition between three species. *SIAM J. App. Math.* **29**: 243–253.

Neill, W. E. 1974. The community matrix and interdependence of the competition coefficients. *Am. Nat.* **108**: 399–408.

Neill, W. E. 1975. Experimental studies of microcrustacean competition, community composition and efficiency of resource utilization. *Ecology* **56**: 809–826.

Nevo, E., G. Gorman, M. Soule, S. Y. Yang, R. Glover, and V. Jovanociv. 1972. Competitive exclusion between insular Lacerta species (Sazuria Lacertidae). Notes on experimental introductions. *Oecologia Berlin* **10**: 183–190.

Pianka, E. R. 1973. The structure of lizard communities. *Annu. Rev. Ecol. Syst.* **4**: 53–74.

Pianka, E. R. 1974. Niche overlap and diffuse competition. *Proc. Nat. Acad. Sci. USA* **71**: 2141–2145.

Pianka, E. R. 1975. Niche relations of desert lizards. In M. Cody and J. Diamond, Eds. *Ecology and evolution of communities*. Cambridge, Harvard University Press, pp. 292–314.

Pielou, E. C. 1972. Niche width and niche overlap: a method of measuring them. *Ecology* **53**: 687–692.

Pulliam, H. R. 1975. Coexistence of sparrows: a test of community theory. *Science*. **189**: 474–476.

Ross, H. H. 1958. Further comments on niches and natural coexistence. *Evolution* **12**: 112–113.

Roughgarden, J. 1974. The fundamental and realized niche of a solitary population. *Am. Nat.* **108**: 232–235.

Roughgarden, J. 1974b. Species packing and the competition function with illustrations from coral reef fish. *Theor. Popul. Biol.* **5**: 163–186.

Roughgarden, J. and M. Feldman. 1975. Species packing and predation pressure. *Ecology* **56**: 489–492.

Schoener, T. W. 1968. The Anolis lizards of Bimini: resource partitioning in a complex fauna. *Ecology* **49**: 704–726.

Schoener, T. W. 1974. Resource partitioning in ecological communities. *Science* **185**: 27–39.

Schoener, T. W. 1976. Competition and the niche. In D. W. Tinkee and W. W. Milstead, Eds. *Biology of the reptilia*. New York: Academic.

Simberloff, Daniel S. 1974. Equilibrium theory of island biogeography and ecology. *Ann. Rev. Ecol. Syst.* **5**: 161–182.

Simberloff, D. S. and E. O. Wilson, 1969. Experimental zoogeography of islands: the colonization of empty islands. *Ecology* **50**: 278–295.

Sutherland, J. P. 1974. Multiple stable points in natural communities. *Am. Nat.* **108**: 859–873.

Terborgh, J. and J. Faaborg. 1973. Turnover and ecological release in the avifauna of Mona Island, Puerto Rico. *Auk.* **90**:759–779.

Vance, R. R. 1972. Competition and mechanism of coexistence in three sympatric species of intertidal hermit crabs. *Ecology* **53**:1062–1074.

Vandermeer, J. H. 1969. The competitive structure of communities: an experimental approach using protozoa. *Ecology* **50**(3):363–371.

Vandermeer, John H. 1970. The community matrix and the number of species in a community. *Am. Nat.* **104**:73–83.

Vandermeer, John H. 1972a. The covariance of the community matrix. *Ecology* **53**:187–189.

Vandermeer, John H. 1972b. Niche theory. *Annu. Rev. Ecol. Syst.* **3**:107–132.

Vuilleumier, F. 1970. Insular biogeography in continental regions: the northern Andes of South America. *Am. Nat.* **104**:373–388.

Williams, W. T. and J. M. Lambert. 1959. Multivariate methods in plant ecology. I. Association-analysis in plant communities. *J. Ecol.* **47**:83–101.

Williamson, M. 1973. Species diversity in ecological communities. In M. S. Bartlett and R. W. Hiorns, Eds. *The Mathematical theory of the dynamics of Biological populations.* New York: Academic, pp. 325–335.

Wittlson, E. O. and D. L. Simberloff. 1969. Experimental zoogeography of islands: defaunation and monitoring techniques. *Ecology* **50**:267–278.

Index

Age distribution vector, 29-32, 36, 78, 104
Age structure/distribution, 11, 27-38, 69-73, 95-98
Aggregation, 141
Allee effect, 51, 209, 212
Ants, 171, 172, 193
Asymptotes, 54, 65, 66, 200, 215
Autonomous systems, 199

Binomial probability, 123
Birth rate, 5-8, 18, 36, 85, 99, 208, 209
  crude, 99, 110, 111
Bristleberry trees, 34, 36
Butterflies, 234

Canonical distribution, 238, 239
Carrying capacity, 7, 9, 38, 50, 166, 171, 184, 207, 278, 285
Centers, 200, 203, 213
Chaos, 57
Clumped distribution, 120, 129-135, 139, 147, 156, 240
Cohort, 69, 85, 97, 98
Community matrix, 167, 186, 187, 204, 274-277, 286
Competition coefficients, 160, 170-172, 275
Competitor, 193, 207
Complexity, 130
Computer simulation, 37
Consumer-resource equations, 277-279, 288
Contagious, 120, 121
Crude birth rate, 99, 110, 111
Crude death rate, 99, 110, 111

Death rate, 5, 8, 18, 19, 71, 85, 95, 99, 208, 209, 227
  crude, 99, 110, 111
Delayed implantation, 12
Density dependence, 6-8, 11, 13, 35-38, 45, 51, 76, 77, 158, 160, 203-212
Doubling time, 5, 16, 17

Eigenvalue, 32, 38, 54, 93, 201-206, 217-226, 276, 277
Equilibrium, 50-57, 159, 161, 167-172, 184, 193, 194, 198-203, 207, 213, 219-225, 229, 266, 270, 273-279

Evenness species, 240, 241, 273
Exponential, 3-13, 33, 35, 48, 51, 60, 75-77, 158, 198
Extinction, 267-274, 285

Fecundity, 27, 33-38, 72-77, 95
Finite rate of increase, 13
Flour beetle, 171
Focus, 200, 203

Gause-Volterra equations, 160
Global analysis, 199
Guppies, 69, 70

Hamsters, 69
Higher order interactions, 161
Horizontal life table, 97, 110
Hump, 210-212, 229
Hyperdispersed, 121, 139

Immigration, 267-274, 285
Individuals curve, 238, 239
Interspecific competition, 158-163, 167, 170-173, 207, 273
Intertidal, 234
Intraspecific competition, 158, 204
Intrinsic rate of increase, 4, 6, 31, 33, 74-77, 93, 95, 111, 159, 171
Island biogeography, 266-274, 279-281
Isocline, 208, 209, 212, 227, 228, 230

Jacobian matrix, 201-205, 225, 226, 275-277

K-vector, 274, 275, 286

Life expectancy, 72
Life tables, 71, 72, 95, 99, 100, 101
  horizontal, 97, 110
  vertical, 97, 98
Limit cycles, 212, 213
Lloyd's method, 132-135
Logarithmic series, 234
Logistic equation, 7-12, 21, 36, 38, 47, 48, 54, 77, 158-161, 165, 207
Lognormal distribution, 234-239, 241-244, 272, 273

Loop analysis, 277
Lotka's equation, 72-78, 93, 95, 102
Lotka-Volterra competition equations, 158, 166, 167, 171, 190, 202, 208, 218, 275, 277
Lotka-Volterra predator-prey equations, 198, 204, 277

Mean, 131, 132, 148
Mean crowding, 133, 134, 147, 149
Mean generation time, 99, 100, 114
Mollusk, 234
Morisita's method, 132-135, 139
Mortality, 27
Multiple species assemblages, 266-291

Nearest neighbor techniques, 140-142, 156
Neighborhood analysis, 56, 199
Nessiestrus, 171
Niche breadth, 130
Niche overlap, 173
Node, 200, 203, 215
N-vector, 274, 275

Octaves, 235-239, 242, 261
Oscillations, 13, 50-55, 65, 66, 199-204, 211-215, 226, 229

Paramecium, 6, 17
Patchiness, 134, 148, 149
Per capita rate, 7, 17, 18, 212, 278
Permanent cycle, 55-57, 229
Poisson distribution, 121-126, 131
Population-projection matrix, 27-37, 47, 72, 78, 93, 96, 98-101
Predator-prey theory, 198, 204-212, 227, 228, 273, 277, 278, 287
Predictability, 130
Ptherus pubis, 171

Random, 119-121, 126, 129-133, 136, 147, 156

Refugium, 212
Replication rate, 4, 5, 15
Rosenzweig and MacArthur method, 206, 212

Saddle point, 200, 203, 215
Shannon-Weiner function, 240, 241, 242
Singular points, 199, 201
Snouter, 28, 29, 30, 36, 39
Spatial pattern, 119, 132, 136
Species-area relationship, 242-244, 262 263, 271
Species curve, 239
Species diversity, 130, 234, 239-241, 273
Species evenness, 240, 241, 273
Species pool, 272
Species richness, 240, 241, 244, 273
Stable age distribution, 30-38, 43, 45, 72, 73, 76, 95-97, 100-105, 110, 111, 113, 116
Stable stage distribution, 35, 77
Stage distribution vector, 35
Stage projection matrix, 33, 34, 95
Stationary age distribution, 38, 44, 45, 95-98
Stochastic input, 11
Superdispersed pattern, 119, 121, 131, 156
Survivorship, 33-38, 48, 69-77, 85, 95-98, 104, 110, 113

Termites, 171, 172, 193, 196
Time lags, 11-13, 36
Tripling time, 5, 17
Tropical forest, 234
Tsetse flies, 71

Utilization functions, 279

Variance, 130-132, 147, 148, 237
Variance-mean ratio, 131, 132, 134, 147
Veil line, 236, 237, 242, 261
Vertical life table, 97, 98